Advances in Experimental Medicine and Biology

Volume 819

For further volumes:
http://www.springer.com/series/5584

Donald Gullberg
Editor

I Domain Integrins

Springer

Editor
Donald Gullberg
University of Bergen
Bergen
Norway

ISSN 0065-2598 ISSN 2214-8019 (electronic)
ISBN 978-94-024-0290-2 ISBN 978-94-017-9153-3 (eBook)
DOI 10.1007/978-94-017-9153-3
Springer Dordrecht Heidelberg New York London

Printed on acid-free paper

Springer is part of Springer Science+Business Media (www.springer.com)

Preface

What Has Been

The integrin family is composed of 24 members [5]. Ten years ago we published a book devoted to the nine αI domain integrin subunits [24]. These are shown in Fig. 1. I am pleased that most of the original authors have been able to contribute to the updated version.

In 2003, the knockout mouse phenotypes for all of the αI domain integrins had not yet been published. They are now. The phenotypes of mouse strain deficient in individual αI integrins are summarized in Table 1.

During the last decade we have learned more about the role of β2 integrins in leukocytes and in leukocyte adhesion deficiencies [29, 27], and the role of β7 integrins in different subsets of immune cells [20]. Much of this knowledge would not have been possible without the use of animal models and have generated results which could not have been predicted from in vitro analyses. Separate from the interesting results in disease models, analyses of αE knockout mice indicate that there is a missing ligand that has not yet been identified for this integrin [20]. Indeed, in human skin and oral mucosa, there is evidence of a ligand for αEβ7 other than E-cadherin [30].

Regarding the role of collagen-binding integrins the knockout phenotypes of mice deficient in integrin α10 and α11, respectively, have now been published [6, 45] and interestingly the enigmatic DDR collagen receptors have recently been shown to affect the function of collagen-binding integrins [1, 53, 62]. In coming years we are likely to learn more about the cross-talk of collagen-binding integrins with other receptor groups. Maybe most surprising in the field of collagen receptors are the relatively mild phenotypes seen in individual knockout strains and the limited role collagen-binding integrins appear to play in classical connective tissue diseases like fibrosis. This is in contrast to the phenotypes observed for different members of the collagen family, where mutants are characterized by major structural defects impacting tissue structure during development and tissue integrity in adult animals [63]. This discrepancy between collagen and collagen receptor-knockout mouse phenotypes is summarized in Table 2. Interestingly, a recent α10 integrin mutation in dogs have indicated that collagen-binding integrins in the muscoskeletal system might have much more severe phenotypes in larger animals/humans compared to the mild integrin phenotypes observed in collagen-binding integrin deficient mice [33].

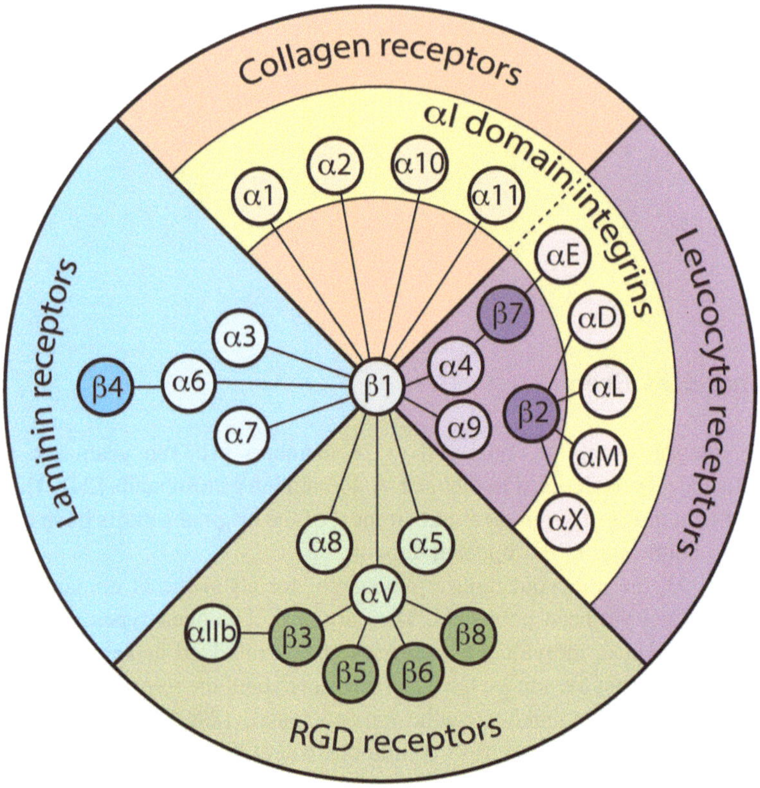

Fig. 1 The integrin family

Table 1 αI integrin knockout phenotypes

Integrin subunit	Distribution	Ligands	Knockout viability	Knockout phenotype
Collagen receptors				
α1	Endothelial cells, smooth muscle cells, fibroblasts, and more cell types [19]	Collagens	+	Normal development [19], hypocellular dermis [18, 47], isolated cells display defect in collagen IV cell attachment
α2	Platelets, epithelial cells, endothelial cells [65], mesenchymal stem cells [44], fibroblasts, and more cell types	Collagens	+	Mild mammary gland phenotype, otherwise normal development [13, 28], cell attachment defect to collagen I of isolated platelets [49], needed for thrombus stabilization [32]
α10	Chondrocytes and subsets of junctional fibroblasts [9, 10]	Collagens	+	Mild cartilage phenotype [6]
α11	Subsets of fibroblasts [46, 55, 57], cancer associated fibroblasts [64], increased levels on myofibroblasts [11], developmental expression in odontoblasts, mesenchymal stem cells [44], induced in cultures of mesenchymally derived cells including myoblasts (do not express α11 in vivo) [25]	Collagens	+	Defective incisor eruption [45], dwarfism [8], increased mortality
Leucocyte receptors				
αD	Macrophages and eosinophils [23, 56]	ICAM-3, VCAM-1	+	Fertile, no gross abnormalities, mild T-cell phenotypic changes [61]
αE	Intraepithelial lymphocytes, some circulating lymphocytes, lamina propria lymphocytes, subsets of CD4+ T-cells, CD8+ T-cells, dendritic cells, mast cells [12, 31, 35]	E-cadherin, uncharacterized ligand	+	Impaired development of gut associated lymphoid tissue [51]
αL	All leucocytes [54]	ICAM-1,-2,-3,-4,-5, JAM-1	+	Splenomegaly and reduced lymph node size [50], increased white blood cells counts [15], reduced lymphocyte homing [7], reduced neutrophil adhesion [15], Treg and NKT cell development affected [42, 59] reduced T-cell proliferation and co-stimulation [21, 50, 52]
αM	Monocytes, macrophages, NK cells, neutrophils, and subsets of T-cells [22, 34, 41]	iC3b, fibrinogen, and more ligands	+	Neutrophil phagocytosis and degranulation reduced [14, 40], impaired mast cell development and function [48], excessive macrophage and dendritic cell toll-like receptor signaling [26, 4], excessive Th17 differentiation [17]
αX	Monocytes, macrophages, dendritic cells, NK cells [41]	iC3b, fibrinogen and more ligands	+	Fertile, no gross abnormalities, affects monocyte firm adhesion [60]

Table 2 Phenotypes of mice deficient in fibrillar collagens and integrin collagen receptors

Ligand			Receptor		
Fibrillar collagen	KO phenotype in mouse	KO phenotype in human	Putative collagen receptor in vivo	Correlation KO phenotypes collagen/ receptor in mouse	KO phenotype in human/dog
I	Mov13 mice [39]: embryonic lethality E12-14, major blood vessel rupture	EDS[a] VIIA, EDS VIIB, OI[b], osteoporosis, joint hypermobility	$\alpha2\beta1$ $\alpha11\beta1$	Not in single integrin mutant strains	?
II	Perinatal lethality [2, 36] short long bones, rudimentary vertebral arches, lack of inter-vertebral discs, notochord defect	Lethal achondrogenesis II, osteochondrodysplasia, osteoarthrosis	$\alpha1\beta1$ $\alpha2\beta1$ $\alpha10\beta1$	$\alpha10$ integrin mutation [6], mild cartilage defect $\beta1$ integrin [3], severe cartilage defect	Chondrodysplasia in dogs, integrin $\alpha10$ mutation [33], severe cartilage phenotype
III	Neonatal lethality [38], 5 % survival with shorter lifespan, intestinal defect, skin lesions, arterial rupture	EDS IV, arterial aneurysms	$\alpha2\beta1$ $\alpha11\beta1$	?	?
V	Embryonic lethality E10-11 [58], cardiovascular insufficiency, lack of collagen fibrillogenesis	EDS I, EDS II	$\alpha2\beta1$ $\alpha11\beta1$	?	?
XI	Cho mice: perinatal lethality by asphyxia [37], weak tracheal cartilage, short snout and mandible, cleft palate, short limbs, externally rotated distal portion of hindlimbs	Schmid chondrodysplasia, non-syndromic hearing loss, osteoarthrosis	$\alpha2\beta1$ $\alpha10\beta1$ $\alpha11\beta1$	$\alpha10$ integrin mutation [6], mild cartilage defect $\beta1$ integrin [3], severe cartilage defect	Chondrodysplasia in dogs, integrin $\alpha10$ mutation [33], severe cartilage phenotype
XXIV	?	?		?	?
XXVII	Mutant transgene [43]: perinatal lethality, lung defect, chondrodysplasia	?	$\alpha2\beta1$ $\alpha11\beta1$	?	?

a Ehlers–Danlos syndrome
b Osteogenesis imperfecta

What Will Come

As in all biological fields, techniques are moving the fields forward as methods become more refined. We now have access to new tools, enabling studies at the nano-scale, and reagents designed to block integrin function can thus be applied to nanoparticles.

In the cancer field, the microenvironment is taking center stage, and here integrins on fibroblasts are predicted to play important roles in paracrine signaling, in regulating tissue stiffness [16] and matrix remodeling.

With exome sequencing of rare genetic diseases becoming more widely used, this will enable new human integrin mutations to be tested in disease models. The development of new molecular techniques to more easily generate mutations in vivo might also contribute to more animal disease models being established.

New technologies, new mouse models in combination with analyses of αI integrins in larger animals/humans are thus predicted to increase our knowledge about this group of receptors. With these things in mind we look forward to another 10 years of research with αI domain integrins.

Acknowledgments

We are grateful for the generous help of Dr. Susanna C. Fagerholm (University of Dundee, UK) with Table 1.

Bergen, March 2014 Donald Gullberg

References

1. Abbonante V, Gruppi C, Rubel D, Gross O, Moratti R, Balduini A (2013) Discoidin domain receptor 1 protein is a novel modulator of megakaryocyte–collagen interactions. J Biol Chem 288:16738–16746
2. Aszodi A, Chan D, Hunziker E, Bateman JF, Fassler R (1998) Collagen II is essential for the removal of the notochord and the formation of intervertebral discs. J Cell Biol 143:1399–1412
3. Aszodi A, Hunziker EB, Brakebusch C, Fassler R (2003) Beta1 integrins regulate chondrocyte rotation, G1 progression, and cytokinesis. Genes Dev 17:2465–2479
4. Bai Y, Qian C, Qian L, Ma F, Hou J, Chen Y et al (2012) Integrin CD11b negatively regulates TLR9-triggered dendritic cell cross-priming by upregulating microRNA-146a. J Immunol 188:5293–5302
5. Barczyk M, Carracedo S, Gullberg D (2010) Integrins. Cell Tissue Res 339:269–280
6. Bengtsson T, Aszodi A, Nicolae C, Hunziker EB, Lundgren-Akerlund E, Fassler R (2005) Loss of alpha10beta1 integrin expression leads to moderate dysfunction of growth plate chondrocytes. J Cell Sci 118:929–936
7. Berlin-Rufenach C, Otto F, Mathies M, Westermann J, Owen MJ, Hamann A et al (1999) Lymphocyte migration in lymphocyte function-associated antigen (LFA)-1-deficient mice. J Exp Med 189:1467–1478
8. Blumbach K, Niehoff A, Belgardt BF, Ehlen HW, Schmitz M, Hallinger R et al (2012) Dwarfism in mice lacking collagen-binding integrins alpha2beta1 and alpha11beta1 Is caused by severely diminished IGF-1 Levels. J Biol Chem 287:6431–6440

9. Camper L, Hellman U, Lundgren-Akerlund E (1998) Isolation, cloning, and sequence analysis of the integrin subunit alpha10, a beta1-associated collagen binding integrin expressed on chondrocytes. J Biol Chem 273:20383–20389
10. Camper L, Holmvall K, Wangnerud C, Aszodi A, Lundgren-Akerlund E (2001) Distribution of the collagen-binding integrin alpha10beta1 during mouse development. Cell Tissue Res. 306:107–116
11. Carracedo S, Lu N, Popova SN, Jonsson R, Eckes B, Gullberg D (2010) The fibroblast integrin alpha11beta1 is induced in a mechanosensitive manner involving activin A and regulates myofibroblast differentiation. J Biol Chem 285:10434–10445
12. Cepek KL, Parker CM, Madara JL, Brenner MB (1993) Integrin alpha E beta 7 mediates adhesion of T lymphocytes to epithelial cells. J Immunol 150:3459–470
13. Chen J, Diacovo TG, Grenache DG, Santoro SA, Zutter MM (2002) The alpha(2) integrin subunit-deficient mouse: a multifaceted phenotype including defects of branching morphogenesis and hemostasis. Am J Pathol 161:337–344
14. Coxon A, Rieu P, Barkalow FJ, Askari S, Sharpe AH, von Andrian UH et al (1996) A novel role for the beta 2 integrin CD11b/CD18 in neutrophil apoptosis: a homeostatic mechanism in inflammation. Immunity 5:653–666
15. Ding ZM, Babensee JE, Simon SI, Lu H, Perrard JL, Bullard DC et al (1999) Relative contribution of LFA-1 and Mac-1 to neutrophil adhesion and migration. J Immunol 163:5029–5038
16. DuFort CC, Paszek MJ, Weaver VM (2011) Balancing forces: architectural control of mechanotransduction. Nat Rev Mol Cell Biol 12:308–319
17. Ehirchiou D, Xiong Y, Xu G, Chen W, Shi Y, Zhang L (2007) CD11b facilitates the development of peripheral tolerance by suppressing Th17 differentiation. J Exp Med 204:1519–1524
18. Gardner H, Broberg A, Pozzi A, Laato M, Heino J (1999) Absence of integrin alpha1beta1 in the mouse causes loss of feedback regulation of collagen synthesis in normal and wounded dermis. J Cell Sci 112 (Pt 3):263–272
19. Gardner H, Kreidberg J, Koteliansky V, Jaenisch R (1996) Deletion of integrin alpha 1 by homologous recombination permits normal murine development but gives rise to a specific deficit in cell adhesion. Dev Biol 175:301–313
20. Gorfu G, Rivera-Nieves J, Ley K (2009) Role of beta7 integrins in intestinal lymphocyte homing and retention. Curr Mol Med 9:836–850
21. Graf B, Bushnell T, Miller J (2007) LFA-1-mediated T cell costimulation through increased localization of TCR/class II complexes to the central supramolecular activation cluster and exclusion of CD45 from the immunological synapse. J Immunol 179:1616–1624
22. Graff JC, Jutila MA (2007) Differential regulation of CD11b on gammadelta T cells and monocytes in response to unripe apple polyphenols. J Leukoc Biol 82:603–607
23. Grayson MH, Van der Vieren M, Sterbinsky SA, Michael Gallatin W, Hoffman PA, Staunton DE et al (1998) Alphadbeta2 integrin is expressed on human eosinophils and functions as an alternative ligand for vascular cell adhesion molecule 1 (VCAM-1). J Exp Med 188:2187–2191
24. Gullberg D (2003) I domains in integrins. vol 1. Landes Bioscience, Georgetown, Texas. pp 1–185
25. Gullberg D, Velling T, Sjoberg G, Sejersen T (1995) Up-regulation of a novel integrin alpha-chain (alpha mt) on human fetal myotubes. Dev Dyn 204:57–65
26. Han C, Jin J, Xu S, Liu H, Li N, Cao X (2010) Integrin CD11b negatively regulates TLR-triggered inflammatory responses by activating Syk and promoting degradation of MyD88 and TRIF via Cbl-b. Nat Immunol 11:734–742
27. Harris ES, Weyrich AS, Zimmerman GA (2013) Lessons from rare maladies: leukocyte adhesion deficiency syndromes. Curr Opin Hematol 20:16–25
28. Holtkotter O, Nieswandt B, Smyth N, Muller W, Hafner M, Schulte V et al (2002) Integrin alpha 2-deficient mice develop normally, are fertile, but display partially defective platelet interaction with collagen. J Biol Chem 277:10789–10794
29. Hu X, Wohler JE, Dugger KJ, Barnum SR (2010) beta2-integrins in demyelinating disease: not adhering to the paradigm. J Leukoc Biol 87:397–403

30. Jenkinson SE, Whawell SA, Swales BM, Corps EM, Kilshaw PJ, Farthing PM (2011) The alphaE(CD103)beta7 integrin interacts with oral and skin keratinocytes in an E-cadherin-independent manner*. Immunology 132:188–196
31. Johansson-Lindbom B, Svensson M, Pabst O, Palmqvist C, Marquez G, Forster R et al (2005) Functional specialization of gut CD103+ dendritic cells in the regulation of tissue-selective T cell homing. J Exp Med 202:1063–1073
32. Kuijpers MJ, Pozgajova M, Cosemans JM, Munnix IC, Eckes B, Nieswandt B et al (2007) Role of murine integrin alpha2beta1 in thrombus stabilization and embolization: contribution of thromboxane A2. Thromb Haemost 98:1072–1080
33. Kyostila K, Lappalainen AK, and Lohi H (2013) Canine chondrodysplasia caused by a truncating mutation in collagen-binding integrin alpha subunit 10. PLoS One 8:e75621
34. Larson RS, Springer TA (1990) Structure and function of leukocyte integrins. Immunol Rev 114:181–217
35. Lehmann J, Huehn J, de la Rosa M, Maszyna F, Kretschmer U, Krenn V et al (2002) Expression of the integrin alpha Ebeta 7 identifies unique subsets of CD25+ as well as CD25-regulatory T cells. Proc Natl Acad Sci USA 99:13031–13036
36. Li SW, Prockop DJ, Helminen H, Fassler R, Lapvetelainen T, Kiraly K et al (1995a) Transgenic mice with targeted inactivation of the Col2 a1 gene for collagen II develop a skeleton with membranous and periosteal bone but no endochondral bone. Genes Dev 9:2821–2830
37. Li Y, Lacerda DA, Warman ML, Beier DR, Yoshioka H, Ninomiya Y et al (1995b) A fibrillar collagen gene, Col11a1, is essential for skeletal morphogenesis. Cell 80:423–430
38. Liu X, Wu H, Byrne M, Krane S, Jaenisch R (1997) Type III collagen is crucial for collagen I fibrillogenesis and for normal cardiovascular development. Proc Natl Acad Sci USA 94:1852–1856
39. Lohler J, Timpl R, Jaenisch R (1984) Embryonic lethal mutation in mouse collagen I gene causes rupture of blood vessels and is associated with erythropoietic and mesenchymal cell death. Cell 38:597–607
40. Lu H, Smith CW, Perrard J, Bullard D, Tang L, Shappell SB et al (1997) LFA-1 is sufficient in mediating neutrophil emigration in Mac-1-deficient mice. J Clin Invest 99:1340–1350
41. O'Doherty U, Peng M, Gezelter S, Swiggard WJ, Betjes M, Bhardwaj N et al (1994) Human blood contains two subsets of dendritic cells, one immunologically mature and the other immature. Immunology 82:487–493
42. Ohteki T, Maki C, Koyasu S, Mak TW, Ohashi PS (1999) Cutting edge: LFA-1 is required for liver NK1.1+TCR alpha beta+ cell development: evidence that liver NK1.1+TCR alpha beta+ cells originate from multiple pathways. J Immunol 162:3753–3756
43. Plumb DA, Ferrara L, Torbica T, Knowles L, Mironov A, Jr., Kadler KE et al (2011) Collagen XXVII organises the pericellular matrix in the growth plate. PLoS One 6:e29422
44. Popov C, Radic T, Haasters F, Prall WC, Aszodi A, Gullberg D et al (2011) Integrins alpha2beta1 and alpha11beta1 regulate the survival of mesenchymal stem cells on collagen I. Cell Death Dis 2:e186
45. Popova SN, Barczyk M, Tiger CF, Beertsen W, Zigrino P, Aszodi A et al (2007) Alpha11 beta1 integrin-dependent regulation of periodontal ligament function in the erupting mouse incisor. Mol Cell Biol 27:4306–4316
46. Popova SN, Rodriguez-Sanchez B, Liden A, Betsholtz C, Van Den Bos T, Gullberg D (2004) The mesenchymal alpha11beta1 integrin attenuates PDGF-BB-stimulated chemotaxis of embryonic fibroblasts on collagens. Dev Biol 270:427–442
47. Pozzi A, Wary KK, Giancotti FG, Gardner HA (1998) Integrin $\alpha 1\beta 1$ mediates a unique collagen-dependent proliferation pathway in vivo. J Cell Biol. 142:587–594
48. Rosenkranz AR, Coxon A, Maurer M, Gurish MF, Austen KF, Friend DS et al (1998) Impaired mast cell development and innate immunity in Mac-1 (CD11b/CD18, CR3)-deficient mice. J Immunol 161:6463–6467

49. Sarratt KL, Chen H, Zutter MM, Santoro SA, Hammer DA, Kahn ML (2005) GPVI and $\alpha 2\beta 1$ play independent critical roles during platelet adhesion and aggregate formation to collagen under flow. Blood 106:1268–7127
50. Schmits R, Kundig TM, Baker DM, Shumaker G, Simard JJ, Duncan G et al (1996) LFA-1-deficient mice show normal CTL responses to virus but fail to reject immunogenic tumor. J Exp Med 183:1415–1426
51. Schon MP, Arya A, Murphy EA, Adams CM, Strauch UG, Agace WW et al (1999) Mucosal T lymphocyte numbers are selectively reduced in integrin alpha E (CD103)-deficient mice. J Immunol 162:6641–6649
52. Shier P, Otulakowski G, Ngo K, Panakos J, Chourmouzis E, Christjansen L et al (1996) Impaired immune responses toward alloantigens and tumor cells but normal thymic selection in mice deficient in the beta2 integrin leukocyte function-associated antigen-1. J Immunol 157:5375–5386
53. Staudinger LA, Spano SJ, Lee W, Coelho N, Rajshankar D, Bendeck MP et al (2013) Interactions between the discoidin domain receptor 1 and beta1 integrin regulate attachment to collagen. Biol Open 2:1148–1159
54. Tan SM (2012) The leucocyte beta2 (CD18) integrins: the structure, functional regulation and signalling properties. Biosci Rep 32:241–269
55. Tiger CF, Fougerousse F, Grundstrom G, Velling T, Gullberg D (2001) alpha11beta1 integrin is a receptor for interstitial collagens involved in cell migration and collagen reorganization on mesenchymal nonmuscle cells. Dev Biol 237:116–129
56. Van der Vieren M, Le Trong H, Wood CL, Moore PF, St John T, Staunton DE et al (1995) A novel leukointegrin, alpha d beta 2, binds preferentially to ICAM-3. Immunity 3:683–690
57. Velling T, Kusche-Gullberg M, Sejersen T, Gullberg D (1999) cDNA cloning and chromosomal localization of human alpha11 integrin. A collagen-binding, I domain-containing, beta1-associated integrin alpha-chain present in muscle tissues. J Biol Chem 274:25735–25742
58. Wenstrup RJ, Florer JB, Brunskill EW, Bell SM, Chervoneva I, Birk DE (2004) Type V collagen controls the initiation of collagen fibril assembly. J Biol Chem 279:53331–52227
59. Wohler J, Bullard D, Schoeb T, Barnum S (2009) LFA-1 is critical for regulatory T cell homeostasis and function. Mol Immunol 46:2424–2428
60. Wu H, Gower RM, Wang H, Perrard XY, Ma R, Bullard DC et al (2009) Functional role of CD11c+ monocytes in atherogenesis associated with hypercholesterolemia. Circulation 119:2708–2717
61. Wu H, Rodgers JR, Perrard XY, Perrard JL, Prince JE, Abe Y et al (2004) Deficiency of CD11b or CD11d results in reduced staphylococcal enterotoxin-induced T cell response and T cell phenotypic changes. J Immunol 173:297–306
62. Xu H, Bihan D, Chang F, Huang PH, Farndale RW, Leitinger B (2012) Discoidin domain receptors promote alpha1beta1- and alpha2beta1-integrin mediated cell adhesion to collagen by enhancing integrin activation. PLoS One 7:e52209
63. Zeltz C, Orgel J, Gullberg D (2014) Molecular composition and function of integrin-based collagen glues -Introducing COLINBRIs. BBA 1840:2533–2548
64. Zhu CQ, Popova SN, Brown ER, Barsyte-Lovejoy D, Navab R, Shih W et al (2007) Integrin alpha 11 regulates IGF2 expression in fibroblasts to enhance tumorigenicity of human non-small-cell lung cancer cells. Proc Natl Acad Sci USA 104: 11754–11759
65. Zutter MM, Santoro SA (1990) Widespread histologic distribution of the alpha 2 beta 1 integrin cell-surface collagen receptor. Am J Pathol 137:113–120

Contents

Evolution of Integrin I Domains

1

Mark S. Johnson and Bhanupratap Singh Chouhan

Abstract

In humans, an ~200-residue "inserted" I domain, a von Willebrand factor A domain (vWFA), buds out from the β-propeller domain in 9 of 18 integrin α subunits. The vWFA domain is not unique to the α subunit as it is an integral part of *all* integrin β subunits and many other proteins. The βI domain has always been a component of integrins but the αI domain makes its appearance relatively late, in early chordates, since it is found in tunicates and later diverging species. The tunicate αI domains are distinct from the human collagen and leukocyte recognizing integrin α subunits, but fragments of integrins from agnathastomes suggest that the human-type αI domains arose in an ancestor of the very first vertebrate species. The rise of integrins with αI domains parallels the enormous changes in body plan and systemic development of the chordate line that began some 550 million or more years ago.

Keywords

Integrin · I-domain · Evolution

1.1 Introduction

Integrins are cell-surface receptors that straddle the plasma membrane by means of a single transmembrane helix in each of two subunits. Integrins in general function to mediate cell-cell and cell-matrix interactions (for a review, see Eble and Kühn [26]); furthermore, integrins are mechanical receptors. Thus, they respond to both external and internal ligands through large changes in receptor conformation that are tightly coupled to function. Integrin signalling is also bi-directional [36], meaning that the presence or absence of molecular interactions in either the cytoplasm or in the extracellular space can modulate the integrin-mediated functions within the other compartment. Thus, integrins are dynamic communicators of both the intracellular wishes of a cell for its extracellular environment as well as mediating environment feeding back to intracellular signalling and downstream intracellular events.

M. S. Johnson (✉) · B. S. Chouhan
Structural Bioinformatics Laboratory, Biochemistry, Department of Biosciences, Åbo Akademi University, Tykistökatu 6A, 20520, Turku, Finland
e-mail: johnson4@abo.fi

D. Gullberg (ed.), *I Domain Integrins*, Advances in Experimental Medicine and Biology 819,
DOI: 10.1007/978-94-017-9153-3_1,

Integrins are $\alpha\beta$ heterodimers and in humans there are 18 α subunits and 8 β subunits that associate to form 24 non-covalently linked heterodimers. The human α subunits range between about 1,050 and 1,190 residues and β subunits between about 770 and 800 residues (with the exception of β4 at over 1,800 amino acids, having a unique additional C-terminal domain). Integrin α and β subunits are known to mutually form a large, N-terminal, extracellular multi-domain "ectodomain" structure, with a globular head and tails, followed by a transmembrane region and relative short C-terminal cytoplasmic sequences. This view was initially based on both electron microscopy data [12, 62] and the analysis of sequences of integrin subunits and the location of a single hydrophobic stretch of residues towards the C-termini of each subunit identified as the presumptive membrane-spanning helical region (e.g. [4, 62]). The complex domain structure [5, 54, 102]; see e.g. Fig. 9.1 in [43] of each subunit appears to be key to the overall dynamic structural changes [55, 58, 103] that are associated with integrin functions and the ability to communicate signals from inside-out and from outside-in.

From the earliest X-ray studies, i.e. αL [72] and αM [52, 51], focus was placed on the human αI domains. Today, structures are also known for I domains of integrin α2 and α1 without (e.g. [27, 76, 64]) and with bound collagen-like triple helical peptides [14, 28], and αX [95]; the latter also within the context of the αXβ2 ectodomain [100]. A region near the N-terminus of the β subunit was predicted to be a von Willebrand Factor type domain based upon the analysis of sequence data [52, 90, 92]. The overall structure of the integrin subunits has thus been detailed by multiple three-dimensional structures of ectodomains and other parts thereof (transmembrane and C-terminal regions) as revealed using structural techniques including X-ray crystallography, NMR spectroscopy, and cryoelectron microscopy. Because the integrins undergo dynamic structural changes, the available structural snapshots provide only a partial description of their full range of functional conformations.

The X-ray structure of the ectodomain of human integrin αVβ3 (PDB code: 1JV2; [102]) was the first reported structure of the extracellular regions of the α and β subunits and their mutual interactions, and those features have been found to be generic features of integrins also observed in the subsequent structures solved for, e.g. the ectodomains of αIIbβ3 [99] and αXβ1 [100] and the N-terminal headpieces (includes the β propeller of the α subunit and β I domain of the β subunit) of α4β7 [105] and α5β1 [60]. Ligand complexes [82, 104] with peptides, having e.g. the "RGD" recognition sequence, pinpointed the narrow binding region suitable for loop recognition, located between the β propeller of the α subunit and the βI domain of the β subunit.

The ectodomain structure of αXβ2 [100] illustrates the relative disposition of the αI domain and revealed the high exposure of the binding site of the αX I domain that would allow integrins to recognize an entirely new class of ligands, such as immunoglobulin fold domains and triple-helical collagens. The αI domain buds out of the N-terminal ~440-residue 7-bladed repeat β-propeller domain within the loop between blades 2 and 3. The β propeller is highly conserved and has been identified in some non-integrin bacterial sequences [15, 40].

In humans, I domains are present in one-half of the integrin α subunits. Four of these nine α subunits, namely α1, α2, α10, and α11, partner with the β1 subunit and are generally referred to as collagen receptors. Five other α subunits, αD, αE, αL, αM and αX, are associated with cells of the immune system, where αE forms heterodimers with β7 but αD, αE, αM, and αL form dimers with the β2 subunit.

The integrins and their evolution were already key topics of study from around the mid-1980's as functional, sequence and structural studies were taking place in multiple laboratories, and the "inserted" I domain [50] or "A" domain [4] was recognized as a novel addition in multiple integrin α subunits. In 2003 (Johnson and Tuckwell), based on the fairly high sequence similarity seen between orthologous human and bony fish integrin subunits with αI domains and the presence of a single αI domain detected in tunicates [57], it was fully expected to find broad coverage of αI domains and even human orthologues throughout the bony fish, sharks and rays, lamprey and hagfish, and even within the invertebrate chordates. Over more recent years, considerable genomic data has become available and in 2014 we have the benefits of all of the accumulated data and their interpretation from many sources, leading to a more coherent view of integrin and especially αI domain evolution. These data not only clarify the likely range of αI domains within extant species, they also help clarify the development of α subunits with I domains orthologous to the types seen in humans and other higher vertebrates.

1.2 Relationships Among Integrin α and β Subunits

In order to set the stage for the appearance and diversification of α subunits containing I domains it is important to recount the evolutionary range of integrin α and β subunits. Complete α and β subunits are present in the earliest metazoans (reviewed in [43]), which suggested that integrins would even predate the first animals. Integrin-like domains and folds-types corresponding to individual integrin domains are also found in some prokaryotes. For example, compared with the integrins the repeats within the β-propeller domain of prokaryotes are even more similar to each other in terms of the loops and calcium binding sites; the latter present in all 7 repeats instead of only 3 or 4 repeats seen with the integrins [15]. But, the domain is present in only some prokaryotes and there is the possibility that they were acquired through non-Darwinian means, i.e. horizontal gene transfer. Moreover, where it was detected the β-propeller domain was present in protein sequences whose function is unknown and without the hallmarks of any other integrin subunits. Similarly, most other domains from which integrins are composed of (e.g. vWFA domain—in both α and β subunits, epidermal growth factor domains—in the β subunit, and immunoglobulin domains—in both α and β subunits) can be located within prokaryotic proteins (reviewed in [43]).

The earliest integrin subunits have been reported in single-cell eukaryotes, the choanozoa, which are the closest relatives of the animals. An integrin-like β subunit fragment was identified in *Ministeria vibrans* [80] and *Capsaspora owczarzaki* contains four α and β integrin subunits each [79, 86], and these sequences can be identified through a simple BLAST search even though, for example, a β subunit shares only between 11 and 21 % sequence identity with a set of known integrin β subunits [43].

Integrin subunits occur across the full range of invertebrates [10] and integrin α and β subunits have long been known to exist within the earliest-appearing animals, e.g. sponges, corals and jellyfish [9, 59, 67, 73]. Similarly, integrin α and β subunits have been identified in the genomes of other early diverging metazoans, including other sponges *Oscarella carmela* (Porifera [63]) and *Amphimedon queenslandica* [84], the placozoan Trichoplax adhaerens [78, 83], the coral *Acropora millepora* (Cnidaria [48] and the sea anemone *Nematostella vectensis* (Cnidaria [48]). Like *C. owczarzaki*, these early metazoans have multiple copies of subunits: For example, *T. adhaerens* has two α subunits and one β subunit; and *N. vectensis* has three or more α and β subunits.

Based on sequence data, evolutionary relationships including multiple phylogenetic representations of the relationships among integrins subunits have been described over the past 25 years [20, 32, 10, 30, 33, 37, 29, 34, 42–43, 87] among others), and they are in close agreement with each other although the naming of individual sequences may have changed and clustered groups may differ somewhat. Hynes and Zhao [37] and Hughes [33] segregated the α subunits with respect to the *Drosophila melanogaster* α subunits into the laminin receptor-like "PS1", the RGD-recognising "PS2" clades and a unique set of duplicated α subunits in the "PS3" clade, containing only invertebrate members (see Bökel and Brown [11] for a review). "Position Specific" antigens from *D. melanogaster* had been known for some time before the full complement of integrin α subunits was defined by the fruit fly

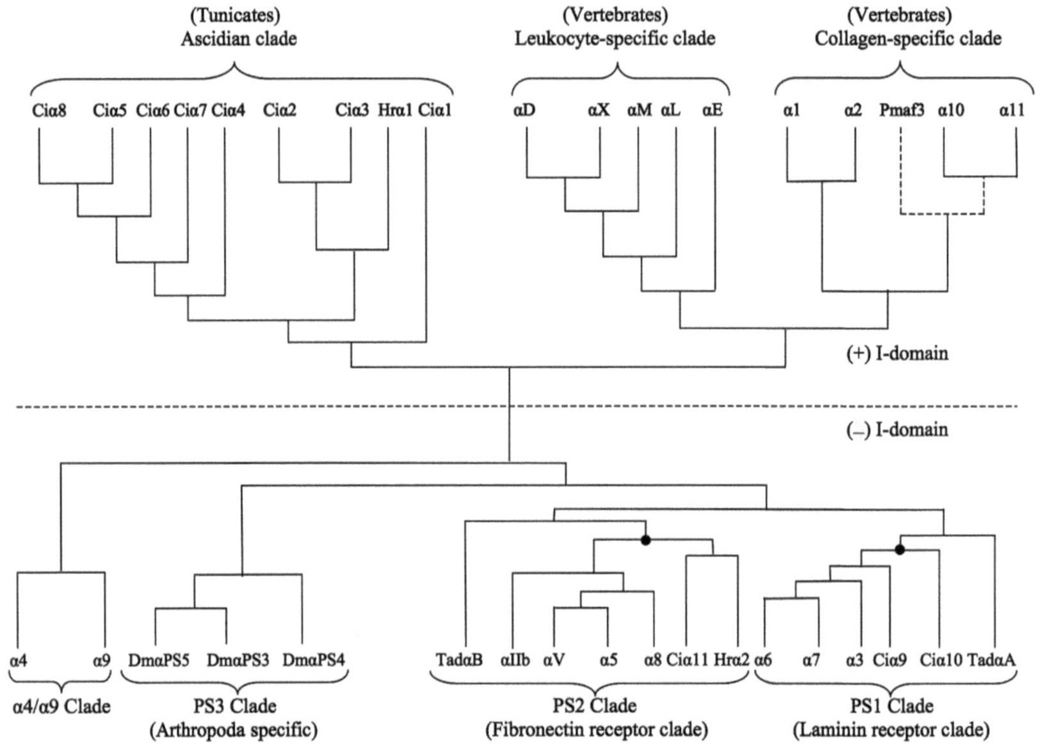

Fig. 1.1 Summary tree for integrin α subunits displaying the phylogenetic relationship among sequences from human, the sea lamprey *P. marinus* (Pmaf3), the tunicates *C. intestinalis* (Ci) and *H. roretzi* (Hr) as well as the arthropod *D. melanogaster* (Dm) and the placozoan *T. adhaerens* (Tad). This summary tree is based on a sequence comparison of 38 vertebrate and invertebrate sequences. The I-domain clade consists of nine representative sequences each from human and the tunicates. The branching patterns for the two *T. adhaerens* sequences have low bootstrap support. The branch lengths in the tree are arbitrary

genomic sequence [1, 75]. The summary tree for integrin α subunits represented in Fig. 1.1 reflects this classification scheme. The two α subunits of the nematode *Caenorhabditis elegans*, Ina-1 and Pat-2, respectively cluster with the PS1 and PS2 clades. In human, α3, α6 and α7 are found in the PS1 group and αIIb, αV, α5 and α8 cluster within the PS2 clade. The αA and αB subunits of the placozoan *T. adhaerens*, depending on the other sequences being compared, cluster respectively with the PS1 group and PS2 group (Fig. 1.1), or together with the PS1 clade [43], but the bootstrap significance in either case is low. In addition to the human members of the PS1 and PS2 clades, there is an α4/α9 clade and clades involving α subunits with inserted I domains (Fig. 1.1).

The β subunits have been clustered into two vertebrate clades, vertebrate A and vertebrate B, as well as an invertebrate clade [10, 34]. Of the human β subunits, β1, β2 and β7 belong to the vertebrate A clade and β3–β6 and β8 belong to the vertebrate B cluster. The β subunits of the vertebrate A clade pair with α subunits having the inserted I domain, but in human β1 is overall the most promiscuous subunit, pairing with the four collagen receptor α subunits plus eight α subunits lacking an I domain. Human β7 clusters with both αE and α4 (which also pairs with β1), the latter lacking the I domain; and human β2 partners only with the immune cell subunits αD, αL, αM and αX. The third major cluster contains invertebrate β subunits, e.g. the two β subunits, βv and βvPS from *D. melanogaster* and pat-3 from *C. elegans* (see Fig. 1.4, [34]). The invertebrate cluster also includes 4 of 5 β subunits in the genome [21] of an early diverging invertebrate chordate—the tunicate *Ciona intestinalis* (Urochordate). A fifth β subunit—Ciβ5—clusters within the vertebrate B

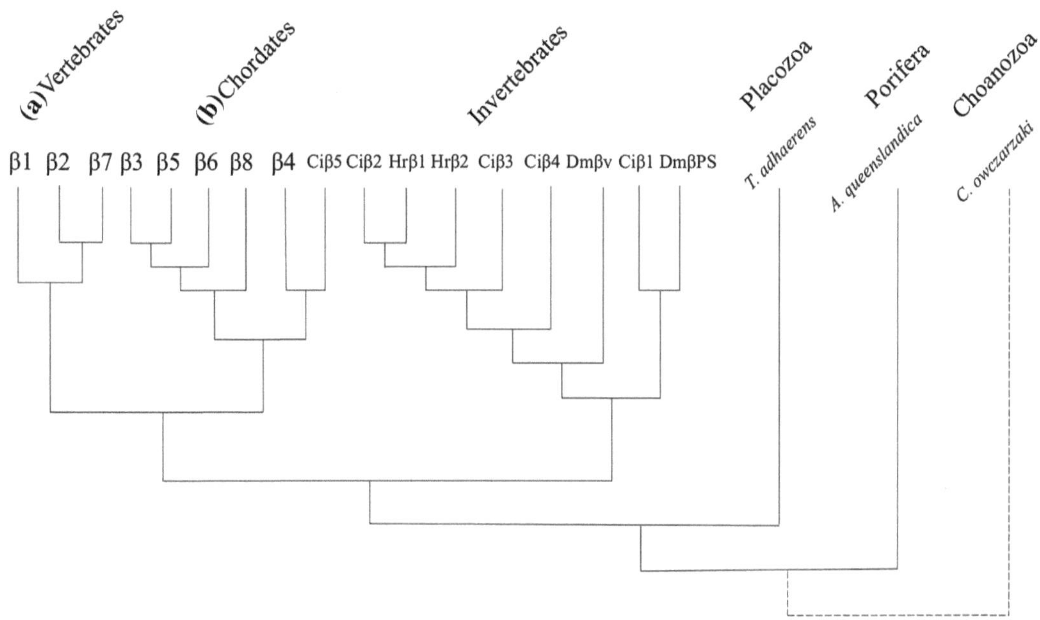

Fig. 1.2 Summary tree for integrin β-subunits displaying the phylogenetic relationship based on 40 vertebrate and invertebrate sequences including sequences from the poriferan *A. queenslandica* and the choanozoan *C. owczarzaki*. The integrin β-subunit sequences derived from *C. owczarzaki* have been utilized to root the tree. The branch lengths in the tree are not to scale

clade [34], so the vertebrate B clade extends at least to one invertebrate species; we have thus labeled it the "Chordate" group to reflect Ciona's inclusion (Fig. 1.2). Representatives of the earliest diverging metazoans, *T. adhaerens* and *A. queenslandica* cluster as outliers of all of the other species when the single-cell eukaryote *C. owczarzaki* is used as an out-group to root the tree (Fig. 1.2).

1.3 Human Integrin α Subunits with the I Domain Have Unique and Common Features

The clustering of the nine human α subunits containing the I domain (Fig. 1.1) follows the phylogeny described by Hughes [33]. When the sequences of full-length integrin α subunits are compared two major clusters are observed, with the collagen-binding cluster ((α1, α2) (α10, α11)), separating cleanly from the integrin α subunits of the immune system: ((((αD, αX)αM)αL)αE).

In contrast to the human integrins, no I domains were found in the α subunits from the genomes of either *C. elegans* or *D. melanogaster* [37], nor in sequences from other invertebrates including very early diverging metazoan species like the placozoa, porifera and cnidarians. Neither are αI domains found among the first diverging deuterostomes—the echinoderms—that directly precede the appearance of the chordate line. Orthologues of the human αI-containing subunits (and other subunits) were, however, identified in sequences from mouse, other mammals, birds and amphibians [33, 42, 96], among others) and likely orthologues in bony fish were identified [42].

One characteristic feature of integrin αI and βI domains (and of many vWFA domains) is the metal ion dependent adhesion site "MIDAS" (Fig. 1.3) where a metal cation, e.g. Mg^{2+}, is bound. Thus, MIDAS is present in all integrins, but the binding modes and the ligand features that can be recognized in the presence or absence of an αI domain are different [88]. The metal ion at MIDAS is bound by conserved

```
                       MIDAS                              αC helix                           Intrinsic
                                                                                              ligand
α2     DVVVVCDESNSIYP......LTNTF......VTDGESH......ILRFGIAVLGYLNRNALDTKNLIKE......FSIEGT
α1     DIVIVLDGSNSIYP......QTMTA......VTDGESH......IQRFSIAILGSYNRGNLSTEKFVEE......FALEAT
α10    DVVIVLDGSNSIYP......ETKTA......VTDGESH......VTRYGIAVLGHYLRRQRDPSSFLRE......FGLEGS
α11    DIVIVLDGSNSIYP......ETRTA......ITDGESH......VTRYAVAVLGYYNRRGINPETFLNE......FSLEGT
Pmaf1  DIVFVLDGSNSIYP......MERGN......VTDGESH......ITRYAIAVLRSYSSNADDVARLINE......FSLEGT
Pmaf2  DIVIVLDGSNSIYP......RTASA......VTDGESH......ITRYAIAVLGYYKRKNIDPSNFISE......FSLEGT
Pmaf3  DIVLVLDGSNSIWP......VTNTA......VTDGESS......ITRFGIAVLDYYISSNMNVEKLQAE......YSLEGT
Cmα11  DIVIVLDGSNSIYP......ETNTA......ITDGESH......ITRYSIAVLGYYNRRGINPTHFLKE......FSLEGT
Cmα1   DIMIVLDGSNSIYP......QTKTA......VTDGESH......ITMFAIAVLGSYNRGNQSTVKFLKE......FALEAT
Cmα2   DIVIVLDGSNSIWP......ETNTA......VTDGESH......IIRFGIAVLGYYNRVGIDTSNLIKE......FSIEGT
αD     DIVFLIDGSGSIDQ......LTFTA......ITDGQKY......IIRYAIGVG-----HAFQGPTARQE......YAVEGT
αX     DIVFLIDGSGSISS......FTYTA......ITDGKKE......IIRYAIGVG-----LAFQNRNSWKE......FAIEGT
αM     DIAFLIDGSGSIIP......RTHTA......ITDGEKF......VIRYVIGVG-----DAFRSEKSRQE......FAIEGT
αL     DLVFLFDGSMSLQP......LTNTF......ITDGEAT......IIRYIIGIG-----KHFQTKES-QE......YVIEAT
αE     EIAIILDGSGSIDP......VTKTA......LTDGGIF......VERFAIGVG-----EEFKSARTARE......ISMEGT
CmαE   EIAIILDGSGSIDA......VTKTA......VTDGEIY......VERFAIGVG-----DATKKPKPVEE......VGIEGT
Ebu_f  DIVVLFDGSRSVTD......GTNAY......ISDGESD......DALN---------------------......------
Hrα1   DVLFVLDGSGSVGK......TTYTG......LTDGQAK......IATFAVGVG----------EYDISE......FVLEGG
Ciα1   DLIFLIDESTSVLE......GTATG......LTDGKSQ......IVMFAIGVG----------KVVMGE......ASLESQ
Ciα2   DMLFVLDGSGSVGK......TTYTA......LTDGLST......ITTFAVGVG----------EANEKE......FVLEGA
Ciα3   DLVYVVDSSNSISD......NTFTS......ITDGKAN......ITVYAIGVA----------LKSDAE......SSGEGQ
Ciα4   DIIILLDGSTSVFP......QTFIH......ITDGEAT......IILTAVGIG-----------SSVNE......------
Ciα5   DIIFVVDESGTVNR......GTYIG......LTDGRAD......IVTVSVGVG-----------DKINE......VKLEGA
Ciα6   DIIFVVDESGSVDV......LTYIG......LTDGAAT......IVTVSVGVG-----------SRVDE......VKLEGD
Ciα7   DIMFVLDDSSSVDD......GTYIS......LTDGGAS......IVLVSVGVG-----------TSVNN......LTARTN
Ciα8   DIIFVVDESGSVDT......LTYIG......LTDGRAT......IVTVSVGVG-----------SGIIE......VKLEGQ
```

Fig. 1.3 Sequence alignment over regions of representative integrin αI domains with key amino acids highlighted: residues of MIDAS involved in binding the metal cation where ligand binding takes place via a glutamate residue as in the X-ray structures of the α2 I domain with bound $GFOGER_3$ collagen-like triple helical peptide (PDB code: 1DZI; [28] and the αL I domain with bound ICAM3 (PDB code: 1T0P; [81]; the diagnostic region referred to as the αC helix, is formed in the ligand-free form of the α2 I domain (PDB code: 1AOX; [27], but the corresponding region is missing from members of the leukocyte αI clade and the ascidian αI domains. The intrinsic ligand, e.g. E^{336} in the α2 I domain, is located at the opposite end of the domain from MIDAS and binds to MIDAS of the βI domain in the β subunit, forming part of the activation mechanism when external ligands bind to MIDAS at the αI domain of the α subunit

residues of the I domain; in the α2 I domain [27, 28] by $^{151}DxSxS^{155}$, T^{221} (and D^{151}) via a water molecule, and D^{254} (Fig. 1.4a). The other coordination positions of the metal ion bind water molecules and the negatively-charged residue from the ligand recognized by the I domain displaces a water molecule and binds directly to the metal ion. The βI domain, in addition to MIDAS, contains the synergistic metal ion-dependent site (SyMBS) and the adjacent to metal ion-dependent adhesion site (ADMIDAS), both of which bind calcium.

In integrins *without* αI domains, the βI domain binds a negatively-charged aspartic acid from the ligand via a positively charged metal ion at MIDAS as seen in the X-ray structures of αVβ3 (cyclic "RGD" peptide, PDB code 1L5G, [104] and αIIbβ3 (fibrinogen γ chain "AGD" peptides: e.g. PDB code 2VDO, and chimeric "RDG" peptide in 2VDR; [82]. In integrins *with* αI domains, an "intrinsic ligand"—a glutamate residue—is conserved across the αI domains (e.g. E^{336} in the α2 I domain; Figs. 1.3, 1.4b) and when ligands bind to MIDAS of αI the glutamate is proposed to bind to MIDAS of the βI domain as a part of the integrin conformational regulatory mechanism [2]; Yang et al. 2004; [44, 100].

In comparison with extrinsic ligand binding to the βI domain, ligands that bind to MIDAS of the integrin αI domain [57] do so through a slightly larger, negatively charged glutamic acid side chain, as illustrated for the two subsets of αI domains in human: The collagen-type, e.g. from the collagen-like triple helical-peptide GFOGER

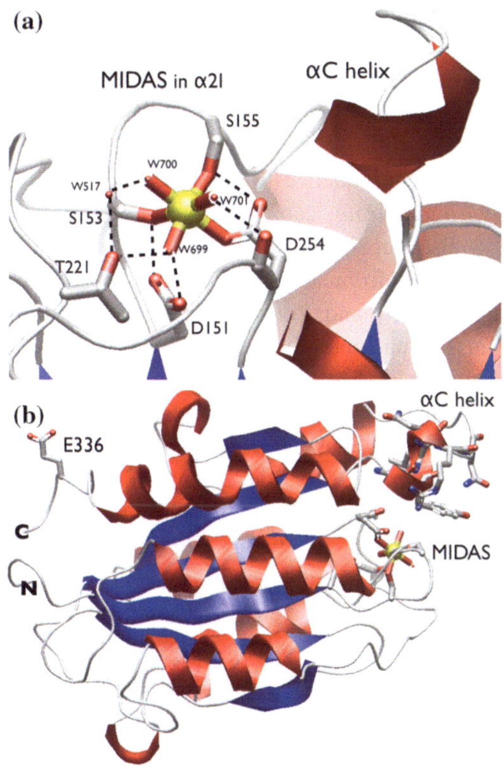

Fig. 1.4 Key features of the αI domain based on the human α2 I domain structure (PDB: 1AOX [27]). **a** The MIDAS site and location of αC helix, including Mg^{2+} (*yellow sphere*), water molecules (*W*) key residues involved in binding directly to Mg^{2+} or via water; dash lines represent likely hydrogen bonds. **b** Relative positions of MIDAS, the αC helix and E336 with respect to the C-terminal (*c*) and N-terminal (*N*) ends of the αI domain, which buds out from the *β* propeller domain of the α subunit. The residues of the αC helix in the α2 I domain are GYLNR, and are shown in ball-and-stick figures. Helices are in *red* and strands in *blue*. The figure was rendered in Bodil [53]

in complex with the α2 I domain (PDB code: 1DZI; [28] and the immune cell recognizing integrin α subunits, e.g. from the immunoglobulin domain of ICAM3 bound to the αL I domain (PDB code: 1T0P; [81]. Note that some "unintended" ligands can bind αI domains in a metal-independent fashion, such as lovastatin to the αL I domain [45], and the "RKKH" motif of a peptide from a snake venom metalloproteinase [38, 68] and echovirus 1 [6, 44, 47, 101] to the human α2 I domain.

Whereas, the MIDAS site is a universal feature of integrin αI domains (Fig. 1.3), and even of the *β*I domain of the *β* subunit and present in some but not all vWFA domains, the collagen-binding αI domains are easily distinguished from those of the immune system and from other vWFA domains by simply examining one key feature of their sequence alignment: Neither the leucocyte-specific αI domains (Fig. 1.3) nor other vWFA domains contain the αC helix present in all of the collagen receptor members. The α2 I-domain structure described by Emsley et al. [27]; PDB code: 1AOX) pinpointed this major difference with the αL and αM I-domain structures—there is an additional helix, αC (Fig. 1.4), in the vicinity of MIDAS whose conformation, along with adjoining regions, changes in response to collagen binding as seen in the α2 I domain bound to a collagen-like triple-helical GFOGER peptide (PDB code: 1DZI; [28]). In the α2 I domain the αC helix corresponds to the sequence ^{284}GYLNR^{288} (GSYNR in α1, GHYLR in α10 and GYYNR in α11). The presence of the αC-helix region is diagnostic of the collagen-binding αI domains, and it is neither found in I domains of the immune system integrins (Fig. 1.3) nor in other vWFA domains and it has been essential for identifying collagen-type integrin αI domains in sequence fragments e.g. from lamprey [16].

1.4 Tunicates and Bony Fish Set Boundaries for Understanding αI Domain Evolution

The answers to two questions posed earlier [31, 42, 40] on the evolution of integrins with αI domains are increasingly being clarified, largely as a result of the genome sequencing studies occurring within chordate species. Firstly, what are the earliest diverging species with integrins having αI domains? And, related to this question—are they orthologues of the human integrins? And, if not, when did the first orthologues arise?

In 2001, Miyazawa et al. identified two integrin α subunits in an early diverging chordate, the ascidian *Halocynthia roretzi* (Urochordate; tunicate). One, αHr2, belongs to the PS2 clade along with the human subunits αIIb, αV, α5, and α8 (see, e.g. [34]), but the second, αHr1, contained an αI domain (Fig. 1.1). This was the first invertebrate I domain to be identified and αHr1 was experimentally associated with the recognition of complement factors in the ascidian immune system [57]. An early phylogenetic reconstruction [42] placed αHr1 as an outlier to both the vertebrate collagen and immune systems clades when the tree was rooted by the position of the non-integrin vWFA3 domain. Furthermore, αHr1 does not have a sequence corresponding to an αC helix (Fig. 1.3).

Soon thereafter, the genome sequence from another ascidian, *C. intestinalis* [21], led to identification of 11 α subunits. Three α subunits lacked the αI domain (Ciα9 and Ciα10 cluster with the PS1 clade and Ciα11 cluster with PS2; Fig. 1.1), but eight others, Ciα1-Ciα8, cluster with αHr1 and separately from both the human leukocyte clade and the collagen receptor clade [29, 34]. The ascidian sequences, individual αI domains or full-length integrins, with the exception of Ciα1 cluster consistently into two main groups using multiple methods (e.g. Bayesian, maximum likelihood, neighbor joining) for phylogeny reconstruction [17], with αHr1 and Ciα2-3 in one group and Ciα6-8 in the other (Fig. 1.1). Based on the phylogenetic reconstructions and closest sequence matches, it is clear that the ascidian integrins are not orthologues of any of the bony vertebrate α subunits having I domains [34, 77]. It was therefore not a surprise that, like αHr1 of *H. roretzi*, none of the *C. intestinalis* sequences have the αC helix characteristic of the collagen-binding αI clade (Fig. 1.3).

Thus, the ascidian data show that some early invertebrate chordates already had integrins with I domains when they diverged within the chordate lineage, but they represent paralogues of the nine human integrins; no orthologous pairs exist. In contrast, at the other end of the spectrum, orthologues of human I domains could be traced back through other mammals, birds, reptiles, amphibians, and even to bony fish [34, 42], suggesting that orthologues of the human integrins with αI domains might be found across the whole range of vertebrate species [42]. Individual integrin sequences from bony fish (e.g. from *Cyprinus carpio*, carp; *Danio rereo*, zebrafish) had also become available as well as genome sequences from e.g. the pufferfish *Takifugu rubripes* [3] and *Tetraodon nigroviridis* [39]. Thus, human integrin subunit orthologues were identified in fish, including α subunits with I domains, and clustered these sequences to the collagen and leukocyte clades (e.g. see [34]). Fish, thought to have undergone an extra round of whole genome duplication in comparison to later diverging vertebrates (see e.g. [22, 89]), also exhibited duplicate isoforms orthologous to the human subunits (e.g. [34]), and today there are orthologous representatives identified in bony fish for nearly all of the human integrin chains and duplicate isoforms are the rule rather than an exception [17].

Thus, the ascidians and the bony fish now provide key and well-established demarcations for integrin αI domain evolution. Urochordates, thought at the time to be the earliest diverging species of the chordate line, have non-human-like integrin subunits with αI domains, whereas human orthologues are present in bony fish. Only three extant groups of organisms were considered to have diverged after the urochordates and prior to the bony fish: The lancelets, the cyclostomes (or agnaths; jawless vertebrates) and the cartilaginous fish. The lancelets, according to established taxonomy, were positioned as the closest living relatives of the vertebrates but this notion has now been challenged by the molecular data.

1.5 Lancelets, the Jawless Vertebrates, the Cartilaginous Fish and the Origin of Vertebrate Orthologues

The last common ancestor of the echinoderms and chordates is estimated to have occurred around 520–550 million years ago according to the fossil record, and considerably earlier when based on sequence comparisons and phylogenetic reconstructions (see e.g. [8, 24]). There are two extant invertebrate chordate groups, the cephalochordates and the urochordates. The presence of an I-domain containing the α subunit in the urochordate *H. roretzi* supported the idea that α subunits containing I-domains would be found in the "later-diverging" cephalochordates (lancelets), and especially within the agnathastomes (cyclostomes; hagfish and lamprey) and chondrichthyes (sharks and rays) [42, 40]. In 2008, the genome of the lancelet *Branchiostoma floridae* (Cephalochordate; amphioxus; [71]) was reported. A search of the genome revealed integrin α and β subunits but α subunits with I domains were not identified [31].

This then led to the following quandary: If the urochordates have αI domains, then how is it possible that they are absent in the later-diverging lancelet? Interestingly, a controversy on the relative divergence times of the two earliest representatives of the chordates was also underway. The lancelets, based on e.g. morphological features (for a considered review of the non-sequence based evidence, see [85], were long thought to have diverged *after* the tunicates—i.e. after the divergence of *H. roretzi* and *C. intestinalis* and other urochordates from the chordate line. In such a case, it would be difficult to reconcile how the lancelet would have "lost" the integrin subunits with αI domains. However, the comparison of 146 genes across 49 species [69], 1,029 concatenated sequences among the deuterostomes [71], and ~40 Mb of expressed sequence tags across 21 phyla [25] all concluded that the urochordates, not the cephalochordates, are the closest extant relatives of the vertebrates. Earlier, a similar controversy led to the reassignment of hagfish and lamprey to a monophyletic group based on the molecular data (see, e.g. [65]) and in contrast to the morphological arguments. Considered in this light, the absence of αI domains in echinoderms and all earlier animals, as well as the lancelet, coupled with the αI domain presence in the ascidians, is congruent and pinpoints the origin of the integrin αI domain integration event to have occurred after the divergence of the cephalochordates and prior to the divergence of the urochordates.

Since the urochordates have integrin αI domains that are not orthologues of the human types, the remaining two groups of extant species diverging prior to the bony fish—namely, the hagfish/lampreys (Agnatha; jawless vertebrates) and the sharks/skates/rays/chimera (Chondrichthyes; cartilaginous fish)—should provide evidence for the origin of the human-type integrin αI domains.

Fragments of sequences have appeared from the genomic sequencing of Petromyzon marinus (sea lamprey) and searches against the ENSEMBL data (http://www.ensembl.org/Petromyzon_marinus/Info/Index) yielded several fragments and a near full-length integrin sequence [16]. Phylogenetic reconstructions with the nearly full-length lamprey integrin and the fragments clearly showed that they are not part of the urochordate cluster, nor are they part of the immune cell recognizing set of αI domains; however, they have the αC-helix (Fig. 1.3; [16]) and they do cluster within the collagen recognizing integrin αI domain set that includes human α1, α2, α10 and α11 (Pmaf3 in Fig. 1.1; Fig. 1.5; [17]). Furthermore, preliminary binding studies on expressed sequences suggest that the lamprey sequences bind different collagens and, unlike the α1I domain of *C. intestinalis* [93], the binding is metal-dependent [17]. Thus, the *P. marinus* sequences are functional members of the collagen recognizing set of integrin αI domains; e.g. Pmaf3 branches as an outlier to both the α10 and α11 I domains (Fig. 1.1) and Pmaf1 and Pmaf2 associate with the collagen recognizing αI domains too (Fig. 1.5). A short sequence fragment, an Expression Sequence Tag from *Eptatretus burgeri*—the inshore hagfish, was also identified in searches and may correspond to the N-terminal portion of an αI domain but the sequence ends just at the junction where the αC helix would have begun if present (Fig. 1.3; [16]. Nonetheless, if that fragment does in fact correspond to part of an αI domain then the sequence is clearly not of the urochordate type, nor is it a member of the collagen-binding clade. Instead, it seems most similar in sequence to the

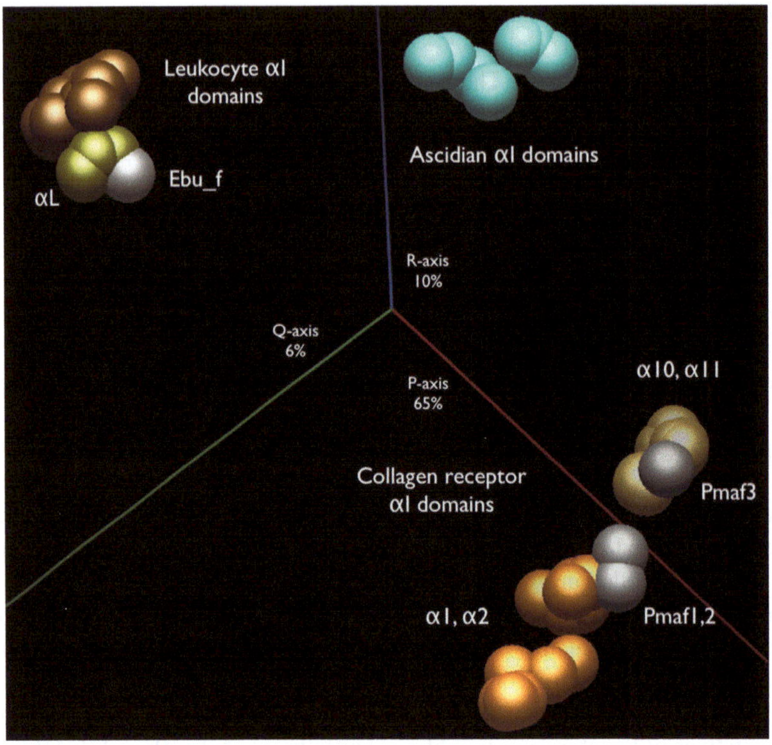

Fig. 1.5 Multivariate analysis showing the mutual relationships among αI domains from human integrins and other vertebrate integrins, ascidians, and fragmentary sequences from lamprey (Pmaf1-3) and hagfish (Ebu_f). Spheres represent individual αI domains and some domains are superimposed at the same location or "behind" other I domains and thus may not be separably visible. Sequences were aligned with Malign [41] and the C program PCA (MS Johnson) was used to compute the three-dimensional projections maximizing the view of the overall variance among the data, generating pseudo-PDB coordinates, displayed and rendered in Bodil [53]. The percentage of the total variance displayed along the P, Q and R axes is shown

leukocyte-binding, human-type integrin αL I domains (Fig. 1.5). Two other C-terminal αI domain fragments were found in searches of sequence data from the genome sequencing of *Callorhinchus milii* (Chondrichthyes; ghost shark, elephant shark). The fragments began at the αC helix junction, having the sequences GSYNR (an α1 orthologue) and GYYNR (an α11 orthologue) of the αC helix, and they correspond closely to human collagen-binding αI domains. Indeed, they represent true orthologues (see Fig. 1.3 for the comparison with the extracts from the full-length sequences) and with the recent published genome data for *C. milii* [94], a large hole in the vertebrate data coverage is plugged; thus, full-length orthologues to the human sequences αE, α1, α2 and α11 are at least present in this cartilaginous fish [17].

1.6 A Nebulous Origin of the Integrin I Domain

The vWFA domain, present as the αI domain is some α subunits and as βI domains in all β subunits, is found in a wide range of other proteins with diverse function (e.g. collagens, complement factors, copines, matrilins, ion channels, protease inhibitors, among others), which are distributed across all of the domains of life [42, 70, 97]. vWFA domains are highly represented in proteins that have roles especially related to processes within the immune and circulatory systems, and are associated with cell-cell and cell-extracellular-matrix (ECM) recognition, being component domains of adhesion molecules at cell membranes as well as proteins of the ECM (see e.g. [18, 19, 97]), among others.

Despite the wealth of information that exists on vWFA domains, it has so far not been possible to establish the likely source for the vWFA domains inserted into the integrin subunits [42, 91]. The branch orders in trees constructed from sequence comparisons of vWFA domains may be robust for similar members within a clade (according to bootstrapping of the sequence alignments), but the relative branching orders among clades are not reliable [42, 91]. The reasons for this is unknown. The dynamic expansion of proteins domains within composite proteins, especially related to extracellular processes, took place in the eukaryotes [61]. One can thus speculate that if multiple vWFA domain duplication events and incorporation into different proteins occurred over a relatively short period of time, for instance with the earliest eukaryotes and perhaps later as the invertebrate chordate line led to the vertebrates, that the similar degree of differences among groups of vWFA domains may make it impossible to resolve the relationships among them because they are all fairly equidistant from each other. Or perhaps the αI domain has arisen from duplication of the βI domain itself, but this appears not to be the case since the βI domain is among the most dissimilar of the vWFA domains in comparison of the αI domain (Fig. 1.6). The vWFA domain is small and even within the closely related human α subunits with I domains, or consider Ciα1 in Fig. 1.6, differences in some branching orders may arise if only the αI domain is compared rather than the full-length integrins having longer sequences and hence higher information content; this effect would likely be magnified when sequences are at even greater distances from each other. Whatever the reason, at the present there is insufficient information in the known sequences to resolve this issue. Interestingly, the sequence "Uncharacterized Protein 2" clusters closely with the ascidian αI domain as does one collagen IV α4 chain (Fig. 1.6).

1.7 A Summary of αI Domain Evolution and the Origin of the Vertebrate αI Domains

The integrin α and β subunits have a long history, likely originating in single-cell eukaryotes and thus predating the rise of the metazoans (Fig. 1.7). Despite the presence of homologues of most of the component domains of integrins within prokaryotes, integrins subunits have not been detected in bacteria. Integrins have also not been identified in non-metazoan multicellular organisms, namely fungi and plants (despite attributing a small protein from *Arabidopsis thaliana* as an integrin-like protein rather than possibly a fibronectin-like domain that might be *recognized by an integrin*; [49].

The earliest observed integrin subunits are in single cell eukaryotes diverging close to the origin of the metazoans [79]. Already in *C. owczarzaki* multiple integrin α and β subunits are observed and this is true for the first metazoans too, e.g. *T. adhaerens* and *A. queenslandica*, suggesting that integrin function assumed multiple roles from the very beginning. In a single species, integrin α subunits of humans frequently outnumber β subunits and thus a single β subunit may have multiple α subunit partners; in humans, for example, the β1 subunit forms dimers with 12 of 18 α subunits.

From the earliest integrins, ligand binding (e.g. [98]) was likely based on interactions at MIDAS of the βI domain and with the β propeller domain of the α subunit. The proteins ligands recognized by these early integrins likely also had short recognition signatures, e.g. RGD, LVD and variants, that were presented on surface loops that could occupy the fairly narrow site between the β propeller of the α subunit and the βI domain of the β subunit. The integrin domain structure then remained quite static in terms of domain structure throughout the invertebrates and into the first invertebrate deuterostomes, e.g. the echinoderms. This is also

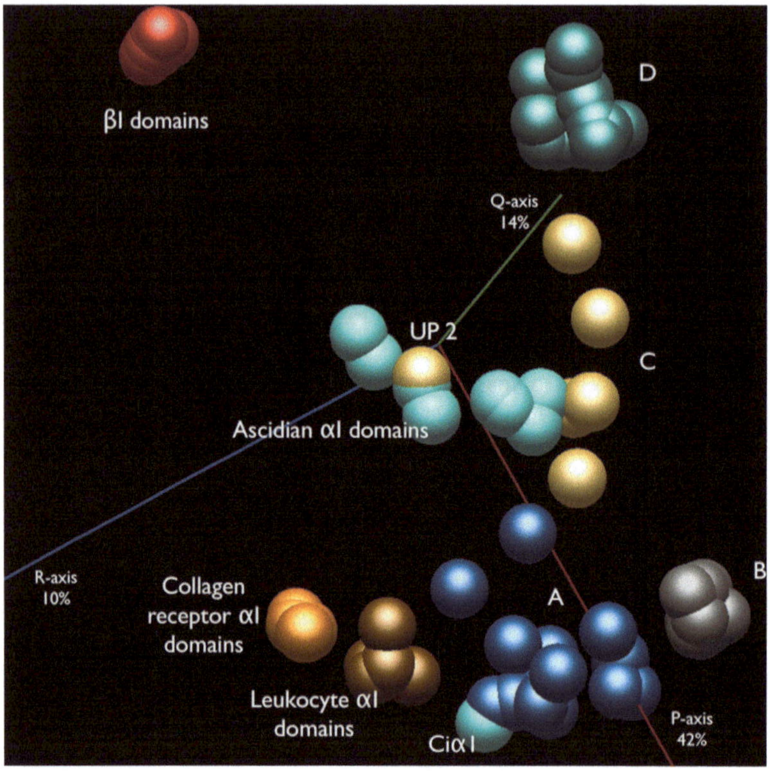

Fig. 1.6 Multivariate analysis showing the mutual relationships among integrin αI and βI domains from human and the ascidians, and other vWFA domains. The integrin βI domains from human and the ascidians cluster together and are the most dissimilar cluster from all others. Ciα1 clusters with a group A, whereas UP2 (Uncharacterized Protein 2) clusters with the Ciona αI domains. *A* Chains from collagen type VI, XII, XXII, UP1, Sushi 2–3, Fibrillins, Cartilage matrix protein. *B* Chains from collagen type XIV, XII; Selectins (E,F,P); Matrillin, Sushi 1,3; Fibrillins; UP3-6; *C* Collagen type VI chain, Anthrax toxin receptor 1, Plasmodium CTRP, Plasmodium micronemal protein. *D* Calcium channels; Trypsin inhibitors; Copines; Midasin: Bacterial Mg^{2+} chelatase; Yeast DNA repair factor; Yeast Proteosome regulatory factor). The comparisons and figure were generated as in Fig. 1.5

true for the lancelets. The cephalochordates are now accepted to have diverged prior to the tunicates and are the earliest diverging extant chordate subphylum.

In an ancestor of the urochordates, a key change in an integrin α subunit gene led to a major alteration of the protein structure and function of some integrin heterodimers: A vWFA-type domain was inserted within a loop at the surface of the β-propeller domain of the α subunit. The urochordates remain the earliest diverging extant species found to have αI domains, with one in *H. roretzi* and eight in *C. intestinalis*. The nine integrins α subunits with I domains for their own clade are clearly not of the collagen or leukocyte types seen in humans and other vertebrates. The α1 I domain of *C. intestinalis* has been tested for collagen binding and, unlike the human collagen binding set, does not bind to fibril forming collagens I–V nor to GFOGER-like peptides, but it does bind strongly to collagen IX through a mechanism that is metal and MIDAS independent [93]. MIDAS-independent binding occurs with other vWFA domains: e.g. vWFA3 [7, 35, 74], known to bind collagen I and III, and vWFA1 [13], which binds platelet glycoprotein 1b alpha.

The αI domain thus appeared early in chordate evolution and within an invertebrate. This inserted domain relocated the integrin external

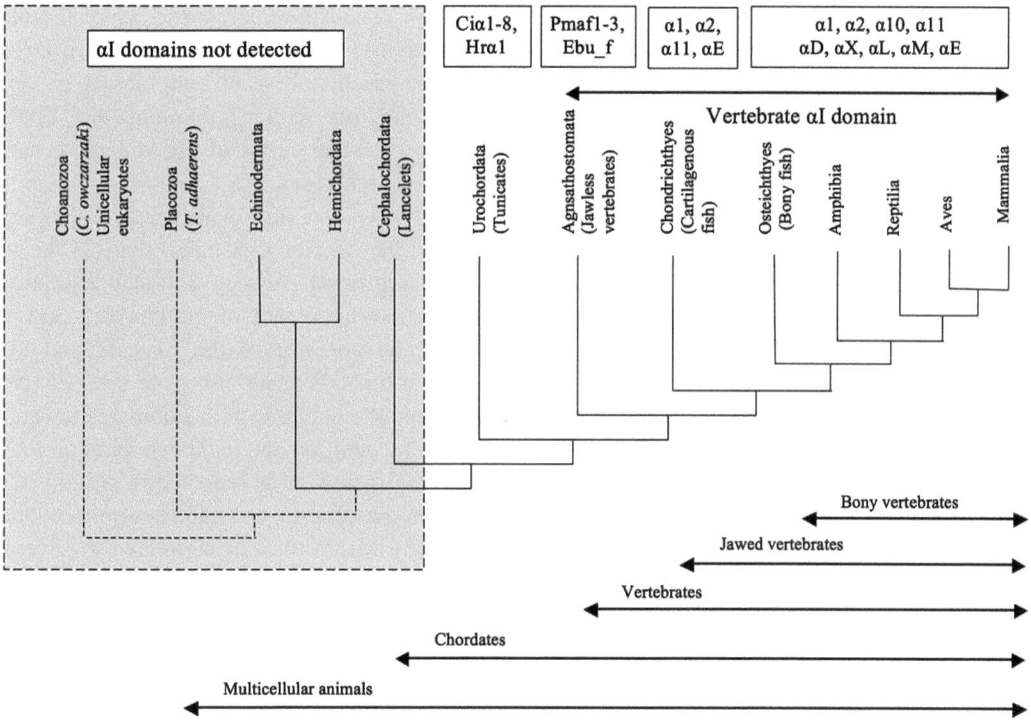

Fig. 1.7 The spectrum of identified integrin and αI domain sequences. The earliest diverging species with identifiable integrin sequences are in the single-cell eukaryotes, phylum choanozoa, e.g. *C. owczarzaki*. Multiple α and β subunits found in the choanozoan are typically found in the metazoans and the number of α and β subunits generally increases for species with later divergence times too, and especially within the chordates. βI domains are an essential part of the heterodimeric structure and are found in all β subunits. αI domains have not been detected among non-deuterostome invertebrates, nor in echinoderms and the earliest diverging chordate invertebrate, the lancelet. The earliest diverging species having integrin α subunits with I domains are the invertebrate urochordates, but these α subunits form a clade distinct from the αI domains found in the vertebrates. Fragments of α subunits with I domains are found in lamprey and possibly hagfish and they appear most similar to sequences from the collagen-binding group and the leukocyte group found in other vertebrates, including the elephant shark, bony fish and other vertebrates through to humans

ligand binding site away from the β propeller—βI domain interfacial cleft to location on the αI domain, which itself budded out from the β propeller. The αI domain in humans make use of a key feature of many vWFA domains, including the βI domain; that is, MIDAS for binding negatively-charged amino acids at the positively-charged metal ion located at the site. But, it is not clear what mechanisms are employed by the ascidian tunicate αI domains for ligand recognition. Nevertheless, MIDAS is conserved and the majority of the known ascidian sequences contain the intrinsic glutamate ligand involved in triggering the receptor activation mechanism (Fig. 1.3). A major advantage of the αI domain in integrins is high solvent exposure near MIDAS, meaning that recognized proteins no longer needed flexible loop regions to snake into a binding cleft. Instead, larger and more bulky surfaces and fibrils could now directly interact with the integrins. Thus, we see that glutamate within collagen-like triple helical peptides bind to MIDAS and, similarly for ICAMs, glutamate extends from the end of a beta-strand at the immunoglobulin fold surface and binds to MIDAS at the αI domain.

In humans there are nine integrin α subunits with I domains; four are within the collagen-binding set and five belong to the leukocyte-specific clade. Orthologues extend across the mammals, birds, reptiles, amphibians, and bony fish—the latter often having duplicate isoforms. The cartilaginous fish also have orthologues of human-type α subunits with I domains. Preliminary data now exist that the earliest diverging extant vertebrates have human-like integrins with αI domains too, since lamprey and perhaps hagfish appear to have integrin αI sequences that are respectively most similar to the collagen binding (i.e. α1, α2, α10 and α11) and immune cell recognizing (i.e. αD, αE, αL, αM, αE) integrin α subunits.

The I domain in the integrin α subunit may have helped to facilitate or accommodate the large scale changes and stresses that accompanied the rise and diversification of chordates, and the added complications of expanded interrelated physiological systems and specific functional organs and tissues. The incorporation of the αI domains into some α subunits, no matter how this occurred, did provide the chordates with a broad spectrum of tools for recognizing the extracellular matrix and other cell types, and allowed cells to deal with a wider range of complications associated with the multicellular complexity of the rapidly expanding vertebrate line.

A key defining feature of the vertebrates is cartilage and bone, and collagen receptors with high avidity may have been necessary for their development as well as for other tissues [31]. The skeletal system is also tightly interconnected to the immune and circulatory system, two other systems where major changes also took place during chordate and especially vertebrate evolution. In the cyclostomes, e.g. lamprey, the circulatory system has fibrinogen-based blood clotting [23], and integrins with αI domains are key receptors involved in e.g. vertebrate platelet aggregation. The vertebrate adaptive immune system relies heavily on I domain containing integrins. In the ascidian H. roretzi, αHr1 is associated with the innate complement system of defense, whereas lamprey and hagfish had developed a unique adaptive immune system preceding that seen in higher vertebrates [66, 106, 46]. The interrelationships between these systems for support, defense, and mediation of nutrition and waste removal from distant tissues would have benefited from a expanded functional set of both classes of vertebrate integrins with αI domains that could recognise a wider array of ECM ligands and cell-surface receptors.

Finally, we can state with relative certainty based on the data now at hand, that [1] the αI domain originated early in chordate evolution but after the divergence of the lancelets and in an common ancestor of the tunicate and the vertebrate lines. We can also state that [2] specialization of α subunits with I domains towards collagen recognition and leukocyte binding took place soon thereafter as reflected in preliminary data from the lamprey and hagfish genomes; and [3] already within the cartilaginous fish several orthologues of the human type integrin α subunits with I domains are identifiable and thus, given the range of observations, the collagen-binding integrin clade and the leukocyte clade are characteristics not just of humans and other later diverging vertebrates, but of the vertebrates as a whole (Fig. 1.7) including the agnathostomes—they are the vertebrate αI domains.

Acknowledgments We thank the agencies that have contributed funding to our studies on integrins, including the Foundation of Åbo Akademi University (Center of Excellence in Cell Stress and Aging), the Sigrid Juselius Foundation, and the Joe, Pentti and Tor Borg Memorial Fund to MSJ; and the National Doctoral Program in Informational and Structural Biology to BC. We gratefully acknowledge computational resources from the Biocenter Finland bioinformatics infrastructure network (Academy of Finland) and CSC IT Center for Science.

References

1. Adams MD, Celniker SE, Holt RA, Evans CA, Gocayne JD, Amanatides PG, Scherer SE, Li PW, Hoskins RA, Galle RF, George RA, Lewis SE, Richards S, Ashburner M, Henderson SN, Sutton GG, Wortman JR, Yandell MD, Zhang Q, Chen LX, Brandon RC, Rogers YH, Blazej RG, Champe M, Pfeiffer BD, Wan KH, Doyle C, Baxter EG, Helt G, Nelson CR, Gabor GL, Abril JF, Agbayani A, An HJ, Andrews-Pfannkoch C, Baldwin D, Ballew RM, Basu A, Baxendale J, Bayraktaroglu L, Beasley EM, Beeson KY, Benos PV, Berman BP, Bhandari D, Bolshakov S, Borkova D, Botchan MR, Bouck J,

Brokstein P, Brottier P, Burtis KC, Busam DA, Butler H, Cadieu E, Center A, Chandra I, Cherry JM, Cawley S, Dahlke C, Davenport LB, Davies P, de Pablos B, Delcher A, Deng Z, Mays AD, Dew I, Dietz SM, Dodson K, Doup LE, Downes M, Dugan-Rocha S, Dunkov BC, Dunn P, Durbin KJ, Evangelista CC, Ferraz C, Ferriera S, Fleischmann W, Fosler C, Gabrielian AE, Garg NS, Gelbart WM, Glasser K, Glodek A, Gong F, Gorrell JH, Gu Z, Guan P, Harris M, Harris NL, Harvey D, Heiman TJ, Hernandez JR, Houck J, Hostin D, Houston KA, Howland TJ, Wei MH, Ibegwam C, Jalali M, Kalush F, Karpen GH, Ke Z, Kennison JA, Ketchum KA, Kimmel BE, Kodira CD, Kraft C, Kravitz S, Kulp D, Lai Z, Lasko P, Lei Y, Levitsky AA, Li J, Li Z, Liang Y, Lin X, Liu X, Mattei B, McIntosh TC, McLeod MP, McPherson D, Merkulov G, Milshina NV, Mobarry C, Morris J, Moshrefi A, Mount SM, Moy M, Murphy B, Murphy L, Muzny DM, Nelson DL, Nelson DR, Nelson KA, Nixon K, Nusskern DR, Pacleb JM, Palazzolo M, Pittman GS, Pan S, Pollard J, Puri V, Reese MG, Reinert K, Remington K, Saunders RD, Scheeler F, Shen H, Shue BC, Sidén-Kiamos I, Simpson M, Skupski MP, Smith T, Spier E, Spradling AC, Stapleton M, Strong R, Sun E, Svirskas R, Tector C, Turner R, Venter E, Wang AH, Wang X, Wang ZY, Wassarman DA, Weinstock GM, Weissenbach J, Williams SM, Woodage T, Worley KC, Wu D, Yang S, Yao QA, Ye J, Yeh RF, Zaveri JS, Zhan M, Zhang G, Zhao Q, Zheng L, Zheng XH, Zhong FN, Zhong W, Zhou X, Zhu S, Zhu X, Smith HO, Gibbs RA, Myers EW, Rubin GM, Venter JC (2002) The genome sequence of Drosophila melanogaster. Science 287:2185–2195

2. Alonso JL, Essafi M, Xiong JP, Stehle T, Arnaout MA (2002) Does the integrin alphaA domain act as a ligand for its betaA domain? Curr Biol 12:R340–R342
3. Aparicio S, Chapman J, Stupka E, Putnam N, Chia JM, Dehal P, Christoffels A, Rash S, Hoon S, Smit A, Gelpke MD, Roach J, Oh T, Ho IY, Wong M, Detter C, Verhoef F, Predki P, Tay A, Lucas S, Richardson P, Smith SF, Clark MS, Edwards YJ, Doggett N, Zharkikh A, Tavtigian SV, Pruss D, Barnstead M, Evans C, Baden H, Powell J, Glusman G, Rowen L, Hood L, Tan YH, Elgar G, Hawkins T, Venkatesh B, Rokhsar D, Brenner S (2002) Whole-genome shotgun assembly and analysis of the genome of Fugu rubripes. Science 297:1301–1310
4. Arnaout MA (1990) Structure and function of the leukocyte adhesion molecules CD11/CD18. Blood 75(5):1037–1050
5. Arnaout MA, Goodman SL, Xiong JP (2007) Structure and mechanics of integrin-based cell adhesion. Curr Opin Cell Biol 19:495–507
6. Bergelson JM, St. John NF, Kawaguchi S, Pasqualini R, Berdichevsky F, Hemier ME, Finberg RW (1994) The I domain is essential for echovirus 1 interaction with VLA-2. Cell Adhes Commun 2:455–464
7. Bienkowska J, Cruz M, Atiemo A, Handin R, Liddington R (1997) The von willebrand factor A3 domain does not contain a metal ion-dependent adhesion site motif. J Biol Chem 272:25162–25167
8. Blair JE, Hedges SB (2005) Molecular phylogeny and divergence times of deuterostome animals. Mol Biol Evol 22(11):2275–2284
9. Brower DL, Brower SM, Hayward DC, Ball EE (1997) Molecular evolution of integrins: genes encoding integrin α subunits from a coral and a sponge. Proc Natl Acad Sci USA 94:9182–9187
10. Burke RD (1999) Invertebrate integrins: structure, function and evolution. Int Rev Cytol 191:257–284
11. Bökel C, Brown NH (2002) Integrins in development: moving on, responding to, and sticking to the extracellular matrix. Dev Cell 3:311–321
12. Carrell NA, Fitzgerald LA, Steiner B, Erickson HP, Phillips DR (1985) Structure of human platelet membrane glycoprotein IIb and IIIa as determined by electron microscopy. J Biol Chem 260(3):1743–1749
13. Celikel R, Varughese KI, Madhusudan, Yoshioka A, Ware J, Ruggeri ZM (1998) Crystal structure of the von Willebrand factor A1 domain in complex with the function blocking NMC-4 Fab. Nat Struct Biol 5:189–194
14. Chin YK, Headley SJ, Mohanty B, Patil R, McEwan PA, Swarbrick JD, Mulhern TD, Emsley J, Simpson JS, Scanion MJ (2013) The structure of integrin α1I domain in complex with collagen-mimetic peptide. J Biol Chem 288(52):36796–36809
15. Chouhan B, Denesyuk A, Heino J, Johnson MS, Denessiouk K (2011) Conservation of the human-type integrin beta propeller domain in bacteria. PLoS 6(10):e25069
16. Chouhan B, Denesyuk A, Heino J, Johnson MS, Denessiouk K (2012) Evolutionary origin of the alpha C helix in integrins. WASET 65:546–549
17. Chouhan B, Käpylä J, Denesyuk A, Denessiouk K, Heino J, Johnson MS. unpublished
18. Colombatti A, Bonaldo P (1991) The superfamily of proteins with von Willebrand factor type A-like domains: one theme common to components of extracellular matrix, hemostasis, cellular adhesion, and defense mechanisms. Blood 77:2305–2315
19. Colombatti A, Bonaldo P, Doliana R (1993) Type A modules: interacting domains found in several non-fibrillar collagens and in other extracellular matrix proteins. Matrix 13:297–306
20. DeSimone DW, Hynes RO (1988) Xenopus laevis integrins. Structure and evolutionary divergence of the β subunits. J Biol Chem 163:5333–5340
21. Dehal P, Satou Y, Campbell RK, Chapman J, Degnan B, De Tomaso A, Davidson B, Di Gregorio A, Gelpke M, Goodstein DM, Harafuji N, Hastings KE, Ho I, Hotta K, Huang W, Kawashima T, Lemaire P, Martinez D, Meinertzhagen IA, Necula S, Nonaka M, Putnam N, Rash S, Saiga H, Satake M, Terry A, Yamada L, Wang HG, Awazu S,

Azumi K, Boore J, Branno M, Chin-Bow S, DeSantis R, Doyle S, Francino P, Keys DN, Haga S, Hayashi H, Hino K, Imai KS, Inaba K, Kano S, Kobayashi K, Kobayashi M, Lee BI, Makabe KW, Manohar C, Matassi G, Medina M, Mochizuki Y, Mount S, Morishita T, Miura S, Nakayama A, Nishizaka S, Nomoto H, Ohta F, Oishi K, Rigoutsos I, Sano M, Sasaki A, Sasakura Y, Shoguchi E, Shin-i T, Spagnuolo A, Stainier D, Suzuki MM, Tassy O, Takatori N, Tokuoka M, Yagi K, Yoshizaki F, Wada S, Zhang C, Hyatt PD, Larimer F, Detter C, Doggett N, Glavina T, Hawkins T, Richardson P, Lucas S, Kohara Y, Levine M, Satoh N, Rokhsar DS (2002) The draft genome of Ciona intestinalis: insights into chordate and vertebrate origins. Science 298:2157–2167

22. Donoghue PCJ, Purnell MA (2005) Genome duplication, extinction and vertebrate evolution. Trends Ecology Evol. 20:312–319
23. Doolittle RF (1976) The evolution of vertebrate fibrinogen. Fed Proc 35(10):2145–2149
24. Doolittle RF, Feng D-F, Tsang S, Cho G, Little E (1996) Determining divergence times of the major kingdoms of living organisms with a protein clock. Science 271:470–477
25. Dunn CW, Hejnol A, Matus DQ, Pang K, Browne WE, Smith SA, Seaver E, Rouse GW, Obst M, Edgecombe GD, Sørensen MV, Haddock SHD, Schmidt-Rhaesa A, Okusu A, Kristensen RM, Wheeler WC, Martindale MQ, Giribet G (2008) Broad phylogenetic sampling improves resolution of the animal tree of life. Nature 452:745–750
26. Eble JA, Kühn K, eds (1997) Integrin-ligand interactions. Chapman and Hall, New York, pp 1–273
27. Emsley J, King SL, Bergelson JM, Liddington R (1997) Crystal structure of the I domain from integrin alpha2beta1. J Biol Chem 272:28512–28517
28. Emsley J, Knight CG, Farndale RW, Barnes MJ, Liddington RC (2000) Structural basis for collagen recognition by integrin $\alpha2\beta1$. Cell 101:47–56
29. Ewan R, Huxley-Jones J, Mould AP, Humphries MJ, Robertson DL, Boot-Handford RP (2005) The integrins of the urochordate Ciona intestinalis provide novel insights into molecular evolution of the vertebrate integrin family. BMC Evol Biol 5:31
30. Flemming JC, Pahl HL, Gonzalez DA, Smith TF, Tenen DG (1993) Structural analysis of the CD11b gene and phylogenetic analysis of the alpha-integrin gene family demonstrate remarkable conservation of genomic organization and suggests early diversification during evolution. J Immunol 150(2):480–490
31. Heino J, Huhtala M, Käpylä J, Johnson MS (2009) Evolution of collagen-based adhesion systems. Int J Biochem Cell Biol (Darwin Special Issue) 41(2):341–348
32. Hughes AL (1992) Coevolution of the vertebrate α- and β-chain genes. Mol Biol Evol 9:216–234
33. Hughes AL (2001) Evolution of the integrin α and β protein families. J Mol Evol 52:63–72
34. Huhtala M, Heino J, Casciari D, de Luise A, Johnson MS (2005) Integrin evolution: insights from ascidian and teleost fish genomes. Matrix Biol 24:83–95
35. Huizinga EG, Martijn van der Plas R, Kroon J, Sixma JJ, Gros P (1997) Crystal structure of the A3 domain of human von Willebrand factor: implications for collagen binding. Structure 5:1147–1156
36. Hynes RO (2002) Integrins: bidirectional, allosteric signaling machines. Cell 110(6):673–687
37. Hynes RO, Zhao Q (2000) The evolution of cell adhesion. J Cell Biol 150:F89–F95
38. Ivaska J, Käpylä J, Pentikäinen O, Hoffren A-M, Hermonen J, Huttunen P, Johnson MS, Heino J (1999) A peptide inhibiting the collagen binding function of integrin alpha2I domain. J Biol Chem 274:3513–3521
39. Jaillon O, Aury JM, Brunet F, Petit JL, Stange-Thomann N, Mauceli E, Bouneau L, Fischer C, Ozouf-Costaz C, Bernot A, Nicaud S, Jaffe D, Fisher S, Lutfalla G, Dossat C, Segurens B, Dasilva C, Salanoubat M, Levy M, Boudet N, Castellano S, Anthouard V, Jubin C, Castelli V, Katinka M, Vacherie B, Biémont C, Skalli Z, Cattolico L, Poulain J, De Berardinis V, Cruaud C, Duprat S, Brottier P, Coutanceau JP, Gouzy J, Parra G, Lardier G, Chapple C, McKernan KJ, McEwan P, Bosak S, Kellis M, Volff JN, Guigó R, Zody MC, Mesirov J, Lindblad-Toh K, Birren B, Nusbaum C, Kahn D, Robinson-Rechavi M, Laudet V, Schachter V, Quétier F, Saurin W, Scarpelli C, Wincker P, Lander ES, Weissenbach J, Roest Crollius H (2004) Genome duplication in the teleost fish Tetraodon nigroviridis reveals the early vertebrate proto-karyotype. Nature 431:946–957
40. Johnson MS, Lu N, Denessiouk K, Heino J, Gullberg D (2009) Integrins during evolution: evolutionary trees and model organisms. BBA 1788(4):779–789
41. Johnson MS, Overington JP (1993) A structural basis for the comparison of sequences: an evaluation of scoring methodologies. J Mol Biol 233:716–738
42. Johnson MS, Tuckwell D (2003) Evolution of integrin I-domains: I domains. In: Gullberg D (ed) Integrins. Landes Bioscience, Texas, pp 1–26
43. Johnson MS, Käpylä J, Dnessiouk K, Airenne T, Heino J (2013) Evolution of cell adhesion to cellular matrix. In: Keeley F, Mecham RP (eds) Evolution of extracellular matrix, Springer, Berlin, Heidelberg
44. Jokinen J, White DJ, Salmela M, Huhtala M, Käpylä J, Sipilä K, Puranen JS, Nissinen L, Kankaanpää P, Marjomäki V, Hyypiä T, Johnson MS, Heino J (2010) Molecular mechanism of $\alpha2\beta1$ integrin interaction with human echovirus 1. EMBO J 29:196–208
45. Kallen J, Welzenbach K, Ramage P, Geyl D, Kriwacki R, Legge G, Cottens S, Weitz-Schmidt G,

Hommel U (1999) Structural basis for LFA-1 inhibition upon lovastatin binding to the CD11a I-domain. J Mol Biol 292:1–9
46. Kasahara M, Sutoh Y (2014) Two forms of adaptive immunity in vertebrates: similarities and differences. Adv Immunol 122:59–90
47. King SL, Kamata T, Cunningham JA, Emsley J, Liddington RC, Takada Y, Bergelson JM (1997) Echovirus 1 interaction with the human very late antigen-2 (integrin alpha2beta1) I domain. Identification of two independent virus contact sites distinct from the metal ion-dependent adhesion site. J Biol Chem 272:28518–28522
48. Knack BA, Iguchi A, Shinzato C, Hayward DC, Ball EE, Miller DJ (2008) Unexpected diversity of cnidarian integrins: expression during coral gastrulation. BMC Evol Biol 8:136
49. Knepper C, Savory EA, Day B (2011) Arabidopsis NDR1 is an integrin-like protein with a roles in fluid loss and plasma membrane-cell wall adhesion. Plant Physiol 156:286–300
50. Larson RS, Corbi AL, Berman L, Springer T (1989) Primary structure of the leukocyte function-associated molecule-1 α subunit: an integrin with an embedded domain defining a protein superfamily. J Cell Biol 108:703–712
51. Lee J-O, Bankston LA, Arnaout MA, Liddington R (1995) C. Two conformations of the integrin A-domain (I-domain): a pathway for activation? Structure 3:1333–1340
52. Lee J-O, Rieu P, Arnaout MA, Liddington R (1995) Crystal structure of the A domain from the alpha subunit of integrin CR3. Cell 80:631–638
53. Lehtonen JV, Still D-J, Rantanen V-V, Ekholm J, Björklund D, Iftikhar Z, Huhtala M, Repo S, Jussila A, Jaakkola J, Pentikäinen O, Nyrönen T, Salminen T, Gyllenberg M, Johnson MS (2004) BODIL: a molecular modeling environment for structure-function analysis and drug design. J Comp Aided Molec Des 18:401–419
54. Luo B-H, Carman CV, Springer TA (2007) Structural basis of integrin regulation and signaling. Annu Rev Immunol 25:619–647
55. Luo B-H, Springer TA (2006) Integrin structures and conformational signaling. Curr Opin Cell Biol 18:579–586
56. Mischishita M, Videm V, Arnaout MA (1993) A novel divalent cation-binding site in the A domain of the β2 integrin CR3 (CD11b/CD18) is essential for ligand binding. Cell 72(6):857–867
57. Miyazawa S, Azumi K, Nonaka M (2001) Cloning and characterization of integrin alpha subunits from the solitary ascidian, *Halocynthia roretzi*. J Immuno 166:1710–1715
58. Mould AP, Humphries MJ (2004) Regulation of integrin function through conformational complexity: not simply a knee-jerk reaction? Curr Opin Cell Biol 16:544–551
59. Müller WE (1997) Origin of metazoan adhesion molecules and adhesion receptors as deduced from cDNA analyses in the marine sponge Geodia cydonium: a review. Cell Tissue Res 289:383–395
60. Nagae M, Re S, Mihara E, Nogi T, Sugita Y, Takagi J (2012) Crystal structure of α5β1 integrin ectodomain: atomic details of the fibronectin receptor. J Cell Biol 197:131–140
61. Nasir A, Kim KM, Caetano-Anollés G (2014) Global patterns of protein domain gain and loss in superkingdoms. PLoS Comput Biol 10(1):e1003452
62. Nermut MV, Green NM, Eason P, Yamada SS, Yamada KM (1988) Electron microscopy and structural model of human fibronectin receptor. EMBO J 7(13):4093–4099
63. Nichols SA, Dirks W, Pearse JS, King N (2006) Early evolution of animal cell signaling and adhesion genes. Proc Natl Acad Sci USA 103(33):12451–12456
64. Nymalm Y, Puranen JS, Nyholm TKM, Käpylä J, Kidron H, Pentikäinen O, Airenne TT, Heino J, Slotte P, Johnson MS, Salminen TA (2004) Jararhagin-derived "RKKH"-peptides induce structural changes in α1 I domain of human integrin α1β1. J Biol Chem 279:7962–7970
65. Ota KG, Kuratani S (2007) Cyclostome embryology and early evolutionary history of vertebrates. Integr Comp Biol 47(3):329–337
66. Pancer Z, Cooper MD (2006) The evolution of adaptive immunity. Annu Rev Immunol 24:497–518
67. Pancer Z, Kruse M, Müller I, Müller WE (1997) On the origin of Metazoan adhesion receptors: cloning of integrin α subunit from the sponge Geodia cydonium. Mol Biol Evol 14:391–398
68. Pentikäinen O, Hoffrén A-M, Ivaska J, Käpylä J, Nyrönen T, Heino J, Johnson MS (1999) RKKH peptides from the snake venom metalloproteinase of Bothrops jararaca bind near the MIDAS site of the human integrin α2I -domain. J Biol Chem 274:31493–31505
69. Philippe H, Lartillot N, Brinkmann H (2005) Multigene analysis of bilaterian animals corroborate the monophyly of Ecdysozoa, Lophotrochozoa, and Protostomia. Mol Biol Evol 22(5):1246–1253
70. Ponting CP, Aravind L, Schultz J, Bork P, Koonin EV (1999) Eukaryotic signalling domain homologues in archaea and bacteria. Ancient ancestry and horizontal gene transfer. J Mol Biol 289:729–745
71. Putnam NH, Butts T, Ferrier DE, Furlong RF, Hellsten U, Kawashima T, Robinson-Rechavi M, Shoguchi E, Terry A, Yu JK, Benito-Gutiérrez EL, Dubchak I, Garcia-Fernández J, Gibson- Brown JJ, Grigoriev IV, Horton AC, de Jong PJ, Jurka J, Kapitonov VV, Kohara Y, Kuroki Y, Lindquist E, Lucas S, Osoegawa K, Pennacchio LA, Salamov AA, Satou Y, Sauka-Spengler T, Schmutz J, Shin-I T, Toyoda A, Bronner-Fraser M, Fujiyama A, Holland LZ, Holland PW, Satoh N, Rokhsar DS (2008) The amphioxus genome and the evolution of the chordate karyotype. Nature 453:1064–1071

72. Qu A, Leahy DJ (1995) Crystal structure of the I-domain from the CD11a/CD18 (LFA-1, alpha L beta 2) integrin. Proc Natl Acad Sci USA 92:10277–10281
73. Reber-Muller S, Studer R, Muller P, Yanze N, Schmid V (2001) Integrin and talin in the jellyfish Podocoryne carnea. Cell Biol Int 25:753–769
74. Romijn RAP, Bouma B, Wuyster W, Gros P, Kroon J, Sixma JJ, Huizinga EG (2001) Identification of the collagen-binding site of the von Willebrand factor A3-domain. J Biol Chem 276:9985–9991
75. Rubin GM, Yandell MD, Wortman JR, Gabor Miklos GL, Nelson CR, Hariharan IK, Fortini ME, Li PW, Apweiler R, Fleischmann W, Cherry JM, Henikoff S, Skupski MP, Misra S, Ashburner M, Birney E, Boguski MS, Brody T, Brokstein P, Celniker SE, Chervitz SA, Coates D, Cravchik A, Gabrielian A, Galle RF, Gelbart WM, George RA, Goldstein LS, Gong F, Guan P, Harris NL, Hay BA, Hoskins RA, Li J, Li Z, Hynes RO, Jones SJ, Kuehl PM, Lemaitre B, Littleton JT, Morrison DK, Mungall C, O'Farrell PH, Pickeral OK, Shue C, Vosshall LB, Zhang J, Zhao Q, Zheng XH, Lewis S (2000) Comparative genomics of the eukaryotes. Science 287:2204–2215
76. Salminen T, Nymalm Y, Kankare J, Käpylä J, Heino J, Johnson MS (1999) Large-scale purification and preliminary x-ray analysis of the human integrin α1I domain. Acta Crystallogr D55:1365–1367
77. Sasakura Y, Shoguchi E, Takatori N, Wada S, Meinertzhagen IA, Satou Y, Satoh N (2003) A genomewide survey of developmentally relevant genes in Ciona intestinalis. X. Genes for cell junctions and extracellular matrix. Dev Genes Evol 213:303–313
78. Schierwater B, Eitel M, Jakob W, Osigus HJ, Hadrys H, Dellaporta SL, Kolokotronis SO, Desalle R (2009) Concatenated analysis sheds light on early metazoan evolution and fuels a modern "urmetazoon" hypothesis. PLoS Biol 7:e20
79. Sebé-Pedrós A, Roger AJ, Lang FB, King N, Ruiz-Trillo I (2010) Ancient origin of the integrin-mediated adhesion and signaling machinery. Proc Natl Acad Sci USA 107:10142–10147
80. Shalchian-Tabrizi K, Minge MA, Espelund M, Orr R, Ruden TA, Jakobsen KS, Cavalier-Smith T (2008) Multigene phylogeny of choanozoa and the origin of animals. PLoS ONE 3(5):e2098
81. Song G, Yang Y, Liu JH, Casasnovas JM, Shimaoka M, Springer TA, Wang JH (2005) An atomic resolution view of ICAM recognition in a complex between the binding domains of ICAM-3 and integrin alphaLbeta2. Proc Natl Acad Sci USA 102:3366–3371
82. Springer TA, Zhu J, Xiao T (2008) Structural basis for distinctive recognition of fibrinogen gammaC peptide by the platelet integrin alphaIIbbeta3. J Cell Biol 182(4):791–800
83. Srivastava M, Begovic E, Chapman J, Putnam NH, Hellsten U, Kawashima T, Kuo A, Mitros T, Salamov A, Carpenter ML, Signorovitch AY, Moreno MA, Kamm K, Grimwood J, Schmutz J, Shapiro H, Grigoriev IV, Buss LW, Schierwater B, Dellaporta SL, Rokhsar DS (2008) The Trichoplax genome and nature of placozoans. Nature 454:955–960
84. Srivastava M, Simakov O, Chapman J, Fahey B, Gauthier ME, Mitros T, Richards GS, Conaco C, Dacre M, Hellsten U, Larroux C, Putnam NH, Stanke M, Adamska M, Darling A, Degnan SM, Oakley TH, Plachetzki DC, Zhai Y, Adamski M, Calcino A, Cummins SF, Goodstein DM, Harris C, Jackson DJ, Leys SP, Shu S, Woodcroft BJ, Vervoort M, Kosik KS, Manning G, Degnan BM, Rokhsar DS (2010) The Amphimedon queenslandica genome and the evolution of animal complexity. Nature 466:720–726
85. Stach T (2008) Chordate phylogeny and evolution: a not so simple three-taxon problem. J Zool 276:117–141
86. Suga H, Chen Z, de Mendoza A, Sebé-Pedrós A, Brown MW, Kramer E, Carr M, Kerner P, Vervoort M, Sánchez-Pons N, Torruella G, Derelle R, Manning G, Lang BF, Russ C, Haas BJ, Roger AJ, Nusbaum C, Ruiz-Trillo I (2013) The Capsaspora genome reveals a complex unicellular prehistory of animals. Nature Commun 4:2325
87. Takada Y, Ye X, Simon S (2007) The integrins. Genome Biol 8:215
88. Takagi J (2007) Structural basis for ligand recognition by integrins. Curr Opin Cell Biol 19:557–564
89. Taylor JS, de Peer YV, Braasch I, Meyer A (2001) Comparative genomics provides evidence for an ancient genome duplication event in fish. Phil Trans R Soc Lond B 356:1661–1679
90. Tozer EC, Liddington RC, Sutcliffe MJ, Smeeton AH, Loftus JC (1996) Ligand binding to integrin alphaIIbbeta3 is dependent on a MIDAS-like domain in the beta3 subunit. J Biol Chem 271:21978–21984
91. Tuckwell D (1999) Evolution of von Willebrand factor A (VWA) domains. Biochem Soc Trans 27:835–840
92. Tuckwell D, Humphries MJ (1997) A structure prediction for the ligand-binding region of the integrin beta subunit: evidence for the presence of a von Willebrand factor A domain. FEBS Lett 400:297–303
93. Tulla M, Huhtala M, Jäälinoja J, Käpylä J, Farndale RW, Ala-Kokko L, Johnson MS, Heino J (2007) Analysis of an ascidian integrin provides new insight into early evolution of collagen recognition. FEBS Lett 581:2434–2440
94. Venkatesh B, Lee AP, Ravi V, Maurya AK, Lian MM, Swann JB, Ohta Y, Flajnik MF, Sutoh Y, Kasahara M, Hoon S, Gangu V, Roy SW, Irimia M,

Korzh V, Kondrychyn I, Lim ZW, Tay BH, Tohari S, Kong KW, Ho S, Lorente-Galdos B, Quilez J, Marques-Bonet T, Raney BJ, Ingham PW, Tay A, Hillier LW, Minx P, Boehm T, Wilson RK, Brenner S, Warren WC (2014) Elephant shark genome provides unique insights into gnathostome evolution. Nature 505:174–179
95. Vorup-Jensen T, Ostermeier C, Shimaoka M, Hommel U, Springer TA (2003) Structure and allosteric regulation of the alpha X beta 2 integrin I domain. Proc Natl Acad Sci USA 100:1873–1878
96. Whittaker CA, DeSimone DW (1993) Integrin alpha subunit mRNAs are differentially expressed in early Xenopus embryos. Development 117:1239–1249
97. Whittaker CA, Hynes RO (2002) Distribution and evolution of von Willebrand/integrin A domains: widely dispersed domains with roles in cell adhesion and elsewhere. Mol Biol Cell 13:3369–3387
98. Wimmer W, Perovic S, Kruse M, Schröder HC, Krasko A, Batel R, Müller WE (1999) Origin of the integrin-mediated signal transduction. Functional studies with cell cultures from the sponge Suberites domuncula. Eur J Biochem 260:156–165
99. Xiao T, Takagi J, Coller BS, Wang JH, Springer TA (2004) Structural basis for allostery in integrins and binding to fibrinogen-mimetic therapeutics. Nature 432:59–67
100. Xie C, Zhu J, Chen X, Mi L, Nishida N, Springer TA (2010) Structure of an integrin with an alpha I domain, complement receptor type 4. EMBO J 29:666–679
101. Xing L, Huhtala M, Pietiäinen V, Käpylä J, Vuorinen K, Marjomäki V, Heino J, Johnson MS, Hyypiä T, Cheng RH (2004) Structural and functional analysis of integrin α2I domain interaction with echovirus 1. J Biol Chem 279:11632–11638
102. Xiong J-P, Stehle T, Diefenbach B, Zhang R, Dunker R, Scott DL, Joachimiak A, Goodman SL, Arnaout MA (2001) Crystal structure of the extracellular segment of integrin $\alpha V\beta 3$. Science 294:339–345
103. Xiong J-P, Stehle T, Goodman SL, Arnaout MA (2003) New insights into the structural basis of integrin activation. Blood 102(4):1155–1159
104. Xiong JP, Stehle T, Zhang R, Joachimiak A, Frech M, Goodman SL, Arnaout MA (2002) Crystal structure of the extracellular segment of integrin $\alpha V\beta 3$ in complex with an Arg-Gly- Asp ligand. Science 296:151–155
105. Yu Y, Zhu J, Mi LZ, Walz T, Sun H, Chen J, Springer TA (2012) Structural specializations of $\alpha(4)\beta(7)$, an integrin that mediates rolling adhesion. J Cell Biol 196:131–146
106. Zapata AG, Torroba M, Vicente A, Varas A, Sacedon R, Jimenez E (1995) The relevance of cell microenvironments for the appearance of lympho-haemopoietic tissues in primitive vertebrates. Histol Histopathol 10:761–778

Integrin $\alpha 1\beta 1$

Humphrey Gardner

2

Abstract

Integrin $\alpha 1\beta 1$ is widely expressed in mesenchyme and the immune system, as well as a minority of epithelial tissues. Signaling through $\alpha 1$ contributes to the regulation of extracellular matrix composition, in addition to supplying in some tissues a proliferative and survival signal that appears to be unique among the collagen binding integrins. $\alpha 1$ provides a tissue retention function for cells of the immune system including monocytes and T cells, where it also contributes to their long-term survival, providing for peripheral T cell memory, and contributing to diseases of autoimmunity. The viability of $\alpha 1$ null mice, as well as the generation of therapeutic monoclonal antibodies against this molecule, have enabled studies of the role of $\alpha 1$ in a wide range of pathophysiological circumstances. The immune functions of $\alpha 1$ make it a rational therapeutic target.

Keywords

Integrin · Collagen · Knockout mouse · Phenotype

2.1 Introduction

The integrin $\alpha 1$ subunit was first discovered by Hemler et al. as the α component of the Very Late Antigen I (VLA1) expressed on a subset of T cells in the joints of patients with rheumatoid arthritis [57], as well as in a subset of lymphocytes after long term in vitro culture. $\alpha 1$ is the largest of the α subunits, with an apparent mw of 190 kDa nonreduced and 210 kDa reduced [60]. $\alpha 1$'s larger size compared to $\alpha 2$ is due to a higher degree of glycosylation [59]. At the C terminus, the intracellular portion of $\alpha 1$ is the shortest of the α subunits, at 13 residues. Functionally, $\alpha 1$ is one of four collagen binding I-domain containing $\beta 1$ partners, along with $\alpha 2$, $\alpha 10$ and $\alpha 11$. None of the four are known to partner with any β subunit other than $\beta 1$. The $\alpha 1$ I domain shows, like $\alpha 2$, 10 and 11, affinity modulation of ligand binding activity in the same way as has been described for αL [89, 133].

H. Gardner (✉)
Translational Medicine, Astrazeneca, 35 Gatehouse Drive, Waltham, MA 02451, USA
e-mail: humphrey.gardner@astrazeneca.com

D. Gullberg (ed.), *I Domain Integrins*, Advances in Experimental Medicine and Biology 819,
DOI: 10.1007/978-94-017-9153-3_2,

2.2 Tissue Distribution and Gene Regulation

$\alpha1\beta1$, like $\alpha11\beta1$, is predominantly present in mesenchyme. In the adult $\alpha1$ is most abundant in vascular and visceral smooth muscle. This smooth muscle expression has been shown, in the chicken, to be due to a unique combination of transcription factors, GATA6, SRF, and Nkx3.2 [101]. The latter is not found in mammals, but similar factors such as Bapx1 and its family members may play the same role. $\alpha1$ expression is switched off during megakaryocytic differentiation and this appears to be due to gene methylation [20]. The regulation of $\alpha1$ baseline expression in other tissues has not been extensively explored. Other sites of $\alpha1$ expression include fibroblasts [136, 142] and, particularly, specialized fibroblast related cells such as hepatic stellate (Ito) cells [112], pericytes [142] and mesangial cells [60, 96]; bone marrow mesenchymal stem cells [36, 54]; chondrocytes [85], in concert with integrin $\alpha10$ [18] and $\alpha2$ [152]; neural cells including undifferentiated Schwann cells [139] and neurons [37]; and many white blood cells [44, 59]. Microvascular endothelium shows abundant $\alpha1$ expression [33], which is upregulated during angiogenesis. Surprisingly, immunoelectron microscopy shows the presence of abundant $\alpha1\beta1$ on the luminal, as well as abluminal, endothelial surface [16], where no canonical $\alpha1$ ligand would be expected to be. $\alpha1$ is generally absent from normal epithelia, other than the endoderm derived hepatocytes [55, 86, 137], retinal pigment epithelium [99], and endometrial glands [10], where it is cyclically expressed.

Although SNPs in *ITGA1* have been associated with osteoporosis in Korean populations [80], these are synonymous and do not have associated expression data to corroborate their relevance.

2.3 Expression During Development

During development, there is abundant and dynamic expression of $\alpha1$ in embryonic tissues. It is first seen at the leading edge of invading trophoblast shortly after implantation [140], and antibody blockade of $\alpha1$ inhibits trophoblast invasion in vitro [32]. During early to mid embryogenesis $\alpha1$ is expressed transiently by neurons of the CNS [37], by maturing skeletal and cardiac muscle [144], in the skin [61], throughout the developing kidney [73], and in neural crest cells as they mature to dorsal root ganglia [37].

2.4 Expression in Malignancy

Dysregulation of $\alpha1$ has been noted in tumors. Some studies of melanoma have shown a correlation of worse clinical behavior with the presence of $\alpha1$ [124, 125], and others with the absence of $\alpha1$ [50]. Leiomyosarcomas often show loss of $\alpha1$ and gain of $\alpha2$ [94]. Bronchoalveolar [75] and gastric [46] carcinomas sometimes show gain of $\alpha1$ expression, as do squamous cell carcinomas of the head and neck [114]. Survey of RNAseq signatures of the GATC database shows that $\alpha1$ is in general reduced in total expression in tumors compared to normal tissues, probably reflecting the increased epithelial to mesenchyme ratio of the tumors, whereas the reverse is seen for the more epithelially expressed $\alpha2$ (Fig. 2.1). The two exceptions to this finding are head and neck SCC, corroborating Ratzinger et al. [114], and clear cell carcinoma of the kidney (Fig. 2.1). Lastly, dermatotrophic T cell lymphomas show expression of $\alpha1$ [138] probably consistent with the ontogeny of their derivation in the T cell lineage. There is no consistent relationship between $\alpha1$ expression and tumor behavior, in contrast to, the well-characterized and functionally significant $\alpha6\beta4$ to $\alpha6\beta1$ switch seen in some epithelial malignancies.

2.5 Integrin $\alpha1$ Ligands

The best-known ligands for $\alpha1$ are the collagens, investigated mostly in fibroblast studies, and laminin 111, investigated primarily in studies of neural cells. Other $\alpha1$ ligands include matrilin-1,

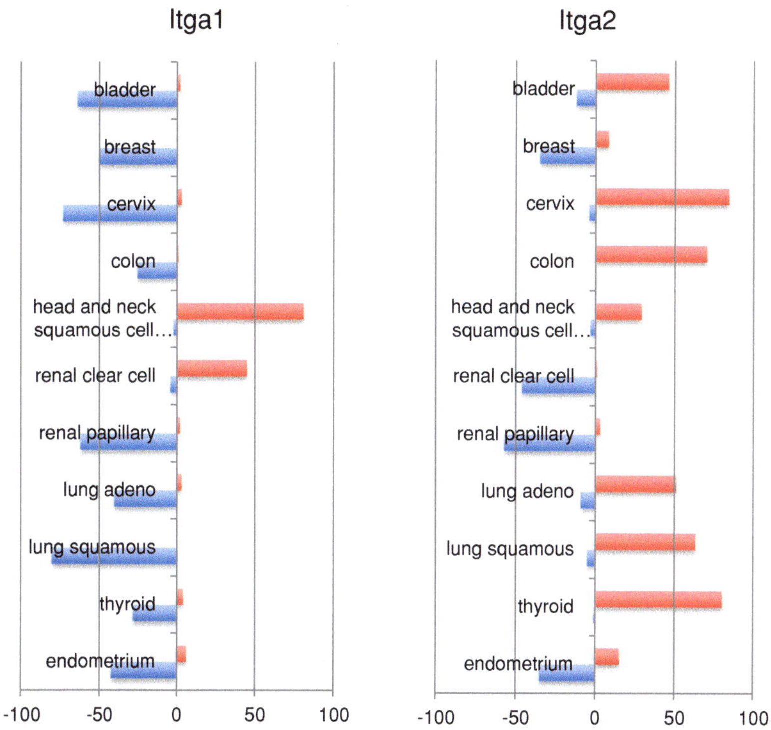

Fig. 2.1 Expression of Integrins alpha1 and 2 in different tumor types. Ratio of RNAseq counts for the gene in tumor versus matched normal was determined. Data taken from TCGA where total evaluable number of samples for the tumor type exceeded 100. Numbers of samples where the ratio of expression exceeded 2 were quantitated. *Red bars* indicate the proportion of cases where tumor expression is twofold or more greater than matched normal, and *blue bars* where tumor expression is twofold or more lower than matched normal. With the exception of renal tumors, Itga2 tended to be increased in expression in tumors in comparison to normal tissue. With the marked exception of clear cell carcinoma of the kidney and head and neck squamous cell carcinoma, Itga1 tended to be downregulated in tumors versus normal tissue

expressed in cartilage, galectins 1, 3 and 8, and the NC1 domain of collagen IV(1), which will be discussed in the context of endothelial regulation. Lastly, semaphorin 7A expressed on macrophages appears to be a counterreceptor to $\alpha1\beta1$ [141]. Ligands are listed in Table 2.1.

2.5.1 Collagens

$\alpha1$ and $\alpha2$ $\beta1$ integrins have collagen binding preferences that are at first glance discordant with their tissue distributions. $\alpha1\beta1$, predominantly expressed on connective tissue, has a higher affinity for collagen type IV than for type I; whereas $\alpha2\beta1$, predominantly on epithelial cells, favors collagen I, which epithelial cells do not normally see, over the collagen IV abundant in epithelial basement membranes. $\alpha1$ and $\alpha2$ (and probably $\alpha10$ and $\alpha11$) bind the triple helical domains of the collagens with highest affinity, and biochemical, cell biological and crystallographic studies show that this binding is contributed to by more than one chain of the triple helix [39, 42]. As such, the binding is dependent on the chains being in register, and would thus be exquisitely sensitive to melting. As collagen melting occur at or below physiological temperatures in a very dynamic fashion [81], it is likely that $\alpha1$ ligand binding, and hence signaling, can be affected by events distal to the receptor along the collagen fibril. This might be especially important in tissue remodeling.

The $\alpha1$ I domain can bind the collagen triple helix at multiple different sites [117, 153], with the relative affinities being divisible into several

Table 2.1 Known ligands of Integrin $\alpha 1\beta 1$

Ligand	Likely cellular context	References
Collagen I	Fibroblasts	[39]
Collagen IV	Fibroblasts, myoepithelium	[39]
Collagen IX	Cartilage	[76]
Collagen XVI	Connetctive tissue	[40]
Arresten (Col4A1 NC1 domain)	Angiogenesis	[25]
Laminin 111	Neural tissue	[143]
Laminin 112	Neural tissue	[143]
Matrilin I	Cartilage	[91]
Galectin 8	T cells	[31]
Galectins 1, 3	Vascular smooth muscle	[98]
Jararhagin	Snake venom	[104]
Obtustatin	Snake venom	[92]
Ross River Virus	Viral infection	[82]
Semaphorin 7A	T cell macrophage interactions	[141]

classes. Among these there are approximately three of the highest binding affinity, with Kds of ~0.25 uM, and about 13 in the next affinity class, with Kds of ~14 uM. The highest binding class regions are adjacent to or overlap with the sites occupied by $\alpha 2$ I domain, and these can be competed off both $\alpha 1$ and $\alpha 2$ I domains by triple helical model molecules containing the core sequence GLOGER or GFOGER, the latter of which was also independently identified as an inhibitor of $\alpha 1$ and $\alpha 2$ binding to collagen I [72] as well as $\alpha 11$ [158]. This core peptide is not effective in blocking $\alpha 1$ binding to collagen IV, but is effective in blocking $\alpha 2$. Recently the peptide GFPGEN was identified as a sequence selective for binding $\alpha 1$ over $\alpha 2$ [130]. The collagen IV binding site for $\alpha 1$ is unique and of higher affinity [17], and has been shown by to require Asp 461 in the $\alpha 1$ chain of collagen IV and Arg 461 in the $\alpha 2$ chain [39]. The binding of integrin $\alpha 1$, $\alpha 2$ and $\alpha 10$ I domains to other collagens has also been explored [103, 147]. More recently $\alpha 1$ has been clearly identified as a receptor for the FACIT collagens IX (predominantly in cartilage) [76] and XVI (predominantly in connective tissue) [40], in a region close to that bound by $\alpha 2$. Mutation of Arg 218 to Asp in $\alpha 1$ causes loss of collagen IV and IX binding, but only partial reduction in collagen I binding [76]. Structural analysis based on modeling from the $\alpha 2$ subunit demonstrates the existence of closed and open states alternately blocking or enabling binding of RKKH type peptides. The two states are energetically very similar, allowing for the possibility of control by inside-out signaling [104]. Another mutation in $\alpha 1$, Glu 317 to Ala, causes increased affinity of the I domain for both collagens and laminin [146], and reveals the possibility that the activated integrin, and ligand bound open integrin, may be slightly different states [77]. Dramatically, this I domain mutation Glu 317 to Ala also causes increased activation of ERK, and enhanced downregulation of collagen synthesis [132], further affirming outside–in signaling and attributing it to the integrin itself. The relationship between the probable affinity modulation of $\alpha 1$, the multiplicity of sites on the collagen fibril along which $\alpha 1$ can bind, and the dynamic instability of the triple helix, suggest a highly dynamic interaction between integrin and collagen. For example, one could see how fibroblast motility along collagen I might be contributed to by detachment and reattachment of the integrin along the fibril. Another possibility is that collagen fibril assembly and extrusion from the fibroblast might be aided by $\alpha 1\beta 1$ protruding from the plasmamembrane surrounding the fibril. Indeed, the $\alpha1$ null mouse has narrower and less well formed collagen fibrils than the wild type animal (Gardner, unpublished). However, the relative importance of $\alpha1$ binding to collagen I versus the basement membrane and facit collagens in vivo has not been established.

2.5.2 Laminins

Laminin 111 and 211 binding by $\alpha1$ is seen in fibroblasts, and is particularly evident on neural cells, for which the pheochromocytoma line PC12 is used as a prototype [143, 159]. These cells show $\alpha1$ dependent adhesion to domain VI of the laminin α chains 1 and 2, at sites adjacent to or congruent with $\alpha2$, at the opposite end of the laminin molecule from the binding regions of the epithelial laminin receptor $\alpha3\beta1$ and the hemidesmosome integrin $\alpha6\beta4$ [24]. This seems reasonable in the context of an epithelial basement membrane, where epithelial cells would bind at one end of the molecule and mesenchymal cells at the other (although binding to laminin 332 by $\alpha1$ is not seen). In vitro, $\alpha1$ has been [13] found to be important for neurite outgrowth on laminin [145] and neural crest cell attachment to collagen [108]. Neural crest cell attachment to laminin can be inhibited by antisense oligonucleotides to $\alpha1$ mRNA [78]. Further studies have shown that neural crest cells migrating on laminin 111 interact, via $\alpha1$, with two distinct sites on the molecule. LN E8—$\alpha1$ interaction drives FAK activation, focal adhesion formation, and migration, while LN E1—$\alpha1$ interaction drives ERK activation and survival [35]. While it is tempting to suggest that this specificity is attributable to subtleties of outside in signaling, the work does not rule out the possibility of some essential coreceptor for one or other interaction. The $\alpha1$ null mouse, however, has normal pigmentation on all genetic backgrounds and appears neurologically and neuroanatomically normal except for a sensitivity to ketamine/xylazine anesthesia (Davidson J, unpublished observations) which may have a neurological basis. Whether $\alpha2$, possibly co-expressed on neurons, provides an adequate alternative for neurite outgrowth, will be seen in the $\alpha1/\alpha2$ double null animal.

2.5.3 Matrilin and Galectin

Matrilin-1 is found in cartilage, and appears to cause increased chondrocyte adhesion to collagen II, via its association with $\alpha1$ [91]. Galectin 8 [31] binds several integrins including $\alpha1$ but not $\alpha2$, and induces Erk phosphorylation independently of cell attachment. Galectins 1 and 3, secreted by vascular smooth muscle, also appear to bind integrin $\alpha1$, the latter in a lactose dependent manner [98]. These glycoproteins, in contrast to matrilin-1, appear to inhibit cell attachment to other matrix components.

2.5.4 Semaphorin 7A

The semaphorins are best known as guidance molecules in the CNS. Interestingly, Sema7A, a subset of semaphorins primarily found in the immune system, appears also to be a component of the immunological synapse in some activated T cells [141], where it interacts specifically with macrophages expressing integrin $\alpha1\beta1$, inducing downstream effects of $\alpha1$ activation. Similarly to $\alpha1$ null mice, sema7A null animals are resistant to encephalitis and DTH models. $\alpha1\beta1$ is widely expressed in the CNS. Whether it interacts with other semaphorins is to be seen.

2.6 Peptide Inhibitors of $\alpha1$

While Jararhagin, a venom protein first noted to bind the alpha2 I domain, also binds the $\alpha1$ I domain [104], Marcinkiewicz and colleagues also identified Obtustatin [92] as a specific inhibitor of $\alpha1$ which does not bind to the I domain. Using blockade of FGF2 driven angiogenesis in the chick CAM model as an assay, they pinned down a specific inhibitory peptide with affinities in the millimolar range, with sequence CWKTSLTSHYC. No further work has been published on this interesting molecule.

2.7 Co-receptors of α1

Many non-I domain containing integrins have been shown to associate in the membrane with other receptors, the best examples being the tetraspanins [8] and Integrin Associated Protein [15]. These may modulate integrin behavior and binding to ligands. α1 has not been shown to associate with such proteins, but this is an area meriting further exploration. On the other hand, α1 is one of a subset of integrins (including α5β1, αvβ3, and α6β4) which associate in the membrane with caveolin and stimulate the Erk pathway via Fyn and Shc [150, 151].

2.8 Integrin α1 Regulation by Cytokines

Most studies of regulation of α1 expression in the adult relate to expression during lymphocyte ontogeny, and in fibroblasts in response to a variety of cytokines. Like integrins α2, α3, α4, and α5β1, α1β1 is upregulated in fibroblast lineages by TGF-β [56], as well as interleukin-1β [123], TNF-α, and interferon gamma [47]. The only cytokine which appears to cause differential regulation of α1 is platelet derived growth factor—BB, which causes downregulation of α1 integrin and upregulation of α5 integrin in fibroblasts [47] and mesangial cells [68].

2.9 The α1 Null Mouse

Aspects of the α1 null mouse will be discussed in subsequent sections. A brief overview is provided here to provide perspective on the known and suspected roles of α1. α1 null mice are viable and fertile, and embryogenesis proceeds normally despite the broad and dynamic expression in trophoblast and developing nervous system. Initially, adult animals are remarkably normal with a mild decrease in weight, normal smooth muscle function, normal rates of wound healing, normal liver function, normal behavior, and no blatant immunodeficiency in a laboratory environment [48]. With ageing, the animals exhibit a series of progressive phenotypes, notably osteoarthritis [156], and retinal degeneration [107], as well as a variety of other vulnerabilities.

2.10 Integrin α1, Signaling, and the Cell Cycle

The potential role of α1 as a cell cycle regulator was suggested by studies showing that α1β1 was a member of a small group of integrins which could activate the adaptor protein Shc, resulting ultimately in MAP kinase activation [150]. Several observations from the α1 null mouse confirmed this, including a reduction in fibroblast proliferation rate in embryonic skin and dermal fibroblast number in the adult, as well as the observations that embryonic fibroblasts from the α1 null failed to activate Shc in response to adhesion to collagen, and that they failed to grow on collagen in conditions of limiting serum whereas growth on the α5 ligand fibronectin or the αv ligand fibrinogen, was normal [109]. As α2 and probably α11 are present on these cells, this suggests that α1 is unique among collagen binding integrins in mesenchyme in being able to stimulate proliferation. Fracture calluses are smaller in α1 null mice, concomitant with a deficiency in bone marrow derived mesenchymal stem cell proliferation [41]. Interestingly, the number and proliferation of mesenchymal stem cell derived hypertrophic chondrocytes in this model is normal—suggesting a specific and transient dependence on α1 for proliferation in the mesenchymal stem cell differentiation pathway. Indeed, α1 has been identified as a very effective tool for the isolation of mesenchymal stem cells [36], and more recently for the selection of the most proliferative subclones of mesenchymal stem cells with the highest multi-differentiation potential [120]. A role for α1 has also been described in osteoblast differentiation [66]. Similar phases of α1 dependence for proliferation appear to be present at some stages of lymphocyte ontogeny [95]. Furthermore, tumor cells derived from Kras transgenic mice are less proliferative on an $\alpha1$ null background [90]. Overall, the subtlety of the proliferative deficit in the $\alpha1$ null mouse must be accounted for by the large number of overlapping proliferative pathways, ligands, and integrins present in the organism.

The $\alpha1$ cytoplasmic domain is very short. It is required for $\alpha1\beta1$ migration into focal adhesions [12], and has a role in binding cytoskeletal components [87, p. 125] FAK, and phospholipase C gamma [149]. A remarkable study by Abair et al. [1], taking advantage of $\alpha1$ null endothelial cells, demonstrated very specific requirements of components of the tail for full activity. The lysine triplet is required for migration and adhesion, and for activation of the Akt and p38 pathways, but not for Erk activation. Furthermore, alanine scanning shows that the most membrane proximal lysine is required for endothelial tubulogenesis, and migration on collagen IV, and that Lys 1151 is required for all functions except for proliferation. It appears that the integrin $\alpha1$ cytoplasmic tail is quite unique among the integrins in being able to bind and activate the small nuclear shuttling phosphatase TCPTP. This phosphatase has many targets, but in the context of collagen ligand binding, TCPTP acts to cause a reduction in EGFR signaling [93], either by dephosphorylating EGFR directly or by reducing the amount of phosphorylated caveolin available to activate EGFR [11]. Whatever the mechanisms, the implication that active ligand binding, which in general would cause Erk activation, can serve to dampen an alternative pro-mitotic signaling pathway is intriguing. The specificity to $\alpha1$ is also intriguing. While the genomic region containing $\alpha1$ is lost in some tumors, and thus $\alpha1$ might be regarded as a candidate tumor suppressor [93], the molecule is not expressed in most epithelial tissues. One physiological site where $\alpha1$ might usefully downregulate EGFR activity is in myoepithelial cells of the breast, where cells express $\alpha1$ [74], as well as EGFR [100], and are juxtaposed to basement membrane.

2.11 Integrin $\alpha1$, Fibroblasts, and Collagen and Collagenase Regulation

Many studies have shown that $\alpha1\beta1$ is a negative feedback regulator of collagen synthesis by fibroblasts. These were initiated by Langholtz et al., who showed that an activating antibody to $\alpha1$ accentuated the normal downregulation of collagen synthesis seen when fibroblasts are suspended in collagen gels [79]. It was also noted that $\alpha1$ levels appeared to be reduced on scleroderma fibroblasts, in conjunction with their upregulation of collagen synthesis [64]. Data from the $\alpha1$ null mouse lent strong support to this role: in vivo the mice show a 20 % increase in the rate of collagen incorporation into the skin, and fibroblasts from these animals are deficient in downregulating synthesis in response to gel suspension [49]. We subsequently examined keloids to determine whether loss of $\alpha1$ could account for the increased collagen expression in these lesions [142]. A high proportion of lesional fibroblasts expressed $\alpha1$ (in contrast to scleroderma lesions), although the levels expressed were somewhat lower than seen in chronic wounds with low collagen production. Thus, absence of $\alpha1$ could not account for the excess collagen production in keloids, but there may be a relative deficiency compared to normal wounds, which show distinct peaks in $\alpha1$ expression at 8 and 30 days [9].

The mechanism for downregulation of collagen synthesis mediated by $\alpha1$ has been extensively dissected. $\alpha1\beta1$ stimulation by ligand activates the MAP kinases Erk1 and 2 via Fyn and Shc [109, 150], and Erk1/2 activation reduces collagen synthesis [116]. Reciprocally, the Erk1/2 inhibitor PD98059 causes upregulation of fibroblast collagen synthesis [1]. This is the reverse of the effect of $\alpha2\beta1$ stimulation, which activates p38 and causes induction of collagen synthesis [65]. Thus, in general, $\alpha1\beta1$ and $\alpha2\beta1$ are opponents in their effect on collagen synthesis, the former inhibitory and the latter activatory. More specific mechanisms of collagen regulation involving reactive oxygen species in mesangial cells will be discussed in the kidney chapter.

The regulation of metalloproteases has similar themes but appears more complex and is probably very cell type specific. Firstly, the structure and function of the mouse and human collagenases is not congruent: MMP1, which is the major fibroblast collagenase in humans, is upregulated by Erk1/2 activation, but the mouse MMP1 structural equivalents, McolA and McolB, are not seen in skin fibroblasts, and have a restricted expression in the placenta and uterus [3]. On the other hand human MMP13, found in chronic ulcers [148], is downregulated by Erk1/2 activation as well as being upregulated by p38 activation [115], and MMP13 is the major fibroblast collagenase in mouse [83]. Although functionally equivalent to human MMP1, mouse MMP13 appears to be regulated like human MMP13, as it is markedly upregulated in $\alpha1$ null mice where there is loss of Erk1/2 signaling but normal $\alpha2$-p38 signaling. The $\alpha1$ null animal shows an increase in expression of several other MMPs, including 7, 9 and 2 in endothelial cells, and 9 and 2 in fibroblasts [49, 111]. For want of other evidence, this may be attributed to reduced Erk1/2 activation. However, whereas $\alpha1$ stimulation is always inhibitory to collagen synthesis, it is sometimes activatory to MMP synthesis. In some systems $\alpha1$ activation by laminin [84] or by collagen IV (Pozzi and Gardner, unpublished) or collagen I [121] causes an increase in MMP synthesis.

In many studies of fibroblast collagen interaction, the complex process of collagen gel contraction is addressed. In dermal fibroblasts integrin $\alpha2\beta1$ is seen to be the dominant player in this process [79], which can be uncoupled from MMP synthesis [14], and is dependent on a functional cytoskeleton. However, in studies of specialized cardiac fibroblasts [19], smooth muscle cells [53], stellate cells [113] and mesangial cells [69], $\alpha1$ blockade has been shown to prevent gel contraction, as has integrin $\alpha v\beta3$ blockade in other cell types [27]. It is possible that whereas $\alpha2$ is structurally more suited to gel contraction (having a far higher affinity for collagen I), $\alpha1$ expression may be required for maintenance of the contractile myofibroblastoid phenotype. It is striking that $\alpha1$ expression is upregulated in vivo in all activated contractile myofibroblastoid cells including myofibroblasts in wound repair, mesangial cells, pericytes, myoepithelial cells [74], and hepatic stellate cells.

In summary, there may be several roles for $\alpha1$ and its interplay with $\alpha2$ in the fibroblast during dermal wound healing and other episodes of mesenchymal repair. $\alpha1$ upregulation in fibroblasts contributes to collagen stimulated cell proliferation, and probably to the myofibroblast transition. $\alpha2$ is the major contributor to the synthetic phenotype, where it contributes the major part of collagen matrix contraction and activates collagen synthesis, as well as activating MMP synthesis for matrix remodeling. $\alpha1$ fine tunes the MMP response, possibly providing general inhibition of MMP release, but allowing for specific activation near the epidermal boundary where there is a greater abundance of the $\alpha1$ high affinity ligand, collagen IV. $\alpha1$ also provides feedback inhibition against excessive collagen synthesis. Consistent with these suggestions, the $\alpha1$ null shows excessive collagen and collagenase synthesis at overlapping phases of wound healing [49], and collagen fibrils are densely aggregated and irregular in the dermis of the $\alpha1$ null, while being individually smaller (Gardner, unpublished observations). Some aspects of this paradigm appear to be different in mesangial cells, which are discussed in the kidney chapter.

2.12 Integrin $\alpha1$ and Angiogenesis

Immunohistochemical analysis of murine and human tissue shows that $\alpha1$ is present on at least some normal microvascular endothelium. $\alpha1$ has also been shown to be upregulated on endothelia in MS lesions [135]. New tumor microvessels appear always to express $\alpha1$, while a smaller proportion of them, predominantly the slightly larger ones, also express $\alpha2$ [111]. Vascular endothelial growth factor (VEGF/VPF) can induce $\alpha1$ on endothelial cells, and as the only collagen receptors expressed, $\alpha1$ and $\alpha2$ are required for endothelial haptotaxis through collagen. Antibodies to $\alpha1$ and $\alpha2$ reduce angiogenesis in response to subcutaneously implanted gels of fibrin or collagen containing VEGF, or to tumor xenografts [128, 129]. However, tumor matrix contains a great variety of alternative integrin ligands. As we have learned from fibroblasts, $\alpha1\beta1$ can activate an Erk1/2 proliferation pathway mediated by Shc. $\alpha2\beta1$ can also positively regulate the progression through the cell cycle in epithelial cells by non overlapping mechanisms [71]. Thus, $\alpha1$ and $\alpha2$ blockade in vivo may cause a simple reduction in endothelial proliferation. With this in mind, there is no deficiency in normal vasculo- and angiogenesis in $\alpha1$ null mice. Analysis of the null mice, however, reveals other, subtler roles for $\alpha1$ in angiogenesis.

Detailed analysis of endothelial cells and tumor vasculature in $\alpha1$ null animals [111] led to independent verification of the significance of plasminogen fragments, the angiostatins [105], in endothelial growth regulation. Pulmonary microvascular endothelial cells from $\alpha1$ null mice grew poorly compared to wild type, regardless of the substratum on which they were grown. This growth deficiency could be completely rescued by frequent media change even if the cells were grown on collagen. Furthermore, media conditioned by $\alpha1$ null endothelial cells was inhibitory to the growth of wild type cells. The growth deficiency in $\alpha1$ null endothelial cells was also corrected by antibodies to angiostatin, or growth in media containing serum from plasminogen null mice (from which no angiostatin could be generated) instead of fetal calf serum. Lastly, the growth deficiency could be rescued by MMP9 blockade. Analysis of conditioned medium from $\alpha1$ null endothelial cells as well as plasma from wounded (but not unwounded) or tumor bearing $\alpha1$ null mice showed an increase in MMP9 and angiostatin compared to wild type animals. These findings in endothelial cells were consistent with the increased MMP expression seen in $\alpha1$ null fibroblasts, due to loss of $\alpha1$-Erk1/2 inhibitory signaling with normal $\alpha2$-p38 activatory signaling. Thus, increased MMP9 released by the $\alpha1$ null cells cleaves plasminogen [106] to yield the endothelial inhibitor, angiostatin.

In vivo, $\alpha1$ null mice, with higher plasma MMP9 and angiostatin levels are less able to vascularize subcutaneous tumors than wild type, but this deficit can be reversed by oral treatment of the animals with the MMP9 inhibitor doxycycline, and consequent reduction of their angiostatin levels [110]. MMP9 levels in the vasculature correlate inversely with tumor vascularization even in wild type mice. These studies have been repeated in several tumor systems with essentially similar results, namely that tumors in the $\alpha1$ null host are smaller and less vascular and the phenotype can be reversed by MMP inhibition [22, 23]. These studies showed that the interplay between $\alpha1$ and $\alpha2$ integrins has significant consequences in the vascular system. Thus, during vascular remodeling, upregulation of endothelial $\alpha1$ and $\alpha2$ occurs, and the balance between them regulates MMP release, and ultimately vessel number.

While plasminogen is an MMP9 target, and its cleavage product angiostatin was entirely responsible for endothelial growth inhibition in vitro, other MMP targets might be of importance in this feedback system in vivo. These include the collagens themselves. In this regard the finding that a collagen NC1 domain is a ligand for $\alpha1$ may be of significance. The NC1 domain of collagen IV $\alpha3$, also known as tumstatin, causes endothelial cytostasis and blocks angiogenesis by binding to integrin $\alpha v\beta3$, and a similar mechanism appears to exist for $\alpha1$ binding to the collagen IV(1) NC1 domain (arresten) [25]. This might be an explanation for the presence of $\alpha1$ on the luminal surface of endothelium, where it could act as a detector of collagen fragments released during remodeling, and provide negative feedback to angiogenesis. While in general $\alpha1$ binding to collagen in fibroblasts causes activation of Erk1/2 via Shc, arresten may provide a growth inhibitory signal. In fact this has been strongly suggested by the work of Nyberg et al. where the arresten—$\alpha1$ interaction appears to mediate an apoptotic response [102], and this interaction has been invoked in the blockade of growth of HSC tongue carcinoma cells [2]. The absence of both signals—collagen IV driving growth and arresten being pro-apoptotic—in the $\alpha 1$ null could explain why normal angiogenesis is unaltered in $\alpha 1$ null mice.

2.13 Smooth Muscle and $\alpha 1$

$\alpha 1$ is extremely abundant on smooth muscle, both visceral and vascular [6], and, in vivo, expresses no other collagen binding integrin (explanted smooth muscle rapidly upregulates $\alpha 2$, complicating studies [134]). Furthermore, smooth muscle basal lamina has abundant collagen IV. There is no upregulation of $\alpha 2$ or $\alpha 10$ in the $\alpha 1$ null smooth muscle in vivo, as assessed by immunostaining [48]. Yet in the $\alpha 1$ null mouse digestion and parturition is entirely normal, and EM studies reveal no alterations in smooth muscle structure. Studies of mesenteric arteries have shown that $\alpha 1$ deficient vessels rupture at lower stresses than wild type, due to a deficiency in the hypertrophic response [88]. Integrin $\alpha 8\beta 1$, a fibronectin receptor, is also abundant in smooth muscle, but the double knockout $\alpha 1/\alpha 8$ animal also had histologically normal smooth muscle (Gardner and Brandenburger, unpublished). Further collagen binding integrin double knockouts may reveal the answer to this mystery.

2.14 Integrin $\alpha 1$ and the Retina and CNS

Retinal pigment epithelium (REPE) cells have been shown to use $\alpha 1$ as one among other receptors for collagen gel contraction [99] but $\alpha 1$ signaling of MAP kinase activation is clearly of unique importance. Peng et al. [107] found that older $\alpha 1$ null mice become blind, with loss of retinal evoked potentials, degeneration of the peripheral retina, irregularities in basal lamina thickness, rod degeneration and synaptic malformation in rod and cone terminals, and failure of transducin α translocation to the outer rod segments upon light exposure.

Frasca et al. [45] have made observations on the role of $\alpha 1$ in contributing to the neurotoxicity of amyloid. This appears to be due to $\alpha 1$-ligand interaction, via Erk activation, being permissive to neuronal entry into the cell cycle after their stimulation by A-beta. Neurons, in contrast to other cell types, appear to meet their demise after cell cycle entry.

2.15 Integrin $\alpha 1$ as a Viral Receptor

Many integrins have been recognized as receptors for viruses. $\alpha 1$ appears to be one of several receptors for Ross River virus, a semliki forest type alphavirus one of whose coat proteins has a region which appears to mimic a collagen fold [82]. There is a possibility that $\alpha 1$ is also a receptor for rotavirus enterotoxin [131].

2.16 Integrin $\alpha 1$ and the Kidney

Expression of $\alpha 1$ by glomerular mesangial cells [30, 60] as well as the developing kidney [73] led to a great deal of interest in the role of this integrin in the kidney. $\alpha 1$ null mice showed no functional or anatomic renal abnormality alterations in $\alpha 1$ null glomeruli in the unperturbed state, but a variety of challenges have exploited the underlying mesangial alterations to create new models of renal disease. Ex vivo studies demonstrate alterations in mesangial homeostasis in the absence of $\alpha 1$, notably an alteration in MMP profile rather different from that seen in cutaneous $\alpha 1$ null fibroblasts [155]. $\alpha 1$ nulls also have poor osmolarity regulation [97]. Streptozocin treated $\alpha 1$ nulls get worse glomerular disease than wild type [157], and the diabetic Akita mouse gets dramatically accelerated renal dysfunction when crossed into an $\alpha 1$ null background [154].

Cross of the $\alpha 1$ null with the collagen IV $\alpha 3$ chain null (COL4A3/Alports) mouse [29] led to unexpected effects. Reduced glomerular basement membrane stiffness in the COL4A3 null leads to a progressive glomerulonephritis with mesangial expansion and secondary tubulointerstitial fibrosis. Surprisingly, the double null animal lived twice as long as the COL4A3 null, due to a delay in the progression of renal failure. This unexpected result appears to be due to several mechanisms. Firstly, in the normal progression of murine Alports, there is a marked influx of monocytes into the interstitium in response to glomerular epithelial damage. $\alpha1$ null monocytes are defective in migrating into the renal interstitium, possibly due to the monocyte requirement for $\alpha1$ to adhere to the collagen XIII generated by endothelium during injury [34], and are therefore reduced in number in the double null kidney. This reduces delivery of TGFβ to the kidney, delaying the onset and progression of interstitial fibrosis [122]. Secondly, mesangial cells are dependent on $\alpha1$ and Rac to invade the glomerular tuft [155], a key process in the initiation of renal repair and injury. In the $\alpha1$/COL4A3 double null, the mesangial expansion is greatly reduced [29]. In another glomerulonephritis model, anti-Thy-1 GN in the rat [69], direct injection of anti-$\alpha1$ in the renal artery caused a marked reduction in mesangial proliferation and matrix accumulation, an important in vivo validation of a series of studies of the role of $\alpha1$ in mesangial cells [67–70]. The role of $\alpha1$ in driving proliferation is complex in mesangial cells. In contrast to studies in most systems which ascribe a pro-proliferative role for $\alpha1$ signaling, overexpression of $\alpha1$ in mesangial cells leads to activation of p27Kip and cell cycle arrest [70]. In fact mesangial cells appear to be an exception in many aspects of $\alpha1$ physiology, in that Erk phosphorylation is upregulated in $\alpha1$ null mesangial cells and p38 is downregulated. Notwithstanding the increased Erk phosphorylation, collagen synthesis is increased, via a reactive oxygen species driven mechanism [21, 28]. This may be due to some kind of integrin crosstalk, where the excess integrin $\alpha1$ activates a pathway normally associated with another integrin [127]. A potential corollary of this is that monomer and polymer collagen have different effects on mesangial cell growth; on the latter substrate growth is inhibited, $\alpha1$ is excluded from focal contacts, and ERK1/2 phosphorylation is diminished [126].

2.17 Integrin $\alpha1$ and the Immune System

Integrin $\alpha1$ was first discovered as a very late antigen on cultured T cells, and being the largest of the α subunits, was named Very Late Antigen 1 (VLA1), a name which persists in immunological studies. Hemler et al. subsequently showed that VLA1 was present on a large proportion of T cells in the joints of rheumatoid arthritics, but was almost absent from the circulation, giving a first clue to a role for $\alpha1$ in tissue migration and T cell activation [60, 58]. More detailed study of the immune system revealed that $\alpha1$ is also expressed on a subset of NK-T cells as well as populations of activated monocytes and NK cells.

$\alpha1$ deficiency generated by knockout or antibody blockade has dramatic consequences in the immune system. $\alpha1$ null mice show no overt immunodeficiency, but they show resistance to many different disease models involving monocyte function or peripheral T cell memory. These include a resistance to anti collagen II antibody induced and mycobacterium induced arthritis [44, 62], colitis [43], DTH, contact hypersensitivity [44], and LCMV induced encephalopathy [7]. Inflamed tissues in these models, as well as the normal gut mucosal epithelium [95], show reduced infiltration by T cells and monocytes. Furthermore, cultured splenocytes from $\alpha1$ null animals show reduced proliferation in response to collagen, and fail to express integrin $\alpha2$ upon long-term culture.

In murine influenza models, $\alpha1$ positive T cells tend to be CD4 and associate with basement membranes, while $\alpha2$ T cells bias to CD4 and an interstitial location. Memory to influenza is maintained by the $\alpha1$ positive T cells, as they are protected from TNF driven apoptosis [119, 118]. Treg cells are VLA1 negative, and stimulated PBMCs can be diverted from generating VLA1 + T effector cells into Treg cells if TNF signaling is blocked [51]. Taken together, the results suggest that $\alpha1$ is needed both for lymphocyte migration into the collagen rich periphery, and for the proliferation of activated

T cells in those locations, or for their long term survival as mediators of peripheral T cell memory [38].

In rheumatoid arthritics, $\alpha 1$ positive T cells are far more abundant and tend to be found in the joints as oligoclonal populations, probably responding to a restricted number of joint antigens [4, 52]. Here they offer an obvious target for therapy. Interestingly, $\alpha 1$ has also been noted to be required for monocyte retention at sites of inflammation in skin [5], and a role for the receptor was similarly shown for T cells in a xenotransplantation model of psoriasis, where epidermal, but not dermal, T cells expressed $\alpha 1$ [26].

2.18 Therapeutics

In the early 2000s Biogen Idec developed a humanized function blocking anti-VLA1 antibody for immune diseases. This has now been taken through a phase 1 single dose escalation study by Santarus, as SAN-300, without remarkable side effects, and with anecdotal demonstration of efficacy in a single rheumatoid arthritis RA patient recruited to the study [63]. The potential for this molecule may be very high in diseases characterized by the persistence of localized pathological effector T cell memory, such as RA and psoriasis.

2.19 Summary and Prospects

Integrin $\alpha 1$ has major roles as a modulator of mesenchymal proliferation and differentiation, matrix turnover, and immune function. Its roles in the immune system make it a clear target for therapy. In its biochemical properties, $\alpha 1$ appears to have a unique role in binding basement membrane collagens, the significance of which in vivo is not yet entirely clear. Like the other collagen binding I domain containing integrins, $\alpha 2$, $\alpha 10$ and $\alpha 11$, its absence is not associated with major structural deficits in the mouse, illustrating the dense interweaving of redundant or partially redundant pathways in tissue morphogenesis.

References

1. Abair TD, Bulus N, Borza C, Sundaramoorthy M, Zent R, Pozzi A (2008) Functional analysis of the cytoplasmic domain of the integrin {alpha}1 subunit in endothelial cells. Blood 112:3242–3254
2. Aikio M, Alahuhta I, Nurmenniemi S, Suojanen J, Palovuori R, Teppo S, Sorsa T, López-Otín C, Pihlajaniemi T, Salo T et al (2012) Arresten, a collagen-derived angiogenesis inhibitor, suppresses invasion of squamous cell carcinoma. PLoS ONE 7:e51044
3. Balbin M, Fueyo A, Knauper V, Lopez JM, Alvarez J, Sanchez LM, Quesada V, Bordallo J, Murphy G, Lopez-Otin C (2001) Identification and enzymatic characterization of two diverging murine counterparts of human interstitial collagenase (MMP-1) expressed at sites of embryo implantation. J Biol Chem 276:10253–10262
4. Bank I, Koltakov A, Goldstein I, Chess L (2002) Lymphocytes expressing alpha1beta1 integrin (very late antigen-1) in peripheral blood of patients with arthritis are a subset of CD45RO(+) T-cells primed for rapid adhesion to collagen IV. Clin Immunol 105:247–258
5. Becker HM, Rullo J, Chen M, Ghazarian M, Bak S, Xiao H, Hay JB, Cybulsky MI (2013) $\alpha 1\beta 1$ integrin-mediated adhesion inhibits macrophage exit from a peripheral inflammatory lesion. J Immunol 190:4305–4314
6. Belkin VM, Koteliansky VE, Belkin AM (1990) Human smooth muscle VLA-1 integrin: purification, substrate specificity, localization in aorta, and expression during development. J Cell Biol 111(5 Pt 1):2159–2170
7. Ben-Horin S, Bank I (2004) The role of very late antigen-1 in immune-mediated inflammation. Clin Immunol 113:119–129
8. Berditchevski F, Bazzoni G, Hemler ME (1995) Specific association of CD63 with the VLA-3 and VLA-6 integrins. J Biol Chem 270:17784–17790
9. Van Beurden HE, Snoek PaM, Von den Hoff JW, Torensma R, Kuijpers-Jagtman A-M (2003) Fibroblast subpopulations in intra-oral wound healing. Wound Repair Regen. 11:55–63
10. Bhat KP, Pezzuto JM (2001) Resveratrol exhibits cytostatic and antiestrogenic properties with human endometrial adenocarcinoma (Ishikawa) cells. Cancer Res 61(16):6137–6144
11. Borza CM, Chen X, Mathew S, Mont S, Sanders CR, Zent R, Pozzi A (2010) Integrin {alpha}1{beta}1 promotes caveolin-1 dephosphorylation by activating T cell protein-tyrosine phosphatase. J Biol Chem 285:40114–40124
12. Briesewitz R, Kern A, Marcantonio EE (1993) Ligand-dependent and -independent integrin focal contact localization: the role of the alpha chain cytoplasmic domain. Mol Biol Cell 4:593–604

13. Briesewitz R Marcantonio EE, Epstein MR (1993) Expression of native and truncated forms of the human integrin alpha 1 subunit. J Biol Chem 268(4):2989–2996
14. Broberg A, Heino J (1996) Integrin alpha2beta1-dependent contraction of floating collagen gels and induction of collagenase are inhibited by tyrosine kinase inhibitors. Exp Cell Res 228:29–35
15. Brown E, Hooper L, Ho T, Gresham H (1990) Integrin-associated protein: a 50-kD plasma membrane antigen physically and functionally associated with integrins. J Cell Biol 111:2785–2794
16. Burggraf D, Trinkl A, Burk J, Martens HK, Dichgans M, Hamann GF (2008) Vascular integrin immunoreactivity is selectively lost on capillaries during rat focal cerebral ischemia and reperfusion. Brain Res 1189:189–197
17. Calderwood DA, Eble J, Kuhn K, Humphries MJ, Tuckwell DS (1997) The integrin alpha1 A-domain is a ligand binding site for collagens and laminin. J Biol Chem 272(19):12311–12317
18. Camper L, Hellman U, Lundgren-Akerlund E (1998) Isolation, cloning, and sequence analysis of the integrin subunit alpha10, a beta1-associated collagen binding integrin expressed on chondrocytes. J Biol Chem 273:20383–20389
19. Carver W, Reaves TA, Borg TK, Terracio L, Molano I (1995) Role of the alpha 1 beta 1 integrin complex in collagen gel contraction in vitro by fibroblasts. J Cell Physiol 165(2):425–437
20. Cheli Y, Kanaji S, Jacquelin B, Chang M, Nugent DJ, Kunicki TJ (2007) Transcriptional and epigenetic regulation of the integrin collagen receptor locus ITGA1-PELO-ITGA2. Biochim Biophys Acta 1769:546–558
21. Chen X, Abair TD, Ibanez MR, Su Y, Frey MR, Dise RS, Polk DB, Singh AB, Harris RC, Zent R et al (2007) Integrin alpha1beta1 controls reactive oxygen species synthesis by negatively regulating epidermal growth factor receptor-mediated Rac activation. Mol Cell Biol 27:3313–3326
22. Chen X, Su Y, Fingleton B, Acuff H, Matrisian LM, Zent R, Pozzi A (2005) An orthotopic model of lung cancer to analyze primary and metastatic NSCLC growth in integrin alpha1-null mice. Clin Exp Metastasis 22:185–193
23. Chen X, Su Y, Fingleton B, Acuff H, Matrisian LM, Zent R, Pozzi A (2005) Increased plasma MMP9 in integrin alpha1-null mice enhances lung metastasis of colon carcinoma cells. Int J Cancer 116:52–61
24. Colognato-Pyke H, Yamada Y, Carbonetto S, Cheng YS, Yurchenco PD, O'Rear JJ (1995) Mapping of network-forming, heparin-binding, and alpha 1 beta 1 integrin-recognition sites within the alpha-chain short arm of laminin-1. J Biol Chem 270(16):9398–9406
25. Colorado PC, Torre A, Kamphaus G, Maeshima Y, Hopfer H, Takahashi K, Volk R, Zamborsky ED, Herman S, Sarkar PK et al (2000) Anti-angiogenic cues from vascular basement membrane collagen. Cancer Res 60:2520–2526
26. Conrad C, Boyman O, Tonel G, Tun-Kyi A, Laggner U, de Fougerolles A, Kotelianski V, Gardner H, Nestle FO (2007) Alpha1beta1 integrin is crucial for accumulation of epidermal T cells and the development of psoriasis. Nat Med 13:836–842
27. Cooke ME, Sakai T, Mosher DF (2000) Contraction of collagen matrices mediated by alpha2beta1A and alpha(v)beta3 integrins. J Cell Sci 113:2375–2383
28. Cosgrove D, Meehan DT, Delimont D, Pozzi A, Chen X, Rodgers KD, Tempero RM, Zallocchi M, Rao VH (2008) Integrin alpha1beta1 regulates matrix metalloproteinases via P38 mitogen-activated protein kinase in mesangial cells: implications for Alport syndrome. Am J Pathol 172:761–773
29. Cosgrove D, Meehan D, Miller C, Bovard K, Gilroy A, Gardner H, Kotelianski V, Gotwals P, Amatucci A, Kalluri R, Rodgers K (2000) Integrin alpha1beta1 and transforming growth factor-beta1 play distinct roles in alport glomerular pathogenesis and serve as dual targets for metabolic therapy. Am J Pathol 157(5):1649–1659
30. Cosio FG, Nahman NS Jr, Sedmak DD (1990) Cellular receptors for matrix proteins in normal human kidney and human mesangial cells. Kidney Int 38(5):886–895
31. Cárcamo C, Pardo E, Oyanadel C, Bravo-Zehnder M, Bull P, Cáceres M, Martínez J, Massardo L, Jacobelli S, González A et al (2006) Galectin-8 binds specific beta1 integrins and induces polarized spreading highlighted by asymmetric lamellipodia in Jurkat T cells. Exp Cell Res 312:374–386
32. Damsky CH, Lim KH, Fitzgerald ML, McMaster MT, Janatpour M, Zhou Y, Logan SK, Fisher SJ, Librach C (1994) Integrin switching regulates normal trophoblast invasion. Development 120(12):3657–3666
33. Defilippi P, Bertolotto A, Rossino P, Silengo L, Tarone G, van Hinsbergh V (1991) Differential distribution and modulation of expression of alpha 1/beta 1 integrin on human endothelial cells 1. J Cell Biol 114(4):855–863
34. Dennis J, Meehan DT, Delimont D, Zallocchi M, Ga Perry, O'Brien S, Tu H, Pihlajaniemi T, Cosgrove D (2010) Collagen XIII induced in vascular endothelium mediates alpha1beta1 integrin-dependent transmigration of monocytes in renal fibrosis. Am J Pathol 177:2527–2540
35. Desban N, Lissitzky JC, Rousselle P, Duband JL (2006) alpha1beta1-integrin engagement to distinct laminin-1 domains orchestrates spreading, migration and survival of neural crest cells through independent signaling pathways. J Cell Sci 119:3206–3218
36. Deschaseaux F, Charbord P (2000) Human marrow stromal precursors are alpha 1 integrin subunit-positive. J Cell Physiol 184(3):319–325

37. Duband JL, Belkin AM, Syfrig J, Thiery JP, Koteliansky VE (1992) Expression of alpha 1 integrin, a laminin-collagen receptor, during myogenesis and neurogenesis in the avian embryo. Development 116:585–600
38. Dustin ML, de Fougerolles AR (2001) Reprogramming T cells: the role of extracellular matrix in coordination of T cell activation and migration. Curr Opin Immunol 13:286–290
39. Eble JA, Golbik R, Mann K, Kuhn K (1993) The alpha 1 beta 1 integrin recognition site of the basement membrane collagen molecule [alpha 1(IV)]2 alpha 2(IV). Embo J 12:4795–4802
40. Eble JA, Kassner A, Niland S, Morgelin M, Grifka J, Grassel S (2006) Collagen XVI harbors an integrin alpha1 beta1 recognition site in its C-terminal domains. J Biol Chem 281:25745–25756
41. Ekholm E, Hankenson KD, Uusitalo H, Hiltunen A, Gardner H, Heino J, Penttinen R (2002) Diminished callus size and cartilage synthesis in alpha 1 beta 1 integrin-deficient mice during bone fracture healing. Am J Pathol 160:1779–1785
42. Emsley J, King SL, Bergelson JM, Liddington RC (1997) Crystal structure of the I domain from integrin alpha2beta1. J Biol Chem 272:28512–28517
43. Fiorucci S, Mencarelli A, Palazzetti B, Sprague AG, Distrutti E, Morelli A, Novobrantseva TI, Cirino G, Koteliansky VE, de Fougerolles AR (2002) Importance of innate immunity and collagen binding integrin alpha1beta1 in TNBS-induced colitis. Immunity 17:769–780
44. De Fougerolles AR, Nickerson-Nutter CL, Chi-Rosso G, Rennert PD, Gardner H, Gotwals PJ, Lobb RR, Koteliansky VE, S.A G (2000) Regulation of inflammation by collagen-binding integrins alpha 1beta1 and alpha2beta1 in models of hypersensitivity and arthritis. J Clin Invest 105(6):721–729
45. Frasca G, Carbonaro V, Merlo S, Copani A, Sortino MA (2008) Integrins mediate beta-amyloid-induced cell-cycle activation and neuronal death. J Neurosci Res 86:350–355
46. Fukuda K, Saikawa Y, Yagi H, Wada N, Takahashi T, Kitagawa Y (2012) Role of integrin alpha1 subunits in gastric cancer patients with peritoneal dissemination. Mol Med Rep 5:336–340
47. Gailit J, Bueller H, Clark RA, Xu J (1996) Platelet-derived growth factor and inflammatory cytokines have differential effects on the expression of integrins alpha 1 beta 1 and alpha 5 beta 1 by human dermal fibroblasts in vitro. J Cell Physiol 169(2):281–289
48. Gardner H, Koteliansky V, Jaenisch R, Kreidberg J (1996) Deletion of integrin alpha 1 by homologous recombination permits normal murine development but gives rise to a specific deficit in cell adhesion. Dev Biol 175(2):301–313
49. Gardner H, Pozzi A, Laato M, Heino J, Broberg A (1999) Absence of integrin alpha1beta1 in the mouse causes loss of feedback regulation of collagen synthesis in normal and wounded dermis. J Cell Sci 112(Pt 3):263–272
50. Gilhar A, Kalish RS, Berkutski T, Azizi E, Bank I, Ullmann Y (1997) Favourable melanoma prognosis associated with the expression of the tumour necrosis factor receptor and the alpha1beta1 integrin: a preliminary report. Melanoma Res 7(6):486–495
51. Goldstein I, Ben-Horin S, Koltakov A, Chermoshnuk H, Polevoy V, Berkun Y, Amariglio N, Bank I (2007) alpha1beta1 Integrin + and regulatory Foxp3 + T cells constitute two functionally distinct human CD4 + T cell subsets oppositely modulated by TNFalpha blockade. J Immunol 178:201–210
52. Goldstein I, Simon AJ, Ben Horin S, Matzri S, Koltakov A, Langevitz P, Rechavi G, Amariglio N, Bank I (2008) Synovial VLA-1 + T cells display an oligoclonal and partly distinct repertoire in rheumatoid and psoriatic arthritis. Clin Immunol 128:75–84
53. Gotwals PJ, Chi-Rosso G, Lindner V, Yang J, Ling L, Fawell SE, Koteliansky VE (1996) The alpha1beta1 integrin is expressed during neointima formation in rat arteries and mediates collagen matrix reorganization. J Clin Invest 97:2469–2477
54. Gronthos S, Graves SE, Robey PG, S.P J (2001) Integrin-mediated interactions between human bone marrow stromal precursor cells and the extracellular matrix. Bone 28(2):174–181
55. Gullberg D, Turner DC, Borg TK, Terracio L, Rubin K (1990) Different beta 1-integrin collagen receptors on rat hepatocytes and cardiac fibroblasts. Exp Cell Res 190:254–264
56. Heino J, Ignotz RA, Hemler ME, Crouse C, Massagué J (1989) Regulation of cell adhesion receptors by transforming growth factor-beta. Concomitant regulation of integrins that share a common beta 1 subunit. J Biol Chem 264(1):380–388
57. Hemler ME, Glass D, Coblyn JS, Jacobson JG (1986) Very late activation antigens on rheumatoid synovial fluid T lymphocytes. Association with stages of T cell activation. J Clin Invest 78:696–702
58. Hemler ME, Jacobson JG, Brenner MB, Mann D, Strominger JL (1985) VLA-1: a T cell surface antigen which defines a novel late stage of human T cell activation. Eur J Immunol 15:502–508
59. Hemler ME, Jacobson JG, Strominger JL (1985) Biochemical characterization of VLA-1 and VLA-2. Cell surface heterodimers on activated T cells. J Biol Chem 260:15246–15252
60. Hemler ME, Sanchez-Madrid F, Flotte TJ, Krensky AM, Burakoff SJ, Bhan AK, Springer TA, Strominger JL (1984) Glycoproteins of 210,000 and 130,000 m.w. on activated T cells: cell distribution and antigenic relation to components on resting cells and T cell lines. J Immunol 132:3011–3018
61. Hertle MD, Adams JC, Watt FM (1991) Integrin expression during human epidermal development in vivo and in vitro. Development 112:193–206

62. Ianaro A, Calignano A, Koteliansky V, Gotwals P, Bucci M, Gerli R, Santucci L, Fiorucci S, Cirino G, Cicala C (2000) Anti-very late antigen-1 monoclonal antibody modulates the development of secondary lesion and T-cell response in experimental arthritis. Lab Invest 80(1):73–80
63. Inderjeeth C, Redfern A, Huang M, Yun H, Grant T, Fritz L, Fuller D (2013) Safety, pharmacokinetics, and pharmacodynamics of SAN-300, a novel monoclonal antibody against very late antigen-1: results of a phase 1 study in healthy volunteers and patients with active rheumatoid arthritis. Arthritis Rheum 65:1439
64. Ivarsson M, McWhirter A, Black CM, Rubin K (1993) Impaired regulation of collagen pro-alpha 1(I) mRNA and change in pattern of collagen-binding integrins on scleroderma fibroblasts. J Invest Dermatol 101:216–221
65. Ivaska J, Reunanen H, Westermarck J, Koivisto L, Kähäri VM, Heino J (1999) Integrin alpha2beta1 mediates isoform-specific activation of p38 and upregulation of collagen gene transcription by a mechanism involving the alpha2 cytoplasmic tail. J Cell Biol 147(2):401–416
66. Jikko A, Chen D, Mendrick DL, Damsky CH, Harris SE (1999) Collagen integrin receptors regulate early osteoblast differentiation induced by BMP-2. J Bone Min. Res 14(7):1075–1083
67. Kagami S, Kondo S, Loster K, Reutter W, Tamaki T, Yoshizumi M, Kuroda Y, Urushihara M (2001) Requirement for tyrosine kinase-ERK1/2 signaling in alpha 1 beta 1 integrin-mediated collagen matrix remodeling by rat mesangial cells. Exp Cell Res 268(2):274–283
68. Kagami S, Loster K, Reutter W, Kuhara T, Yasutomo K, Kuroda Y, Kondo S (1999) Alpha1beta1 integrin-mediated collagen matrix remodeling by rat mesangial cells is differentially regulated by transforming growth factor-beta and platelet-derived growth factor-BB. J Am Soc Nephrol 10(4):779–789
69. Kagami S, Urushihara M, Kondo S, Hayashi T, Yamano H, Loster K, Vossmeyer D, Reutter W, Kuroda Y (2002) Effects of anti-alpha1 integrin subunit antibody on anti-Thy-1 glomerulonephritis. Lab Invest 82:1219–1227
70. Kagami S, Urushihara M, Loster K, Reutter W, Saijo T, Kitamura A, Kobayashi S, Kuroda Y, Kondo S (2000) Overexpression of alpha1beta1 integrin directly affects rat mesangial cell behavior. Kidney Int 58(3):1088–1097
71. Klekotka PA, Santoro SA, Ho A, Dowdy SF, Zutter MM (2001) Mammary epithelial cell-cycle progression via the alpha(2)beta(1) integrin: unique and synergistic roles of the alpha(2) cytoplasmic domain. Am J Pathol 159:983–992
72. Knight CG, Peachey AR, Tuckwell DS, Farndale RW, Barnes MJ, Morton LF (2000) The collagen-binding A-domains of integrins alpha(1)beta(1) and alpha(2)beta(1) recognize the same specific amino acid sequence, GFOGER, in native (triple-helical) collagens. J Biol Chem 275(1):35–40
73. Korhonen M, Ylanne J, Laitinen L, Virtanen I (1990) The alpha 1-alpha 6 subunits of integrins are characteristically expressed in distinct segments of developing and adult human nephron. J Cell Biol 111:1245–1254
74. Koukoulis GK, Virtanen I, Korhonen M, Laitinen L, Quaranta V, Gould VE (1991) Immunohistochemical localization of integrins in the normal, hyperplastic, and neoplastic breast. Correlations with their functions as receptors and cell adhesion molecules. Am J Pathol 139:787–799
75. Koukoulis GK, Virtanen I, Gould VE, Warren WH (1997) Immunolocalization of integrins in the normal lung and in pulmonary carcinomas. Hum Pathol 28(9):1018–1025 (Erratum in: Hum Pathol 1997 28(12):1442)
76. Käpylä J, Jäälinoja J, Tulla M, Ylöstalo J, Nissinen L, Viitasalo T, Vehviläinen P, Marjomäki V, Nykvist P, Säämänen A-M et al (2004) The fibril-associated collagen IX provides a novel mechanism for cell adhesion to cartilaginous matrix. J Biol Chem 279:51677–51687
77. Lahti M, Bligt E, Niskanen H, Parkash V, Brandt AM, Jokinen J, Patrikainen P, Kapyla J, Heino J, Salminen TA (2011) Structure of collagen receptor integrin alpha(1)I domain carrying the activating mutation E317A. J Biol Chem 286:43343–43351
78. Lallier, T., and Bronner-Fraser, M. (1993). Inhibition of neural crest cell attachment by integrin antisense oligonucleotides. *Science (80-.).* **259**, 692–695
79. Langholz O, Mauch C, Kozlowska E, Bank I, Krieg T, Eckes B, Rockel D (1995) Collagen and collagenase gene expression in three-dimensional collagen lattices are differentially regulated by alpha 1 beta 1 and alpha 2 beta 1 integrins. J Cell Biol 131(6 Pt 2):1903–1915
80. Lee HJ, Kim SY, Koh JM, Bok J, Kim KJ, Kim KS, Park MH, Shin HD, Park BL, Kim TH et al (2007) Polymorphisms and haplotypes of integrinalpha1 (ITGA1) are associated with bone mineral density and fracture risk in postmenopausal Koreans. Bone 41:979–986
81. Leikina E, Mertts MV, Kuznetsova N, Leikin S (2002) Type I collagen is thermally unstable at body temperature. Proc Natl Acad Sci USA 99:1314–1318
82. La Linn M, Eble JA, Lubken C, Slade RW, Heino J, Davies J, Suhrbier A (2005) An arthritogenic alphavirus uses the alpha1beta1 integrin collagen receptor. Virology 336:229–239
83. Liu X, Wu H, Byrne M, Jeffrey J, Krane S, Jaenisch R (1995) A targeted mutation at the known collagenase cleavage site in mouse type I collagen impairs tissue remodeling. J Cell Biol 130:227–237
84. Lochter A, Werb Z, Bissell MJ, Navre M (1999) alpha1 and alpha2 integrins mediate invasive activity of mouse mammary carcinoma cells

through regulation of stromelysin-1 expression. Mol Biol Cell 10(2):271–282
85. Loeser RF, Tan L, Goldring MB, Sadiev S (2000) Integrin expression by primary and immortalized human chondrocytes: evidence of a differential role for alpha1beta1 and alpha2beta1 integrins in mediating chondrocyte adhesion to types II and VI collagen. Osteoarthr Cartil 8(2):96–105
86. Loster K, Heidrich C, Hofmann W, Reutter W, Voigt S (1994) Cell-collagen adhesion is inhibited by monoclonal antibody 33.4 against the rat alpha 1-integrin subunit. Exp Cell Res 212(1):155–160
87. Loster K, Hofmann W, Reutter W, Danker K, Vossmeyer D (2001) alpha1 Integrin cytoplasmic domain is involved in focal adhesion formation via association with intracellular proteins. Biochem J 356(Pt 1):233–240
88. Louis H, Kakou A, Regnault V, Labat C, Bressenot A, Gao-Li J, Gardner H, Thornton SN, Challande P, Li Z et al (2007) Role of alpha1beta1-integrin in arterial stiffness and angiotensin-induced arterial wall hypertrophy in mice. Am J Physiol Hear. Circ Physiol 293:H2597–H2604
89. Ma Q, Shimaoka M, Lu C, Jing H, Carman CV, Springer TA (2002) Activation-induced conformational changes in the I domain region of lymphocyte function-associated antigen 1. J Biol Chem 277:10638–10641
90. Macias-Perez I, Borza C, Chen X, Yan X, Ibanez R, Mernaugh G, Matrisian LM, Zent R, Pozzi A (2008) Loss of integrin alpha1beta1 ameliorates Kras-induced lung cancer. Cancer Res 68:6127–6135
91. Makihira S, Ohno S, Kawamoto T, Fujimoto K, Okimura A, Yoshida E, Noshiro M, Hamada T, Kato Y, Yan W (1999) Enhancement of cell adhesion and spreading by a cartilage-specific noncollagenous protein, cartilage matrix protein (CMP/Matrilin-1), via integrin alpha1beta1. J Biol Chem 274(16):11417–11423
92. Marcinkiewicz C, Weinreb PH, Calvete JJ, Kisiel DG, Mousa SA, Tuszynski GP, Lobb RR (2003) Obtustatin: a potent selective inhibitor of $\alpha 1 \beta 1$ integrin in vitro and angiogenesis in vivo advances in brief. 2020–2023
93. Mattila E, Pellinen T, Nevo J, Vuoriluoto K, Arjonen A, Ivaska J (2005) Negative regulation of EGFR signalling through integrin-alpha1beta1-mediated activation of protein tyrosine phosphatase TCPTP. Nat Cell Biol 7:78–85
94. Mechtersheimer G, Barth T, Quentmeier A, Moller P (1994) Differential expression of beta 1 integrins in nonneoplastic smooth and striated muscle cells and in tumors derived from these cells. Am J Pathol 144:1172–1182
95. Meharra EJ, Hassett D, Parker C, Havran W, Gardner H, Schon M (2000) Reduced gut intraepithelial lymphocytes in VLA1 null mice. Cell Immunol 201(1):1–5
96. Mendrick DL, duMont SS, Sandstrom DJ, Kelly DM (1995) Glomerular epithelial and mesangial cells differentially modulate the binding specificities of VLA-1 and VLA-2. Lab Invest 72(3):367–375
97. Moeckel GW, Zhang L, Chen X, Rossini M, Zent R, Pozzi A (2006) Role of integrin alpha1beta1 in the regulation of renal medullary osmolyte concentration. Am J Physiol Ren. Physiol 290:F223–F231
98. Moiseeva EP, Baron JH, de Bono DP, Spring EL (1999) Galectin 1 modulates attachment, spreading and migration of cultured vascular smooth muscle cells via interactions with cellular receptors and components of extracellular matrix. J Vasc Res 36(1):47–58
99. Morales SA, Mareninov S, Prasad P, Wadehra M, Braun J, Gordon LK (2007) Collagen gel contraction by ARPE-19 cells is mediated by a FAK-Src dependent pathway. Exp Eye Res 85:790–798
100. Moumen M, Chiche A, Cagnet S, Petit V, Raymond K, Faraldo MM, Deugnier M-A, Glukhova MA (2011) The mammary myoepithelial cell. Int J Dev Biol 55:763–771
101. Nishida W, Mori S, Takahashi M, Ohkawa Y, Tadokoro S, Yoshida K, Hiwada K, Hayashi K, Sobue K, Nakamura M (2002) A triad of serum response factor and the GATA and NK families governs the transcription of smooth and cardiac muscle genes. J Biol Chem 277(9):7308–7317
102. Nyberg P, Xie L, Sugimoto H, Colorado P, Sund M, Holthaus K, Sudhakar A, Salo T, Kalluri R (2008) Characterization of the anti-angiogenic properties of arresten, an alpha1beta1 integrin-dependent collagen-derived tumor suppressor. Exp Cell Res 314:3292–3305
103. Nykvist P, Ivaska J, Kapyla J, Pihlajaniemi T, Heino J, Tu H (2000) Distinct recognition of collagen subtypes by alpha(1)beta(1) and alpha(2)beta(1) integrins. Alpha(1)beta(1) mediates cell adhesion to type XIII collagen. J Biol Chem 275(11):8255–8261
104. Nymalm Y, Puranen JS, Nyholm TKM, Käpylä J, Kidron H, Pentikäinen OT, Airenne TT, Heino J, Slotte JP, Johnson MS et al (2004) Jararhagin-derived RKKH peptides induce structural changes in alpha1I domain of human integrin alpha1beta1. J Biol Chem 279:7962–7970
105. O'Reilly MS, Holmgren L, Shing Y, Chen C, Rosenthal RA, Moses M, Lane WS, Cao Y, Sage EH, Folkman J (1994) Angiostatin: a novel angiogenesis inhibitor that mediates the suppression of metastases by a Lewis lung carcinoma. Cell 79:315–328
106. Patterson BC, Sang QA (1997) Angiostatin-converting enzyme activities of human matrilysin (MMP-7) and gelatinase B/type IV collagenase (MMP-9). J Biol Chem 272:28823–28825

107. Peng YW, Zallocchi M, Meehan DT, Delimont D, Chang B, Hawes N, Wang W, Cosgrove D (2008) Progressive morphological and functional defects in retinas from alpha1 integrin-null mice. Invest Ophthalmol Vis Sci 49:4647–4654
108. Perris R, Syfrig J, Paulsson M, Bronner-Fraser M (1993) Molecular mechanisms of neural crest cell attachment and migration on types I and IV collagen. J Cell Sci 106:1357–1368
109. Pozzi A, Giancotti FG, Gardner HA, Wary KK (1998) Integrin alpha1beta1 mediates a unique collagen-dependent proliferation pathway in vivo. J Cell Biol 142:587–594
110. Pozzi A, LeVine WF, Gardner HA (2002) Low plasma levels of matrix metalloproteinase 9 permit increased tumor angiogenesis. Oncogene 21:272–281
111. Pozzi A, Miles LA, Wagner S, Soloway P, Gardner HA, Moberg PE (2000) Elevated matrix metalloprotease and angiostatin levels in integrin alpha 1 knockout mice cause reduced tumor vascularization. Proc Natl Acad Sci USA 97(5):2202–2207
112. Racine-Samson L, Bissell DM, Rockey DC (1997a) The role of alpha1beta1 integrin in wound contraction. A quantitative analysis of liver myofibroblasts in vivo and in primary culture. J Biol Chem 272:30911–30917
113. Racine-Samson L, Rockey DC, Bissell DM (1997b) The role of alpha1beta1 integrin in wound contraction. A quantitative analysis of liver myofibroblasts in vivo and in primary culture. J Biol Chem 272:30911–30917
114. Ratzinger S, Grassel S, Dowejko A, Reichert TE, Bauer RJ (2011) Induction of type XVI collagen expression facilitates proliferation of oral cancer cells. Matrix Biol 30:118–125
115. Ravanti L, Lopez-Otin C, Kahari VM, Heino J (1999) Induction of collagenase-3 (MMP-13) expression in human skin fibroblasts by three-dimensional collagen is mediated by p38 mitogen-activated protein kinase. J Biol Chem 274(4):2446–2455
116. Reunanen N, Foschi M, Han J, Kahari VM (2000) Activation of extracellular signal-regulated kinase 1/2 inhibits type I collagen expression by human skin fibroblasts. J Biol Chem 275:34634–34639
117. Rich RL, Owens RT, Carson M, Hook A, Moore D, Symersky J, Yang VW, Narayana SV, Hook M, Deivanayagam CC (1999) Trench-shaped binding sites promote multiple classes of interactions between collagen and the adherence receptors, alpha(1)beta(1) integrin and Staphylococcus aureus cna MSCRAMM. J Biol Chem 274(35):24906–24913
118. Richter M, Ray SJ, Chapman TJ, Austin SJ, Rebhahn J, Mosmann TR, Gardner H, Kotelianski V, deFougerolles AR, Topham DJ (2007) Collagen distribution and expression of collagen-binding alpha1beta1 (VLA-1) and alpha2beta1 (VLA-2) integrins on CD4 and CD8 T cells during influenza infection. J Immunol 178:4506–4516
119. Richter MV, Topham DJ (2007) The alpha1beta1 integrin and TNF receptor II protect airway CD8 + effector T cells from apoptosis during influenza infection. J Immunol 179:5054–5063
120. Rider DA, Nalathamby T, Nurcombe V, Cool SM (2007) Selection using the alpha-1 integrin (CD49a) enhances the multipotentiality of the mesenchymal stem cell population from heterogeneous bone marrow stromal cells. J Mol Histol 38:449–458
121. Ronzière M-C, Aubert-Foucher E, Gouttenoire J, Bernaud J, Herbage D, Mallein-Gerin F (2005) Integrin alpha1beta1 mediates collagen induction of MMP-13 expression in MC615 chondrocytes. Biochim Biophys Acta 1746:55–64
122. Sampson NS, Enke DA, Cosgrove D, Koteliansky V, Gotwals P, Ryan ST (2001) Global gene expression analysis reveals a role for the alpha 1 integrin in renal pathogenesis. J Biol Chem 276(36):34182–34188
123. Santala P, Heino J (1991) Regulation of integrin-type cell adhesion receptors by cytokines. J Biol Chem 266(34):23505–23509
124. Schadendorf D, Haney U, Ostmeier H, Suter L, Czarnetzki BM, Gawlik C (1993) Tumour progression and metastatic behaviour in vivo correlates with integrin expression on melanocytic tumours. J Pathol 170(4):429–434
125. Schadendorf D, Makki A, Alijagic S, Kupper M, Mrowietz U, Henz BM, Fichtner I (1996) Metastatic potential of human melanoma cells in nude mice–characterisation of phenotype, cytokine secretion and tumour-associated antigens. Br J Cancer 74(2):194–199
126. Schocklmann HO, Kralewski M, Hartner A, Ludke A, Sterzel RB, Lang S (2000) Distinct structural forms of type I collagen modulate cell cycle regulatory proteins in mesangial cells. Kidney Int 58(3):1108–1120
127. Schwartz MA, Ginsberg MH (2002) Networks and crosstalk: integrin signalling spreads. Nat Cell Biol 4:E65–E68
128. Senger DR, Benes JE, Perruzzi CA, Sergiou AP, Detmar M, Claffey KP (1997) Angiogenesis promoted by vascular endothelial growth factor: regulation through alpha1beta1 and alpha2beta1 integrins. Proc Natl Acad Sci USA 94(25):13612–13617
129. Senger DR, Perruzzi CA, Streit M, Koteliansky VE, de Fougerolles AR, Detmar M (2002) The alpha(1)beta(1) and alpha(2)beta(1) integrins provide critical support for vascular endothelial growth factor signaling, endothelial cell migration, and tumor angiogenesis. Am J Pathol 160:195–204
130. Seo N, Russell BH, Rivera JJ, Liang X, Xu X, Afshar-Kharghan V, Hook M (2010) An engineered alpha1 integrin-binding collagenous sequence. J Biol Chem 285:31046–31054
131. Seo NS, Zeng CQ, Hyser JM, Utama B, Crawford SE, Kim KJ, Hook M, Estes MK (2008) Integrins alpha1beta1 and alpha2beta1 are receptors for the

rotavirus enterotoxin. Proc Natl Acad Sci USA 105:8811–8818
132. Shi M, Pedchenko V, Greer BH, Van Horn WD, Santoro Sa, Sanders CR, Hudson BG, Eichman BF, Zent R, Pozzi A (2012) Enhancing integrin $\alpha 1$ inserted (I) domain affinity to ligand potentiates integrin $\alpha 1\beta 1$-mediated down-regulation of collagen synthesis. J Biol Chem 287:35139–35152
133. Shimaoka M, Takagi J, Springer TA (2002) Conformational regulation of integrin structure and function. Annu Rev Biophys Biomol Struct 31:485–516
134. Skinner MP, Raines EW, Ross R (1994) Dynamic expression of alpha 1 beta 1 and alpha 2 beta 1 integrin receptors by human vascular smooth muscle cells. Alpha 2 beta 1 integrin is required for chemotaxis across type I collagen-coated membranes. Am J Pathol 145:1070–1081
135. Sobel RA, Maeda A, Chen M, Hinojoza JR (1998) Endothelial cell integrin laminin receptor expression in multiple sclerosis lesions. Am J Pathol 153(2):405–415
136. Soligo D, Luksch R, Manara G, Quirici N, Parravicini C, Lambertenghi Deliliers G, Schirò R (1990) Expression of integrins in human bone marrow. Br J Haematol 76(3):323–332
137. Stamatoglou SC, Johansson S, Forsberg E, Hughes RC, Bawumia S (1991) Affinity of integrin alpha 1 beta 1 from liver sinusoidal membranes for type IV collagen. *1*. FEBS Lett 288(1–2):241–243
138. Sterry W, Konter U, Kellner I, Boehncke WH, Mielke U (1992) Role of beta 1-integrins in epidermotropism of malignant T cells. Am J Pathol 141(4):855–860
139. Stewart HJ, Jessen KR, Mirsky R, Turner D (1997) Expression and regulation of alpha1beta1 integrin in Schwann cells. J Neurobiol 33(7):914–928
140. Sutherland AE, Calarco PG, Damsky CH (1993) Developmental regulation of integrin expression at the time of implantation in the mouse embryo. Development 119:1175–1186
141. Suzuki K, Okuno T, Yamamoto M, Pasterkamp RJ, Takegahara N, Takamatsu H, Kitao et al (2007) Semaphorin 7A initiates T-cell-mediated inflammatory responses through alpha1beta1 integrin. Nature 446(7136):680–684
142. Szulgit G, Wandel A, Tenenhaus M, Panos R, Gardner H, Rudolph R (2002) Alterations in fibroblast alpha1beta1 integrin collagen receptor expression in keloids and hypertrophic scars. J Invest Dermatol 118(3):409–415
143. Tawil NJ, Blacher R, Esch F, Reichardt LF, Turner DC, Carbonetto S, Houde M (1990) Alpha 1 beta 1 integrin heterodimer functions as a dual laminin/collagen receptor in neural cells. Biochem 29(27):6540–6544
144. Terracio L, Rubin K, Gullberg D, Balog E, Carver W, Jyring R, Borg TK (1991) Expression of collagen binding integrins during cardiac development and hypertrophy. Circ Res 68:734–744
145. Tomaselli KJ, Doherty P, Emmett CJ, Damsky CH, Walsh FS, Reichardt LF (1993) Expression of beta 1 integrins in sensory neurons of the dorsal root ganglion and their functions in neurite outgrowth on two laminin isoforms. J Neurosci 13:4880–4888
146. Tulla M, Lahti M, Puranen JS, Brandt AM, Kapyla J, Domogatskaya A, Salminen TA, Tryggvason K, Johnson MS, Heino J (2008) Effects of conformational activation of integrin alpha 1I and alpha 2I domains on selective recognition of laminin and collagen subtypes. Exp Cell Res 314:1734–1743
147. Tulla M, Pentikäinen OT, Viitasalo T, Käpylä J, Impola U, Nykvist P, Nissinen L, Johnson MS, Heino J (2001) Selective binding of collagen subtypes by integrin alpha 1I, alpha 2I, and alpha 10I domains. J Biol Chem 276:48206–48212
148. Vaalamo M, Mattila L, Johansson N, Kariniemi AL, Karjalainen-Lindsberg ML, Kahari VM, Saarialho-Kere U (1997) Distinct populations of stromal cells express collagenase-3 (MMP-13) and collagenase-1 (MMP-1) in chronic ulcers but not in normally healing wounds. J Invest Dermatol 109:96–101
149. Vossmeyer D, Loster K, Reutter W, Danker K, Hofmann W (2002) Phospholipase Cgamma binds alpha1beta1 integrin and modulates alpha1beta1 integrin-specific adhesion. J Biol Chem 277(7):4636–4643
150. Wary KK, Mainiero F, Isakoff SJ, Marcantonio EE, Giancotti FG (1996) The adaptor protein Shc couples a class of integrins to the control of cell cycle progression. Cell 87:733–743
151. Wary KK, Mariotti A, Zurzolo C, Giancotti FG (1998) A requirement for caveolin-1 and associated kinase Fyn in integrin signaling and anchorage-dependent cell growth. Cell 94:625–634
152. Wu JE, Santoro SA (1994) Complex patterns of expression suggest extensive roles for the alpha 2 beta 1 integrin in murine development. Dev Dyn 199:292–314
153. Xu Y, Gurusiddappa S, Rich RL, Owens RT, Keene DR, Mayne R, Höök a, Höök M (2000) Multiple binding sites in collagen type I for the integrins alpha1beta1 and alpha2beta1. J Biol Chem 275:38981–38989
154. Yu L, Su Y, Paueksakon P, Cheng H, Chen X, Wang H, Harris RC, Zent R, Pozzi A (2012) Integrin alpha1/Akita double-knockout mice on a Balb/c background develop advanced features of human diabetic nephropathy. Kidney Int 81:1086–1097
155. Zallocchi M, Johnson BM, Meehan DT, Delimont D, Cosgrove D (2013) alpha1beta1 Integrin/Rac1-Dependent Mesangial Invasion of Glomerular Capillaries in Alport Syndrome. Am J Pathol 183:1269–1280
156. Zemmyo M, Meharra EJ, Kuhn K, Creighton-Achermann L, Lotz M (2003) Accelerated, aging-dependent development of osteoarthritis in alpha1 integrin-deficient mice. Arthritis Rheum 48: 2873–2880

157. Zent R, Yan X, Su Y, Hudson BG, Borza DB, Moeckel GW, Qi Z, Sado Y, Breyer MD, Voziyan P et al (2006) Glomerular injury is exacerbated in diabetic integrin alpha1-null mice. Kidney Int 70:460–470
158. Zhang WM, Kapyla J, Puranen JS, Knight CG, Tiger CF, Pentikainen OT et al (2003) alpha11beta1 integrin recognizes the GFOGER sequence in interstitial collagens. J Biol Chem 278:7270–7277
159. Zhang Z, Turner DC, Tarone G (1993) Expression of integrin alpha 1 beta 1 is regulated by nerve growth factor and dexamethasone in PC12 cells. Functional consequences for adhesion and neurite outgrowth 1. J Biol Chem 268(8):5557–5565

$\alpha2\beta1$ Integrin

3

Aasakiran Madamanchi, Samuel A. Santoro,
and Mary M. Zutter

Abstract

The $\alpha2\beta1$ integrin, also known as VLA-2, GPIa-IIa, CD49b, was first identified as an extracellular matrix receptor for collagens and/or laminins [55, 56]. It is now recognized that the $\alpha2\beta1$ integrin serves as a receptor for many matrix and nonmatrix molecules [35, 79, 128]. Extensive analyses have clearly elucidated the $\alpha2$ I domain structural motifs required for ligand binding, and also defined distinct conformations that lead to inactive, partially active or highly active ligand binding [3, 37, 66, 123, 136, 137, 140]. The mechanisms by which the $\alpha2\beta1$ integrin plays a critical role in platelet function and homeostasis have been carefully defined via in vitro and in vivo experiments [76, 104, 117, 125]. Genetic and epidemiologic studies have confirmed human physiology and disease states mediated by this receptor in immunity, cancer, and development [6, 20, 21, 32, 43, 90]. The role of the $\alpha2\beta1$ integrin in these multiple complex biologic processes will be discussed in the chapter.

Keywords

$\alpha2\beta1$ integrin · Collagen · Disease models

3.1 Collagen Receptors-Structure and Ligand Binding

The $\alpha2\beta1$ integrin consists of an obligate heterodimer formed from the $\alpha2$ integrin subunit non-covalently associated with the $\beta1$ subunit. It is one of four 'I domain' integrins, named for the presence of a highly conserved, extracellular, (inserted) I domain, which mediates specific binding of ligands including, most prominently, collagens [30]. The $\alpha2$ subunit I domain is an autonomously folding domain of approximately 220 amino acids [30]. The I domain found in the collagen receptors is shared with the alpha subunits of the leukocyte $\beta2$ integrins and is highly homologous to the A domain found in Von Willebrand factor, in cartilage matrix protein, in some collagen subtypes and in components of the complement system. The crystal

A. Madamanchi · S. A. Santoro · M. M. Zutter (✉)
Department of Pathology, Microbiology and Immunology, Vanderbilt University School of Medicine, Nashville, TN, USA
e-mail: mary.zutter@Vanderbilt.Edu

D. Gullberg (ed.), *I Domain Integrins*, Advances in Experimental Medicine and Biology 819,
DOI: 10.1007/978-94-017-9153-3_3,

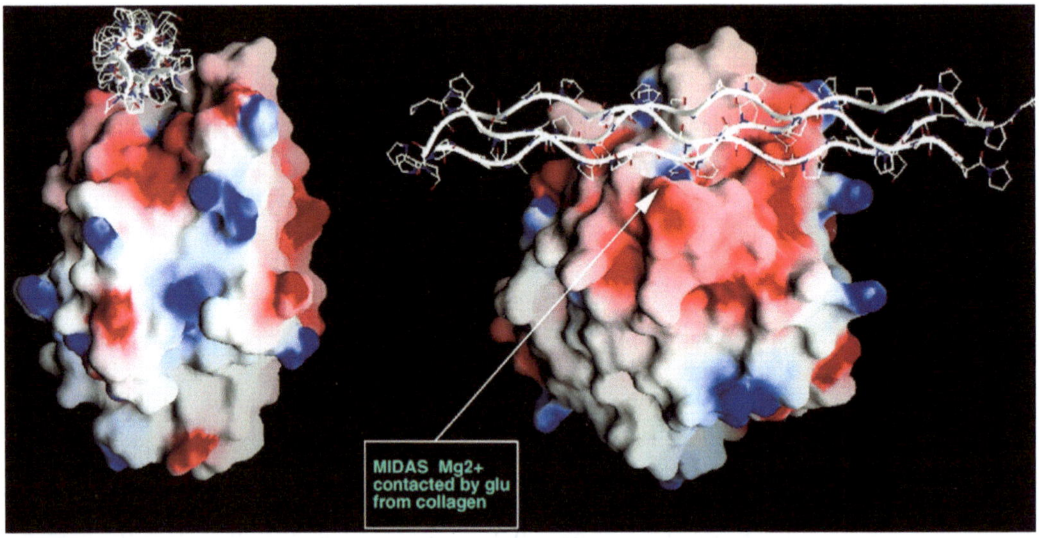

Fig. 3.1 A hypothetical model of an I-domain·collagen complex. A collagen triple helix (*white spiral*) is shown in a possible fit a groove on the MIDAS face. A glutamate side chain from the collagen coordinating the metal ion as indicated by arrow. The I domain is colored according to surface charge distribution (*blue* positive, *red* negative, *white* neutral. Two orthogonal views are shown (Reprinted from Fig. 5, Emsley et al. 1997)

structure of the α2 integrin I domain was first defined in 1997 (Fig. 3.1) [20]. The α2 subunit shares many similarities in structure and ligand binding with the other I domain integrins, including the Mg2+ dependence for binding, and enhancement of integrin function by Mn2+ [36, 60, 116, 118]. The I domain contains a conserved cation binding site, the metal ion-dependent adhesion site (MIDAS) with clear preference for Mg2+/Mn2+. The MIDAS motif is critical for collagen recognition [69].

Structural and other studies of the α2 I domain have identified an inactive or closed conformation, an intermediate or low-affinity conformation, and an active or high-affinity conformation [3, 37, 66, 123, 136, 137, 140]. Experimental approaches have characterized the role that distinct I domain residues play in receptor conformation and ligand binding capability. Mutation of the Mg2+ binding site at T221 disrupts the MIDAS site and inactivates I domain function [112, 135]. Insertion of a disulfide bridge between helices locks the I domain into a high affinity conformation [124]. Within the α2 integrin I domain, amino acid E318 forms a salt bridge with amino acid R288, thereby maintaining the α2 integrin I domain in a closed conformation. Recent reports by Carafoli et al. indicate that mutation of E318 to alanine causes disruption of this salt bridge and promotes the transition to the open, high affinity conformation which enhances α2 integrin I domain binding to low-affinity ligands [19].

Crystal structures of the active α2 I domain E318W complexed with the GFOGER peptides revealed two domains bound to a single triple helix [19], suggesting that a single GxOGER motif in the heterotrimeric collagen V or the FACIT collagen IX, may support binding of the activated integrin. Similarly, a crystal structure of the analogous E317A mutant of α1 I domain also resulted in an opening of the helices [89], and modelling of a similar peptide, GLOGEN, onto E317A [25] allows similar conclusions to be drawn for α1β1.

The α2β1 integrin has high affinity for collagen Type I. Evaluation of the role of the α2β1 integrin structure and function has led to the identification of a number of novel ligands. The other ligands can be subdivided into other collagens, non-collagenous molecules with collagen-like triple helical structures, laminin and molecules with laminin domains, proteoglycans, as well as infectious organisms, primarily viruses, and other potential non-matrix ligands.

Among collagens, the $\alpha 2\beta 1$ integrin preferentially binds fibrillar isoforms (I-III, V and XI). Integrin $\alpha 2\beta 1$ also recognizes the network forming collagen IV [78], the beaded-filament forming collagen VI, and the transmembrane collagen XIII when in an active, high-affinity conformation [67]. Modulation of integrin conformation by cytoplasmic signals provides an integrin-specific mechanism for adjusting ligand affinity known as 'inside-out' signaling. However, the binding of purified recombinant $\alpha 2$ integrin I domain to collagen type I or IV reflects the same relative affinity for the ligand as does the parent integrin; indicating that differences in the integrin-binding motifs of these isoforms most likely account for the differential recognition by the integrin [18]. The development of overlapping sets of collagen-derived peptides, termed Toolkits, facilitated systematic mapping of motifs for integrin binding and identified the collagen sequence GFOGER as the major high-affinity binding motif for the $\alpha 2\beta 1$ integrin [82, 83, 112]. The GFOGER motif, found in Type I, II and XI, is uniquely able to bind platelet integrin $\alpha 2\beta 1$ without prior activation [124], suggesting the ability to induce the active conformation without the inside-out signals needed for lower-affinity motifs.

More recently, other collagens were defined as $\alpha 2\beta 1$ integrin ligands. Collagen XVI, a member of the fibril-associated collagens with interrupted triple helices (FACITs), binds to the $\alpha 2\beta 1$ integrin, as well as to the $\alpha 1\beta 1$ integrin [33]. The $\alpha 2\beta 1$ integrin ligand, collagen XXIII, a transmembrane collagen, has been reported as the primary apical binding partner for the integrin in keratinocyte adhesion in the epidermis [47, 53, 141].

Many molecules of the immune system contain segments of a collagen triple helix, including C1q. As discussed below, our laboratory showed that $\alpha 2\beta 1$ integrin-mediated stimulation of an innate immune response required $\alpha 2\beta 1$ integrin dependent-adhesion to C1q in an immune complex [34]. The full length $\alpha 2\beta 1$ integrin and the $\alpha 2$ integrin I domain adhere to C1q as well as to members of the collectin family of proteins, including surfactant protein A and mannose binding lectin. The $\alpha 2$ integrin I domain adheres to C1q in the absence of activation. However, the activated E318A mutant of $\alpha 2$ I domain bound to C1q with higher affinity than wild type $\alpha 2$ integrin I domain.

As with collagens, adhesion to laminin isoforms is mediated by the $\alpha 2$ integrin I domain, however laminin binding only occurs in the active, high-affinity conformation [18, 22, 36]. Isolated full-length $\alpha 2$ integrin subunit has been shown to bind to laminin-111 (previously laminin-1) and laminin-332 (previously laminin-5). Netrin-4, a member of the netrin family of guidance signals, demonstrates high homology to the beta 1 chain of laminins and binds to the $\alpha 2\beta 1$ integrin and to the $\alpha 3\beta 1$ integrin [148]. To date, an extensive and detailed molecular analysis to identify the recognition site/s on laminin has not been performed. Laminin-binding has proven to occur constitutively in some cell types, and inducibly in others. However, the role of these adhesive events is not well understood.

Perlecan, a heparin sulfate proteoglycan, and its C-terminal fragment, endorepellin, bind the $\alpha 2\beta 1$ integrin [45, 46]. The terminal globular domain of endorepellin, LG3, interacts directly with the $\alpha 2$ I domain. This interaction has been studied in the context of angiogenesis and shown to be important for $\alpha 2\beta 1$ integrin-dependent angiogenesis.

Decorin, another small leucine-rich proteoglycan modulates $\alpha 2\beta 1$ integrin matrix interactions by playing an important role in regulating extracellular matrix assembly as well as directly interacting with the integrin [13, 40, 52, 143]. Decorin binding to collagen has been shown to affect fibril formation by initially delaying lateral fibril growth and reducing average fibril diameter [142]. Additionally, decorin interacts with $\alpha 2\beta 1$, but not $\alpha 1\beta 1$ integrin, at a site distinct from the collagen-binding domain. Adhesive interaction between decorin and the $\alpha 2\beta 1$ integrin was first identified in platelets, and later discovered to be important in angiogenesis.

Single nucleotide polymorphisms in the integrin $\alpha 2$ gene, as discussed later in more detail, have an important role in the predisposition of patients to cardiovascular disease. One such minor allele difference (rs1801106; G1600A) has now been shown to attenuate adhesion of platelets to decorin but not to collagen and is associated with increased risk for recurrence of stroke [87]. The non-conservative amino acid substitution E534K, is the basis of the human platelet alloantigen system HPA-5, providing the first evidence of a functional effect of HPA-5 alleles.

The $\alpha 2\beta 1$ integrin serves as a receptor for many different infectious organisms. In many cases the organisms usurp $\alpha 2\beta 1$ integrin's routine biology for attachment, cell entry and transmission throughout the body. The best studied interaction of $\alpha 2\beta 1$ integrin is with echovirus (EV1) [10–12, 31]. EV1, is a human RNA virus which binds directly to the I domain of human $\alpha 2\beta 1$ integrin. Unlike most viruses that exploit integrin receptors, EV1 does not undergo clathrin-mediated endocytosis, but instead clusters on caveosomes and is internalized via a clathrin- and caveolin-independent macropinocytosis-like mechanism [73, 93]. Additionally, EV1 binding has been demonstrated to activate PKCα, while inhibition of PKCα signaling blocks EV1 internalization [138]. Interestingly, EV1, unlike other $\alpha 2\beta 1$ integrin ligands, preferentially binds the inactive, closed conformation of the integrin over the active, high affinity conformation [68].

Not only do infectious organisms utilize the integrin as a receptor, lectins that recognize high mannose glycans on viruses are produced from bacteria, algae, plants and animals and bind the $\alpha 2\beta 1$ integrin. A recently characterized anti-HIV lectin from Pseudomonas fluorescens Pf0-1 exhibited potent antiviral activity against influenza [121]. The lectin induced loss of cell adhesion and viral death that was dependent on binding to the $\alpha 2\beta 1$ integrin. Following lectin binding to the $\alpha 2\beta 1$ integrin, the complex was internalized to the perinuclear region and not recycled. The process resembled that described for echovirus mediated cell entry and death.

3.2 Signaling

The $\alpha 2\beta 1$ integrin plays a unique contribution in regulating cell migration, proliferation and survival. The $\alpha 2$, but not the $\alpha 1$, integrin cytoplasmic domain mediates p38 MAP kinase pathway activation and a migratory phenotype [80, 81]. Expression of the constitutively active small G protein Rac1 augmented p38 MAP kinase phosphorylation and migration in mammary epithelial cell expressing full length $\alpha 2$ subunit. The role of the $\alpha 2$-cytoplasmic domain in activation of the p38 MAP kinase pathway was also established in fibroblasts. Fibroblasts grown in three-dimensional collagen gels require the $\alpha 2$-cytoplasmic domain for p38 MAP kinase activation that leads to $\alpha 2\beta 1$ integrin-mediated up-regulation of collagen gene expression [62]. Together these results support an important and specific role for the $\alpha 2$-cytoplasmic domain in mediating p38 MAP kinase activation. Similarly, the cytoplasmic domain of the $\alpha 2$ integrin subunit specifically supports insulin-mediated S-phase entry [81]. The $\alpha 2$, but not the $\alpha 1$, cytoplasmic domain mediated activation of the cyclin E/cdk2 complex, which allows entry into S-phase in the absence of growth factors other than insulin. These results suggest that the $\alpha 2$ integrin cytoplasmic domain and the insulin receptor synergize to regulate cell cycle progression.

More recently, Ivaska et al. suggested that the $\alpha 2\beta 1$ integrin induced protein serine/threonine phosphatase 2A (PP2A) activity in a collagen-specific manner [63]. In their studies, collagen-induced PP2A activation and resulting dephosphorylation of Akt and glycogen synthase kinase 3β (GSK3β) in Saos-2 cells was $\alpha 2\beta 1$ integrin-dependent. PP2A is a master regulator of a diverse set of cellular signaling pathways, so its interaction with $\alpha 2\beta 1$ integrin has the potential to dramatically increase the scope of the signaling activities of the integrin. Careful investigation of these putative signaling mechanisms is necessary for a clearer understanding of the role for the integrin in various cell types.

3.3 The α2β1 Integrin: Expression and Function

In addition to differences in collagen recognition, expression of the integrin is dependent on cell type and stage of differentiation. The α2β1 integrin is primarily expressed in vivo by epithelial cells, platelets/megakaryocytes, and fibroblasts [146]. In addition, α2β1 integrin expression on T-cells and endothelial cells varies depending on differentiation and the state of activation [29, 55, 56, 144]. The roles and functions of the integrin are therefore highly dependent not only on cell type but on signals from other cells and the associated microenvironment.

The majority of earlier work defined the role and function of the α2β1 integrin by studies of human platelets and in vitro models. These early studies implicated the α2β1 integrin in a wide range of biologic and pathobiologic functions including platelet adhesion required for hemostasis and thrombosis, epithelial differentiation and branching morphogenesis, tumor biology, wound healing, angiogenesis, and inflammation and immunity. Much has been learned over the last 10 years since development of state of the art inhibitory antibodies and gene silencing approaches, novel in vitro culture systems, and new animal models including the global α2 integrin-subunit deficient and the more recent tissue-specific α2 integrin-subunit deficient mouse. These studies and their impact on our understanding of the integrin in human biology and disease will be reviewed.

3.4 Platelet α2β1 Integrin in Ligand Binding

Patient studies first established the link between α2β1 integrin and platelet function. In 1985 Nieuwenhuis identified a deficiency of platelet glycoprotein 1a (α2 subunit) in a patient with abnormal bleeding [106, 107]. Later other patients with either reduced levels of platelet expression of the α2β1 integrin or the presence of autoantibodies to the integrin were also described to exhibit impaired platelet activation by collagen but not by other agonists.

Studies using purified human platelets established the α2β1 integrin-dependent adhesion to collagens I-VIII in a Mg2+-dependent manner. Although the α2β1 integrin is expressed at relatively low copy number on platelets (2000–4000 copies per platelet), the integrin is required for firm attachment of platelets to collagen in the subendothelium after vascular injury [56, 85, 118]. Experiments with purified platelets from genetically modified α2-deficient mice confirmed these results. Platelets from α2-deficient animals fail to adhere to type I collagen under both static and flow conditions [24]. Platelets from animals heterozygous for the α2-null allele adhere to type I collagen to a lesser degree than platelets from wild type animals, consistent with a gene dosage effect.

Platelets however have not one, but two major collagen receptors: the high affinity α2β1 integrin and the lower affinity glycoprotein VI (GPVI)/Fc receptor γ-chain (FcRγ) complex [65, 102, 105]. Despite the significant evidence supporting the role of α2β1 integrin in platelet adhesion to collagen, the relative contribution and precise roles of α2β1 integrin and GPVI/FcRγ in collagen-induced platelet adhesion and activation is still a focus on experimental inquiry. The Santoro group originally proposed a two-step, two-site model of platelet adhesion and activation to collagen, in which the higher affinity α2β1 integrin supports the initial rapid platelet-collagen interaction that mediates platelet adhesion to vessel wall under conditions of flow [103, 116, 118, 128, 134]. This allowed the subsequent engagement of a lower affinity, signal-transducing co-receptor GPVI to bind collagen and mediate collagen-induced platelet activation and aggregation. GPVI, a member of the immunoglobulin superfamily noncovalently and constitutively associates with the FcRγ chain to form a multimeric signaling complex. In this model, the α2β1 integrin mediates strong adhesion but does not contribute to platelet activation.

Other work raised question about the two-step, two-site model. Studies using a variety of agonists and inhibitors, defined the contributions and mechanisms leading to conformational changes resulting from integrin activation and provided evidence that the $\alpha2\beta1$ integrin can mediate GPVI-independent, collagen-induced platelet activation [59, 70, 75, 131]. Collagen-induced phosphorylation of PLCγ2 and Syk was inhibited by antibodies that block $\alpha2\beta1$ integrin adhesion to collagen or by selective proteases that cleave the $\beta1$ integrin subunit of the $\alpha2\beta1$ integrin. In other studies collagen-induced phosphorylation of c-Src was mediated by the $\alpha2\beta1$ integrin [61]. Platelet adhesion to intact collagen stimulated a different response than adhesion to GPVI-mimetics, further supporting distinct signaling from the $\alpha2\beta1$ integrin and GPVI/FcRγ [57, 70].

New work attempted to reconcile these conflicting stories. Auger et al. used flourescence video microscopy to monitor increases in intracellular free Ca2+ concentration ([Ca2+]i), an early stage in GPVI/FcRγ-mediated platelet activation, upon platelet adhesion to collagen under flow conditions [5]. In both human and mouse platelets under flow conditions, they identified a population of platelets that displayed an immediate increase in [Ca2 +]i upon collagen contact, as well as a second population of platelets that exhibited a delayed increase in [Ca2 +]i (1–30 s after adhering to collagen). The first population was unaffected by anti-$\alpha2\beta1$ integrin antibody blockade suggesting a GPVI/FcRγ-centric mechanism for both adhesion and activation as suggested by Nieswandt et al. The second population conformed to the traditional two-step model. The authors speculated that the apparently heterogeneous mechanism would allow for optimal response to different types of vascular injury. A similar study by Mazzucato et al. used inhibitory antibody-treated human platelets as well as mouse platelets from null animals to link short-lasting α-like and long-lasting γ-like [Ca2+]i oscillation peaks to $\alpha2\beta1$ integrin and GPVI signaling, respectively [97]. Interestingly, they found that $\alpha2\beta1$ integrin-mediated α-like calcium oscillations occur even in GPVI-null backgrounds indicating that inside-out priming of the integrin may also come from non-GPVI sources. Indeed Majoram et al. reported a role for platelet GPCRs, including protease activated receptor 1 and 4 (PAR1 and PAR4), in PLC-mediated $\alpha2\beta1$ integrin activation [94].

Together these studies demonstrated greater synergy between $\alpha2\beta1$ integrin and GPVI/FcRγ in mediating these processes than was previously understood. Resting platelets express the integrin in a low-affinity conformation. Activation, downstream of activation of GPVI, PAR1 or PAR4, or another pathway, leads to a conformational change to a high-affinity state which enhances adhesion to Type I collagen and promotes a more permissive binding to other ligands including Type IV collagen and laminin.

3.5 The $\alpha2\beta1$ Integrin: Genetic Risk for Hemostasis and Thrombosis and Much More

There is substantial variation in the baseline expression of $\alpha2\beta1$ integrin in the population; quantitative measurements of platelet surface membrane $\alpha2\beta1$ integrin expression indicate as much as a 10 fold difference among normal patients [64]. The mechanism of genetic regulation of the gene encoding the $\alpha2$ integrin subunit has been best delineated. The variation is genetically determined and associated with three alleles of the $\alpha2$ integrin subunit gene, *ITGA2* [84, 86]. The three alleles have been defined by 8 nucleotide polymorphisms in the coding region of *ITGA2* gene at nucleotide 807(C or T) and 873(G or A). Individuals carrying the 807T/873A allele express high levels of platelet $\alpha2\beta1$ integrin, whereas individuals carrying the 807C/873G allele exhibit low levels of $\alpha2\beta1$ integrin expression. Cheli et al. described another variant in CA repeat length in the *ITGA2* gene promoter that demonstrated linkage disequilibrium with variants in the coding region

[23]. Expression of α2β1 integrin may be similarly regulated in other cell types.

Genetic regulation of α2β1 integrin expression has meaningful biological implications, which have been most widely appreciated in the area of hemostasis and thrombosis. Kunicki et al. reported functional significance of α2β1 integrin expression levels by demonstrating that the number of α2β1 integrin molecules per platelet correlated with the ability of platelets to adhere to Type I collagen [85]. Clinical and epidemiologic studies based on genetic polymorphism analysis demonstrated direct clinical significance of allelic differences in levels of α2β1 integrin expression. The alleles associated with high levels of α2β1 integrin expression were associated with nonfatal myocardial infarction in individuals less than a mean age of 62 years, with an increased risk of developing diabetic retinopathy in patients with Type II diabetes mellitus, and with an increased risk of stroke [95, 119].

The original assumption was that increased integrin expression led to increased platelet adhesion to collagen and subsequent risk of thrombosis. Recently an alternative mechanism for the association was suggested. The level of α2β1 integrin expression correlated with mean platelet volume in humans and during megakaryocyte differentiation and proplatelet formation in mice [88, 126]. Surprisingly, platelet specific deletion of the integrin using the platelet factor 4 promoter-Cre construct and mice with a floxed ITGA2 gene demonstrated that mice lacking platelet-specific α2β1 integrin showed decreased megakaryocyte differentiation, diminished proplatelet formation and decreased mean platelet volume [49]. Since mice with global deletion of *ITGA2* failed to show altered megakaryocytic/platelet differentiation, compensation by alternative integrins, cell types, or pathways was sufficient to prevent this additional phenotype. Epidemiologic data linking levels of the α2β1 integrin expression with risk of pathologic thrombosis and other cardiovascular complications underscore the importance of further clarifying the role for α2β1 in platelet function.

3.6 The α2β1 Integrin During Wound Healing and Fibrosis

Early in vitro studies suggested that the α2β1 integrin was required for wound healing. Studies using skin explants ex vivo showed that keratinocyte-specific α2β1 integrin expression was re-oriented from the basal cell area to the forward-basal aspect of migrating keratinocytes where the integrin is in contact with type I collagen [114]. Keratinocyte migration into the wound was inhibited by antibodies against the α2β1 integrin [110].

In the late phase of wound healing after reepithelialization, tissue contraction of collagen fibers results in a strengthened scar. The scar is the result of extensive fibrosis, a process of tissue replacement by dense extracellular matrix composed of abundant collagen I. The α2β1 and the α1β1 integrins, both expressed by fibroblasts, are key regulators of collagen turnover in the skin, and other organs including the kidney [58, 62]. After binding to collagen, the α1β1 integrin activates a pathway that down-regulates collagen synthesis. In contrast, activation of the α2β1 integrin promotes collagen synthesis [99]. The alignment of the collagen fibers that occurs in healing wounds is recapitulated in three-dimensional collagen gels. The in vitro models provided evidence supporting critical roles for the α2β1 integrin wound healing and fibrosis.

Surprisingly, despite the results of in vitro and explant studies of wound healing, α2-deficient mice demonstrated no defect or delay in wound repair compared to wild-type animals [47, 152]. The morphology of the wounds also failed to demonstrate any difference in keratinocyte migration over exposed dermis at the wound site, suggesting that α2β1 integrin does not play an obligatory role in wound healing. No differences in scar formation or strength were noted.

Differences between the in vitro experiments and α2-null mouse model systems have several possible explanations. First, human and genetically altered mouse models may not be mechanistically equivalent. Acute loss-of-function as observed with use of inhibitory antibodies may have different effects than the germ-line deletion of α2β1. In addition, antibodies that inhibit integrin binding may produce 'negative signaling' which is distinct from the absence of integrin signaling in the null context.

Interestingly, Zweers et al. and Grenache et al. both reported increased neoangiogenesis in the wound microenvironment of α2-null mice, providing in vivo evidence for an anti-angiogenic role for α2β1 integrin [47, 152]. The increased angiogenesis in the wound healing model was quite surprising. Many studies have focused on understanding the role of the integrin in vascular development and angiogenesis, as discussed below.

Fibrosis also occurs in other tissues; the involvement of α2β1 integrin is particularly well studied in the kidney [16]. Glomerulosclerosis, characterized by excessive collagen deposition in the glomerulus is the most common cause of end stage kidney disease. The specific role of α2β1 integrin in regulating glomerulosclerosis is somewhat controversial. Mesangial cells and podocytes express the α2β1 integrin. One report studying α2-null mice on the C57Bl/6 background suggested that the integrin protected from glomerular injury [44]. In contrast, a study in which α2-null mice were crossed with the COL4A3-null mice, a model of Alport disease demonstrated that α2β1 integrin expression exacerbates glomerular injury, decreased survival, and reduced glomerular matrix deposition and scarring [48].

Consistent with a role for the integrin in promoting collagen synthesis, Miller et al. showed that inhibition of integrin α2β1, using a high-affinity small-molecular weight inhibitor protects mice from glomerular injury [100]. The anti-α2β1 inhibitor also reduced collagen synthesis in wild type but not α2-null mesangial cells, consistent with the α2β1 integrin-dependence of its antifibrotic effect.

In contrast to the kidney, the α2β1 integrin appears to have an anti-fibrotic role in the lung. Xia et al. reported that in idiopathic pulmonary fibrosis (IPF), reduced fibroblast α2β1 integrin levels allowed escape from anti-proliferative signals that normally limit fibroproliferation after tissue injury [147]. Fibroblastic foci in IPF patients were shown to be characterized by low fibroblast α2β1 integrin expression. IPF fibroblasts demonstrated decreased α2β1 integrin-mediated PP2A phosphatase activity. Downstream increases in activity of GSK-3β and β catenin provided the proliferative signals that mark the pathological IPF fibroblast phenotype. Although this work provided an elegant model for how α2β1 integrin downregulation may contribute to the pathogenesis of IPF; the relevant mechanisms for α2β1 integrin loss remain uninvestigated. Additionally, it is unclear how the established role for α2β1 integrin in promoting collagen biosynthesis and ROS production may be involved. Are the disparate elements of α2β1 integrin function somehow context or tissue-specific? Reconciliation of the pro-fibrotic and anti-fibrotic properties of the α2β1 integrin demands further study in light of its potential clinical relevance.

3.7 The α2β1 Integrin and Angiogenesis/Vasculogenesis

Angiogenesis is coordinated by a complex interplay between endothelial cells and their microenvironment. During VEGF-induced angiogenesis in vivo expression of α2β1 integrin is up-regulated and α2β1 integrin expression has been observed on the sprouting tips of neonatal blood vessels [38, 122]. Together these results suggested an important function for α2β1 in angiogenesis, however the precise nature of the integrin's role is still incompletely understood.

The earliest investigations into the functional role of α2β1 in angiogenesis employed inhibitory antibodies during in vitro studies. Early reports from Gamble et al. indicated that anti-α2β1 antibodies inhibited endothelial cell proliferation on collagen [41]. Soon after, Davis reported that anti-α2 inhibited lumen and tube formation by HUVECs in a 3D collagen matrix [28]. Later studies using planar type I collagen gel angiogenesis assays, confirmed that inhibition of α2β1 integrins with function blocking antibodies disrupted tube formation [132]. Senger et al. demonstrated in vivo using subcutaneous matrigel plug angiogenesis assays in mice, that inhibition of $\alpha2\beta1$ and $\alpha1\beta1$ in combination decreased new vessel growth in the implanted plugs. Together these results suggested a pro-angiogenic function for the $\alpha2\beta1$ integrin [122].

Studies from $\alpha2$-deficient mice have yielded contradictory results. Several labs, including our own, reported not only normal developmental angiogenesis, but also increased neoangiogenesis during wound healing in genetically-altered $\alpha2\beta1$ integrin-null mice [47, 149]. Similarly, our lab demonstrated that $\alpha2\beta1$ integrin-deletion increased tumor angiogenesis in a growth factor-dependent manner via modulation of VEGFR-1 signaling. Additionally studies in the diet-induced obesity model also showed increased angiogenesis in $\alpha2$-null mice compared to wild type mice [71]. The contradiction between the evidence for pro and anti-angiogenic functions for $\alpha2\beta1$ integrin are not totally based of differences in mouse and human endothelial cells or in vivo compared to in vitro models. Cailleteau et al. used an $\alpha2$ siRNA approach to alter integrin expression in HUVECs. These studies showed that $\alpha2\beta1$ integrin engagement by laminin promoted endothelial cell cycle arrest and quiescence [17]. Additionally, $\alpha2\beta1$ integrin binding to endorepellin in both human and mouse endothelial cells mediated the angiostatic effects [14, 46, 145].

Based on these inhibitory studies pharmacological inhibitors of $\alpha2\beta1$ may have potential anti-angiogenic drug effects (see therapy section). Small molecule inhibitors (SMI) of $\alpha2\beta1$ blocked both endothelial tube-formation in vitro and sprouting angiogenesis in zebrafish [115]. A more thorough understanding of the role for $\alpha2\beta1$ in angiogenesis promises novel insight into clinical application of $\alpha2\beta1$ integrin targeting compounds. Recent studies implicating the $\alpha2\beta1$ integrin in notch signaling offer an alternative paradigm for understanding $\alpha2\beta1$ integrin in angiogenesis [17, 39, 129]. The notch pathway coordinates sprouting angiogenesis by organizing endothelial cells into migratory 'tip' and proliferative 'stalk' cell conformations with differential capacity to respond to VEGF stimulation [54, 109]. Estrach et al. reported that $\alpha2\beta1$-mediated laminin signaling is necessary but not sufficient for induction of the tip cell determinant, Dll4 [39]. Clarifying the functional relationship between $\alpha2\beta1$ integrin and notch signaling in the endothelium is a promising avenue of future study.

3.8 The $\alpha2\beta1$ Integrin in the Innate and Acquired Immune Response

The $\alpha2\beta1$ integrin was initially identified as an integrin expressed at very late stages of T cell activation, thus the designation very late activation antigen-2 (VLA-2)(CD49b) [55, 56]. The $\alpha2\beta1$ integrin was then noted on a variety of cells of the inflammatory and hematopoietic system, including activated T cells, but not naïve T cells in chronic inflammatory settings. Early studies showed that $\alpha2\beta1$-dependent adhesion to collagen enhanced T cell receptor mediated T cell proliferation and cytokine secretion [120]. Boisvert et al. defined one possible mechanism; they reported that collagen I-stimulated, $\alpha2\beta1$ integrin-mediated both activation-independent and T cell receptor-dependent interferon γ expression via the ERK and JNK MAPKs and PI3K/AKT signaling pathways [15].

The $\alpha2\beta1$ integrin also influenced T cell activation by inhibiting fas ligand expression and apoptosis in effector T cells in a collagen I dependent manner [2, 42]. In animals, inhibitory monoclonal antibodies directed against the $\alpha2\beta1$ integrin significantly inhibited the effector phase of both contact and delayed type hypersensitivity. These early results established a role for the $\alpha2\beta1$ integrin in T cell mediated function. The role of the $\alpha2\beta1$ integrin in the innate and acquired immune response has been an area of active investigation.

To better the define the role of the $\alpha2\beta1$ integrin in T cell function, expression of the $\alpha2\beta1$ integrin on T cell subsets and in response to antigenic challenges was investigated. Kassiotis et al. reported that expression of $\alpha2\beta1$ integrin defined two functionally distinct subsets of memory T cells that played a role in the response to infection and immunization [74]. $\alpha2\beta1$ integrin expression was stably induced by antigen on approximately 50 % of memory T cells with helper function and stimulated production of tumor necrosis factor-α. The $\alpha2\beta1$ integrin expressing, CD49b+, memory Th cells demonstrated enhanced ability to mediate macrophage activation and to kill of intracellular bacteria.

Sasaki et al. demonstrated that mature Th1 and Th2 cells exhibited distinct $\alpha2\beta1$ integrin expression profiles [120]. Although naive Th cells did not express $\alpha2\beta1$ integrin, Th1 cells acquired high levels of $\alpha2\beta1$ integrin expression during maturation in an interferon-γ (IFN-γ) and interleukin (IL)-12-independent manner. This study suggested that high level $\alpha2\beta1$ integrin expression on Th1, but not Th2, cells was functionally important, because stimulation of Th1 or Th2 cells with $\alpha2\beta1$ integrin ligands caused selective activation of Th1 cells to produce interferon-γ after long-term culture.

Richter et al. studied $\alpha2\beta1$ integrin expression during influenza infection in the lung [113]. During the acute phase of infection, the $\alpha2\beta1$ integrin was expressed by a significant proportion of both CD4+ and CD8+ T cells in the lung; however, the integrin was expressed less frequently on memory cells, particularly CD8+ T cells. A similar expression pattern for the $\alpha2\beta1$ integrin in the spleen was found in a model of lymphocytic choriomeningitis viral infection [1]. The data suggested that $\alpha2\beta1$ integrin expression directed localization of CD4+ and CD8+ T cell subsets within the lung and promoted T cell migration within extralymphoid spaces, particularly during the acute phase of infection.

A role for $\alpha2\beta1$ integrin expression by Th17 cells has been described. Boisvert et al. showed that human naïve CD4 T cells stimulated toward Th17 polarization preferentially upregulate $\alpha2\beta1$ integrin [15]. Th17 cells adhered to collagens I and II, but not IVin an $\alpha2\beta1$ integrin-dependent manner. $\alpha2\beta1$ integrin-dependent adhesion combined with anti-CD3 antibody co-stimulated the production of IL-17A, IL-17F and IFN-γ by human Th17 cells.

The importance of $\alpha2\beta1$ integrin to T cell memory has remained controversial. Work by several groups suggested that professional memory CD4 cells reside and rest in the bone marrow. Recently, Hanazawa et al demonstrated that memory CD4 cells expressed high levels of $\alpha2\beta1$ integrin and that antibody-mediated inhibition of $\alpha2\beta1$ integrin of memory CD4 cell precursors caused failure to transmigrate from blood through sinusoidal endothelial cells into the bone marrow [50]. These results suggested that the $\alpha2\beta1$ integrin was required for the migration of memory CD4 cell precursors into their survival niches of the bone marrow.

In addition to its expression on activated T cells, the $\alpha2\beta1$ integrin is expressed at high levels on almost all NK cells and mast cells, and on subpopulations of monocytes and neutrophils [4, 133]. Arase et al. identified the NK cell recognition epitope of the widely used DX5 pan-NK cell monoclonal antibody as CD49b or the $\alpha2\beta1$ integrin. These investigators demonstrated that $\alpha2\beta1$-expressing and nonexpressing subsets of NK cells are present in the mouse spleen and raised the possibility that $\alpha2\beta1$ integrin expression is important in NK cell function. The role of the $\alpha2\beta1$ integrin on subsets of neutrophils and monocytes has also been studied. One study found expression of the $\alpha2\beta1$ integrin on extravasated neutrophils in human skin blister chambers and in the rat peritoneal cavity following chemotactic stimulation [144]. These studies, as well as others, suggested that the $\alpha2\beta1$ integrin on neutrophils is involved in neutrophil migration from the vasculature into extravascular tissue in response to cytokine induction.

Work from our lab has clarified the function of the $\alpha2\beta1$ integrin in mast call activation. We initially observed decreased inflammatory responses to Listeria monocytogenes in $\alpha2$-null mice [34]. This innate immunity defect was determined to arise from a requirement for $\alpha2\beta1$ integrin activation on peritoneal mast cells (PMCs) for mast-cell activation and cytokine release in vivo. We also identified C1q complement protein and collectin family members, including mannose binding lectin and surfactant protein A, as novel ligands for the integrin in mast cell activation in vitro in response to Listeria. Since ligation of the $\alpha 2\beta 1$ integrin alone was insufficient to activate cytokine secretion, we hypothesized that an additional signal emanating from a co-receptor was required to activate mast-cell cytokine secretion. We identified the required co-receptor as hepatocyte growth factor (HGF-R)/c-met [98]. We demonstrated that Listeria induced mast cell activation and cytokine secretion requires costimulatory signals from $\alpha 2\beta 1$ integrin ligation to either type I collagen or C1q as well as c-met activation. The synergistic signal from the two coreceptors resulted in mast cell release of the proinflammatory cytokine IL-6 to trigger the early innate immune response.

3.9 $\alpha 2\beta 1$ in Epithelial Biology

The $\alpha 2\beta 1$ integrin is expressed at high levels on numerous epithelial cells including not only the squamous epithelium, but also ciliated columnar epithelium of the respiratory tract, the epithelial cells of the gastrointestinal tract and urinary tract, and the glandular epithelium of the breast [24]. In contrast to the high $\alpha 2\beta 1$ integrin expression in the normal breast epithelium, markedly reduced or undetectable levels of $\alpha 2\beta 1$ integrin were seen in poorly-differentiated carcinomas. Expression of $\alpha 2\beta 1$-integrin was diminished or lost in a manner that correlated with a loss of epithelial differentiation and tumor progression in mammary carcinoma as well as other adenocarcinomas, including those of the prostate, lung, pancreas, and skin.

Our group's early studies focused on understanding the correlation between $\alpha 2\beta 1$ integrin expression and a differentiated epithelial phenotype and conversely, whether dysregulated $\alpha 2\beta 1$ integrin expression contributed to the malignant behavior of cancer cells. Gain of function and loss of function models in vitro suggested that $\alpha 2\beta 1$ integrin expression contributed to the differentiated epithelial phenotype and branching morphogenesis of mammary and other epithelial cells [130, 150, 151]. These observations were supported by findings from other laboratories. Using a primary human nonmalignant, but immortalized, mammary epithelial cell line, Berdichevsky et al. and D'Souza et al. demonstrated that branching morphogenesis can be blocked by inhibitory monoclonal antibodies directed against the $\alpha 2$ integrin subunit or by altered $\alpha 2\beta 1$-integrin expression mediated by the expression of the c-erbB2 proto-oncogene, respectively [9, 26, 27].

The development of genetically engineered mice with global deletion of *ITGA2* permitted further analysis of the role for $\alpha 2\beta 1$ integrin in vivo. The major changes in branching morphogenesis in vitro were not fully recapitulated in vivo. The $\alpha 2$-null mice have only modest defects in mammary morphology. The in vitro experiments were designed to study a single integrin interaction on epithelial cells with only a small number of matrix molecules. Mammary gland in vivo consists of epithelial cells, fibroblasts, endothelial cells, and immune cells embedded in a complex matrix. The complexity in in vivo systems and compensatory mechanisms may both mitigate the consequences of $\alpha 2\beta 1$ integrin-deficiency.

3.10 The $\alpha 2\beta 1$ integrin Plays a Role in Cancer Progression

Interest in $\alpha 2\beta 1$ integrin in breast cancer began with the observation of a strong correlation between diminished $\alpha 2\beta 1$ integrin expression and a less differentiated phenotype. The $\alpha 2\beta 1$ integrin–deficient mouse model provided our laboratory the opportunity to investigate a role for integrin in the development and progression of breast cancer in vivo. Our group demonstrated that in the spontaneous MMTV-neu mouse model of breast cancer, $\alpha 2\beta 1$ integrin-deletion did not significantly alter the incidence of tumor development or tumor growth, but markedly increased hematogenous metastasis [111]. Increased metastasis in this model resulted in part from increased capacity for cancer cell intravasation.

Detailed in silico examination of publically available data from breast cancer patients supported this finding; expression of the $\alpha 2$ integrin subunit, but not $\alpha1$ or $\beta1$ integrin subunits, was a prognostic indicator of decreased metastasis and better patient outcomes (Fig. 3.2). Similarly, retrospective analysis of lymph node-negative patients from the Wang cohort who relapsed with metastatic disease, revealed an inverse correlation between $\alpha2\beta1$ integrin expression and the occurrence of brain lesions; patients with greater than twice the average $\alpha2\beta1$ integrin expression suffered no brain metastasis whereas all nearly one third of all other patients suffered brain metastasis ($P = 0.0049$).

Expression of the $\alpha2\beta1$ integrin in prostate cancer was also predictive of metastasis and survival. The mouse and human studies supported the in vitro experimental analyses and the reported epidemiologic linkage between the single nucleotide polymorphisms regulating $\alpha2\beta1$ integrin expression and poor prognosis in patients with breast cancer [90]. Together these data suggested that $\alpha2\beta1$ integrin is a valuable biomarker for risk of metastasis in breast cancer.

Our data clearly showed in an animal model of breast cancer and human breast and prostate cancer that the integrin behaved as a metastasis suppressor. Data from other laboratories suggest that $\alpha2\beta1$ integrin's role in prostate and perhaps other cancers may be more complicated. In vitro, $\alpha2\beta1$ integrin was required but not sufficient for survival and metastasis of LNCaP prostate cancer cells to bone [91]. $\alpha2\beta1$ integrin protein and mRNA expression was enhanced in bone metastases to the level observed in normal, nonmalignant prostate tissue and significantly higher than primary prostate cancer lesions or metastasis to other sites such as lymph nodes [127]. Similarly, $\alpha2\beta1$ integrin expression accelerated experimental metastasis or tumor dissemination of melanoma and rhabdomyosarcoma or melanoma, gastric and colon cancer, respectively [7, 8, 51, 92, 96, 139].

Therefore, despite this progress several important questions remain concerning the role of the $\alpha2\beta1$ integrin in cancer biology. What is the precise molecular mechanism through which $\alpha2\beta1$ integrin loss enables increased intravasation? How does integrin down-regulation during breast cancer progression occur? Many other cancers including prostate, colon and lung cancer also appear to have $\alpha2\beta1$ integrin loss associated with cancer progression and metastasis. However, some cancers are associated with high $\alpha2\beta1$ integrin expression levels. Answers to each of these questions will provide novel insight into tumor biology, as well as suggesting new avenues for clinical application of the $\alpha2\beta1$ integrin as a biomarker or therapeutic target.

3.11 Therapies

Over the past several years there has been increased interest in pharmacological targeting of the $\alpha2\beta1$ integrin for treatment of thrombosis and angiogenesis [72]. The $\alpha2\beta1$ integrin is viewed as a safe target because although overexpression was associated with pathological clot formations, mice with integrin deletion lack severe bleeding defects, and inhibition causes only minimal increases in bleeding time. Compound 15, a nonpeptide inhibitor of the integrin, has been demonstrated to block platelet adhesion to collagen I under both static and flow conditions [16]. The inhibitor was originally designed to inhibit $\alpha2\beta1$ on platelets by locking the integrin $\alpha2\beta1$ in the inactive low-affinity conformation [100]. Additionally, in vivo, the compound inhibited thrombus formation in a mouse model and inhibited angiogenesis in a zebrafish model. Other $\alpha2\beta1$ inhibitors have shown similar effects; BTT-3016, a sulfonamide derivative prevented platelet aggregation and reduced thrombus formation in a vascular injury model [108]. Another sulfonamide derivative that targets $\alpha2\beta1$, E7820, is currently in phase II clinical trials as an adjuvant therapy for metastatic colon cancer [77, 101]. The clinical impact of pharmacological targeting the $\alpha2\beta1$ integrin will require further time and experimentation.

3.12 Summary and New Directions

It is increasingly clear that the $\alpha2\beta1$ integrin plays a nuanced but important role in critical cell functions in many different cell types. Several

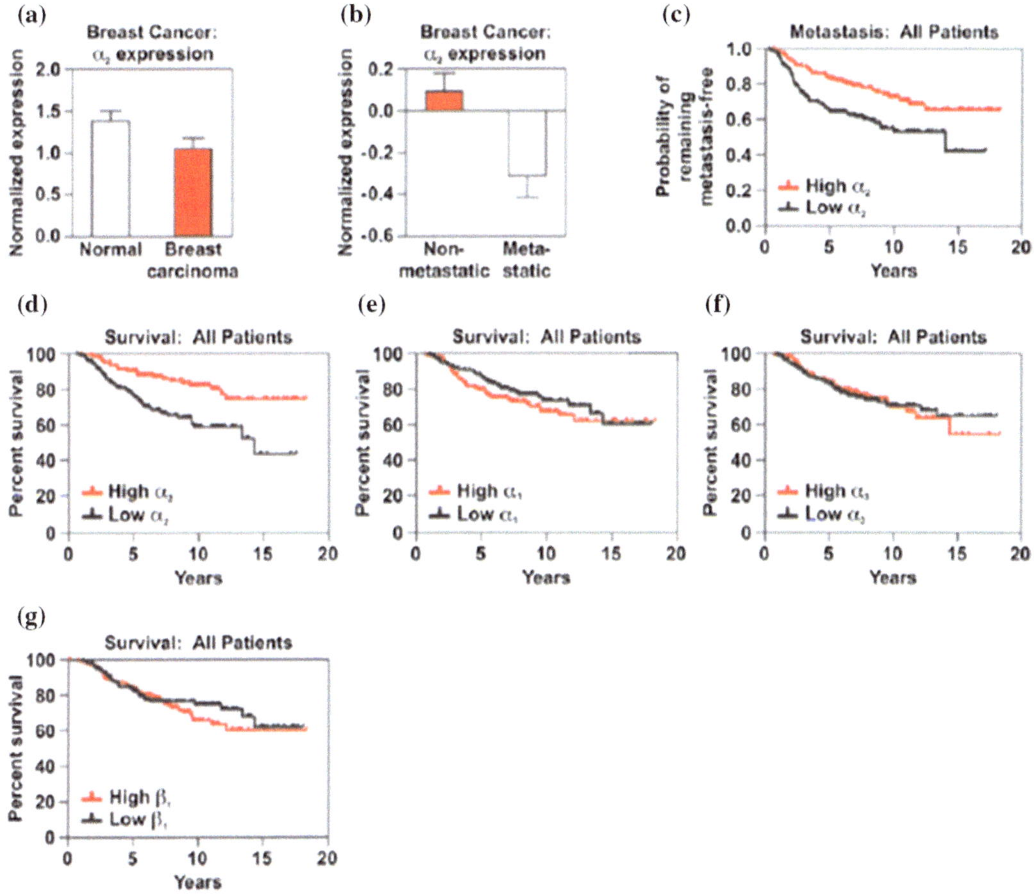

Fig. 3.2 Decreased α2β1 integrin mRNA expression predicts metastasis and decreased survival in breast cancer patients. **a** Expression of the α2 integrin was significantly decreased in breast carcinomas (n = 40) compared with normal breast tissue (n = 7) (P = 0.038). (**b–g**) Analysis of the NKI-295 cohort correlates expression of α2 integrin, but not other integrins, with metastasis (**b** and **c**) and patient survival (**d–g**). The α2 integrin expression was substantially reduced in patients with metastases (n = 101) when compared with non-metastatic patients (n = 194, P = 0.0038) (**b**). Log-rank analysis demonstrates that high-level α2 integrin expression correlates with the probability of both remaining metastasis-free (**c**, P = 0.0022) and with improved long-term survival (**d**, P < 0.0001). In contrast, expression of the α1 (**e**, P = 0.2639), α3 (**f**, P = 0.9509), and β1 (**g**, P = 0.5) integrin subunits did not correlate with patient survival (Reprinted from Fig. 6, [111])

new studies have suggested previously undocumented roles for the integrin in diseases ranging from type 2 diabetes, to dwarfism. In platelets, the combination of animal and in vitro studies have slowly revealed a more nuanced yet equally important role for the integrin than had previously been imagined. The recent development of tissue-specific α2-null mice promises to bring similar clarity and complexity to our understanding of α2β1 integrin function in inflammation, angiogenesis and tumor biology in the years ahead.

References

1. Andreasen SØ, Thomsen AR, Koteliansky VE, Novobrantseva TI, Sprague AG, de Fougerolles AR, Christensen JP (2003) Expression and functional importance of collagen-binding integrins, alpha 1 beta 1 and alpha 2 beta 1, on virus-activated T cells. J Immunol Baltim Md 1950 171:2804–2811
2. Aoudjit F, Vuori K (2000) Engagement of the alpha2beta1 integrin inhibits Fas ligand expression and activation-induced cell death in T cells in a focal adhesion kinase-dependent manner. Blood 95:2044–2051

3. Aquilina A, Korda M, Bergelson JM, Humphries MJ, Farndale RW, Tuckwell D (2002) A novel gain-of-function mutation of the integrin alpha2 VWFA domain. Eur J Biochem 269:1136–1144
4. Arase H, Saito T, Phillips JH, Lanier LL (2001) Cutting edge: the mouse NK cell-associated antigen recognized by DX5 monoclonal antibody is CD49b (alpha 2 integrin, very late antigen-2). J Immunol Baltim Md 1950 167:1141–1144
5. Auger JM, Kuijpers MJ, Senis YA, Watson SP, Heemskerk JW (2005) Adhesion of human and mouse platelets to collagen under shear: a unifying model. Faseb J 19:825–827
6. Ayala F, Corral J, Gonzalez-Conejero R, Sanchez I, Moraleda JM, Vicente V (2003) Genetic polymorphisms of platelet adhesive molecules: association with breast cancer risk and clinical presentation. Breast Cancer Res Treat 80:145–154
7. Baronas-Lowell D, Lauer-Fields JL, Borgia JA, Sferrazza GF, Al-Ghoul M, Minond D, Fields GB (2004) Differential modulation of human melanoma cell metalloproteinase expression by alpha2beta1 integrin and CD44 triple-helical ligands derived from type IV collagen. J Biol Chem 279: 43503–43513
8. Bartolomé RA, Barderas R, Torres S, Fernandez-Aceñero MJ, Mendes M, García-Foncillas J, Lopez-Lucendo M, Casal JI (2013) Cadherin-17 interacts with $\alpha 2\beta 1$ integrin to regulate cell proliferation and adhesion in colorectal cancer cells causing liver metastasis. Oncogene 33:1658–1669
9. Berdichevsky F, Alford D, D'Souza B, Taylor-Papadimitriou J (1994) Branching morphogenesis of human mammary epithelial cells in collagen gels. J Cell Sci 107(Pt 12):3557–3568
10. Bergelson JM, Shepley MP, Chan BM, Hemler ME, Finberg RW (1992) Identification of the integrin VLA-2 as a receptor for echovirus 1. Science 255:1718–1720
11. Bergelson JM, Chan BM, Finberg RW, Hemler ME (1993) The integrin VLA-2 binds echovirus 1 and extracellular matrix ligands by different mechanisms. J Clin Invest 92:232–239
12. Bergelson JM, St John NF, Kawaguchi S, Pasqualini R, Berdichevsky F, Hemler ME, Finberg RW (1994) The I domain is essential for echovirus 1 interaction with VLA-2. Cell Adhes Commun 2:455–464
13. Bidanset DJ, Guidry C, Rosenberg LC, Choi HU, Timpl R, Hook M (1992) Binding of the proteoglycan decorin to collagen type VI. J Biol Chem 267:5250–5256
14. Bix G, Fu J, Gonzalez EM, Macro L, Barker A, Campbell S, Zutter MM, Santoro SA, Kim JK, Hook M et al (2004) Endorepellin causes endothelial cell disassembly of actin cytoskeleton and focal adhesions through alpha2beta1 integrin. J Cell Biol 166:97–109
15. Boisvert M, Gendron S, Chetoui N, Aoudjit F (2007) Alpha2 beta1 integrin signaling augments T cell receptor-dependent production of interferon-gamma in human T cells. Mol Immunol 44:3732–3740
16. Borza CM, Su Y, Chen X, Yu L, Mont S, Chetyrkin S, Voziyan P, Hudson BG, Billings PC, Jo H et al (2012) Inhibition of integrin alpha2beta1 ameliorates glomerular injury. J Am Soc Nephrol 23:1027–1038
17. Cailleteau L, Estrach S, Thyss R, Boyer L, Doye A, Domange B, Johnsson N, Rubinstein E, Boucheix C, Ebrahimian T et al (2010) alpha2beta1 integrin controls association of Rac with the membrane and triggers quiescence of endothelial cells. J Cell Sci 123:2491–2501
18. Calderwood DA, Tuckwell DS, Eble J, Kuhn K, Humphries MJ (1997) The integrin alpha1 A-domain is a ligand binding site for collagens and laminin. J Biol Chem 272:12311–12317
19. Carafoli F, Hamaia SW, Bihan D, Hohenester E, Farndale RW (2013) An activating mutation reveals a second binding mode of the integrin alpha2 I domain to the GFOGER motif in collagens. PLoS One 8:e69833
20. Carlsson LE, Santoso S, Spitzer C, Kessler C, Greinacher A (1999) The alpha2 gene coding sequence T807/A873 of the platelet collagen receptor integrin alpha2beta1 might be a genetic risk factor for the development of stroke in younger patients. Blood 93:3583–3586
21. Casorelli I, De Stefano V, Leone AM, Chiusolo P, Burzotta F, Paciaroni K, Rossi E, Andreotti F, Leone G, Maseri A (2001) The C807T/G873A polymorphism in the platelet glycoprotein Ia gene and the risk of acute coronary syndrome in the Italian population. Br J Haematol 114:150–154
22. Chan BM, Hemler ME (1993) Multiple functional forms of the integrin VLA-2 can be derived from a single alpha 2 cDNA clone: interconversion of forms induced by an anti-beta 1 antibody. J Cell Biol 120:537–543
23. Cheli Y, Williams SA, Ballotti R, Nugent DJ, Kunicki TJ (2010) Enhanced binding of poly(ADP-ribose)polymerase-1 and Ku80/70 to the ITGA2 promoter via an extended cytosine-adenosine repeat. PLoS One 5:e8743
24. Chen J, Diacovo TG, Grenache DG, Santoro SA, Zutter MM (2002) The alpha(2) integrin subunit-deficient mouse: a multifaceted phenotype including defects of branching morphogenesis and hemostasis. Am J Pathol 161:337–344
25. Chin YK-Y, Headey SJ, Mohanty B, Patil R, McEwan PA, Swarbrick JD, Mulhern TD, Emsley J, Simpson JS, Scanlon MJ (2013) The Structure of Integrin α1I Domain in Complex with a Collagen-mimetic Peptide. J Biol Chem 288:36796–36809
26. D'souza B, Taylor-Papadimitriou J (1994) Overexpression of ERBB2 in human mammary epithelial cells signals inhibition of transcription of the E-cadherin gene. Proc Natl Acad Sci USA 91:7202–7206

27. D'Souza B, Berdichevsky F, Kyprianou N, Taylor-Papadimitriou J (1993) Collagen-induced morphogenesis and expression of the alpha 2-integrin subunit is inhibited in c-erbB2-transfected human mammary epithelial cells. Oncogene 8:1797–1806
28. Davis GE, Black SM, Bayless KJ (2000) Capillary morphogenesis during human endothelial cell invasion of three-dimensional collagen matrices. Vitro Cell Dev Biol Anim 36:513–519
29. De Fougerolles AR, Sprague AG, Nickerson-Nutter CL, Chi-Rosso G, Rennert PD, Gardner H, Gotwals PJ, Lobb RR, Koteliansky VE (2000) Regulation of inflammation by collagen-binding integrins alpha1beta1 and alpha2beta1 in models of hypersensitivity and arthritis. J Clin Invest 105:721–729
30. Dickeson SK, Walsh JJ, Santoro SA (1998) Binding of the alpha 2 integrin I domain to extracellular matrix ligands: structural and mechanistic differences between collagen and laminin binding. Cell Adhes Commun 5:273–281
31. Dickeson SK, Mathis NL, Rahman M, Bergelson JM, Santoro SA (1999) Determinants of ligand binding specificity of the alpha(1)beta(1) and alpha(2)beta(1) integrins. J Biol Chem 274: 32182–32191
32. Dodson PM, Haynes J, Starczynski J, Farmer J, Shigdar S, Fegan G, Johnson RJ, Fegan C (2003) The platelet glycoprotein Ia/IIa gene polymorphism C807T/G873A: a novel risk factor for retinal vein occlusion. Eye Lond 17:772–777
33. Eble JA, Kassner A, Niland S, Morgelin M, Grifka J, Grassel S (2006) Collagen XVI harbors an integrin alpha1 beta1 recognition site in its C-terminal domains. J Biol Chem 281:25745–25756
34. Edelson BT, Stricker TP, Li Z, Dickeson SK, Shepherd VL, Santoro SA, Zutter MM (2006) Novel collectin/C1q receptor mediates mast cell activation and innate immunity. Blood 107:143–150
35. Elices MJ, Hemler ME (1989) The human integrin VLA-2 is a collagen receptor on some cells and a collagen/laminin receptor on others. Proc Natl Acad Sci USA 86:9906–9910
36. Elices MJ, Urry LA, Hemler ME (1991) Receptor functions for the integrin VLA-3: fibronectin, collagen, and laminin binding are differentially influenced by Arg-Gly-Asp peptide and by divalent cations. J Cell Biol 112:169–181
37. Emsley J, Knight CG, Farndale RW, Barnes MJ, Liddington RC (2000) Structural basis of collagen recognition by integrin alpha2beta1. Cell 101:47–56
38. Enenstein J, Kramer RH (1994) Confocal microscopic analysis of integrin expression on the microvasculature and its sprouts in the neonatal foreskin. J Invest Dermatol 103:381–386
39. Estrach S, Cailleteau L, Franco CA, Gerhardt H, Stefani C, Lemichez E, Gagnoux-Palacios L, Meneguzzi G, Mettouchi A (2011) Laminin-binding integrins induce Dll4 expression and Notch signaling in endothelial cells. Circ Res 109:172–182
40. Fleischmajer R, Fisher LW, MacDonald ED, Jacobs L Jr, Perlish JS, Termine JD (1991) Decorin interacts with fibrillar collagen of embryonic and adult human skin. J Struct Biol 106:82–90
41. Gamble J, Meyer G, Noack L, Furze J, Matthias L, Kovach N, Harlant J, Vadas M (1999) B1 integrin activation inhibits in vitro tube formation: effects on cell migration, vacuole coalescence and lumen formation. Endothelium 7:23–34
42. Gendron S, Couture J, Aoudjit F (2003) Integrin alpha2beta1 inhibits Fas-mediated apoptosis in T lymphocytes by protein phosphatase 2A-dependent activation of the MAPK/ERK pathway. J Biol Chem 278:48633–48643
43. Gerger A, Hofmann G, Langsenlehner U, Renner W, Weitzer W, Wehrschutz M, Wascher T, Samonigg H, Krippl P (2009) Integrin alpha-2 and beta-3 gene polymorphisms and colorectal cancer risk. Int J Colorectal Dis 24:159–163
44. Girgert R, Martin M, Kruegel J, Miosge N, Temme J, Eckes B, Muller GA, Gross O (2010) Integrin alpha2-deficient mice provide insights into specific functions of collagen receptors in the kidney. Fibrogenesis Tissue Repair 3:19
45. Goyal A, Pal N, Concannon M, Paul M, Doran M, Poluzzi C, Sekiguchi K, Whitelock JM, Neill T, Iozzo RV (2011) Endorepellin, the angiostatic module of perlecan, interacts with both the alpha2beta1 integrin and vascular endothelial growth factor receptor 2 (VEGFR2): a dual receptor antagonism. J Biol Chem 286:25947–25962
46. Goyal A, Poluzzi C, Willis CD, Smythies J, Shellard A, Neill T, Iozzo RV (2012) Endorepellin affects angiogenesis by antagonizing diverse vascular endothelial growth factor receptor 2 (VEGFR2)-evoked signaling pathways: transcriptional repression of hypoxia-inducible factor 1alpha and VEGFA and concurrent inhibition of nuclear factor of activated T cell 1 (NFAT1) activation. J Biol Chem 287:43543–43556
47. Grenache DG, Zhang Z, Wells LE, Santoro SA, Davidson JM, Zutter MM (2007) Wound healing in the alpha2beta1 integrin-deficient mouse: altered keratinocyte biology and dysregulated matrix metalloproteinase expression. J Invest Dermatol 127:455–466
48. Gross O, Beirowski B, Koepke M-L, Kuck J, Reiner M, Addicks K, Smyth N, Schulze-Lohoff E, Weber M (2003) Preemptive ramipril therapy delays renal failure and reduces renal fibrosis in COL4A3-knockout mice with Alport syndrome. Kidney Int 63:438–446
49. Habart D, Cheli Y, Nugent DJ, Ruggeri ZM, Kunicki TJ (2013) Conditional knockout of integrin alpha2beta1 in murine megakaryocytes leads to reduced mean platelet volume. PLoS One 8:e55094
50. Hanazawa A, Hayashizaki K, Shinoda K, Yagita H, Okumura K, Lohning M, Hara T, Tani-ichi S, Ikuta K, Eckes B et al (2013) CD49b-dependent

establishment of T helper cell memory. Immunol Cell Biol 91:524–531
51. Hangan D, Uniyal S, Morris VL, MacDonald IC, von Ballestrem C, Chau T, Schmidt EE, Chambers AF, Groom AC, Chan BM (1996) Integrin VLA-2 (alpha2beta1) function in postextravasation movement of human rhabdomyosarcoma RD cells in the liver. Cancer Res 56:3142–3149
52. Hedbom E, Heinegard D (1989) Interaction of a 59-kDa connective tissue matrix protein with collagen I and collagen II. J Biol Chem 264:6898–6905
53. Heikkinen A, Tu H, Pihlajaniemi T (2012) Collagen XIII: a type II transmembrane protein with relevance to musculoskeletal tissues, microvessels and inflammation. Int J Biochem Cell Biol 44:714–717
54. Hellstrom M, Phng LK, Hofmann JJ, Wallgard E, Coultas L, Lindblom P, Alva J, Nilsson AK, Karlsson L, Gaiano N et al (2007) Dll4 signalling through Notch1 regulates formation of tip cells during angiogenesis. Nature 445:776–780
55. Hemler ME (1990) VLA proteins in the integrin family: structures, functions, and their role on leukocytes. Annu Rev Immunol 8:365–400
56. Hemler ME, Jacobson JG, Brenner MB, Mann D, Strominger JL (1985) VLA-1: a T cell surface antigen which defines a novel late stage of human T cell activation. Eur J Immunol 15:502–508
57. Hers I, Berlanga O, Tiekstra MJ, Kamiguti AS, Theakston RD, Watson SP (2000) Evidence against a direct role of the integrin alpha2beta1 in collagen-induced tyrosine phosphorylation in human platelets. Eur J Biochem 267:2088–2097
58. Honore S, Kovacic H, Pichard V, Briand C, Rognoni JB (2003) Alpha2beta1-integrin signaling by itself controls G1/S transition in a human adenocarcinoma cell line (Caco-2): implication of NADPH oxidase-dependent production of ROS. Exp Cell Res 285:59–71
59. Ichinohe T, Takayama H, Ezumi Y, Arai M, Yamamoto N, Takahashi H, Okuma M (1997) Collagen-stimulated activation of Syk but not c-Src is severely compromised in human platelets lacking membrane glycoprotein VI. J Biol Chem 272:63–68
60. Ignatius MJ, Reichardt LF (1988) Identification of a neuronal laminin receptor: an Mr 200 K/120 K integrin heterodimer that binds laminin in a divalent cation-dependent manner. Neuron 1:713–725
61. Inoue O, Suzuki-Inoue K, Dean WL, Frampton J, Watson SP (2003) Integrin alpha2beta1 mediates outside-in regulation of platelet spreading on collagen through activation of Src kinases and PLCgamma2. J Cell Biol 160:769–780
62. Ivaska J, Reunanen H, Westermarck J, Koivisto L, Kahari VM, Heino J (1999) Integrin alpha2beta1 mediates isoform-specific activation of p38 and upregulation of collagen gene transcription by a mechanism involving the alpha2 cytoplasmic tail. J Cell Biol 147:401–416
63. Ivaska J, Nissinen L, Immonen N, Eriksson JE, Kahari VM, Heino J (2002) Integrin alpha 2 beta 1 promotes activation of protein phosphatase 2A and dephosphorylation of Akt and glycogen synthase kinase 3 beta. Mol Cell Biol 22:1352–1359
64. Jacquelin B, Tarantino MD, Kritzik M, Rozenshteyn D, Koziol JA, Nurden AT, Kunicki TJ (2001) Allele-dependent transcriptional regulation of the human integrin alpha2 gene. Blood 97:1721–1726
65. Jandrot-Perrus M, Busfield S, Lagrue AH, Xiong X, Debili N, Chickering T, Le Couedic JP, Goodearl A, Dussault B, Fraser C et al (2000) Cloning, characterization, and functional studies of human and mouse glycoprotein VI: a platelet-specific collagen receptor from the immunoglobulin superfamily. Blood 96:1798–1807
66. Jin M, Andricioaei I, Springer TA (2004) Conversion between three conformational states of integrin I domains with a C-terminal pull spring studied with molecular dynamics. Structure 12:2137–2147
67. Jokinen J, Dadu E, Nykvist P, Kapyla J, White DJ, Ivaska J, Vehvilainen P, Reunanen H, Larjava H, Hakkinen L et al (2004) Integrin-mediated cell adhesion to type I collagen fibrils. J Biol Chem 279:31956–31963
68. Jokinen J, White DJ, Salmela M, Huhtala M, Kapyla J, Sipila K, Puranen JS, Nissinen L, Kankaanpaa P, Marjomaki V et al (2010) Molecular mechanism of alpha2beta1 integrin interaction with human echovirus 1. EMBO J 29:196–208
69. Kamata T, Liddington RC, Takada Y (1999) Interaction between collagen and the alpha(2) I-domain of integrin alpha(2)beta(1). Critical role of conserved residues in the metal ion-dependent adhesion site (MIDAS) region. J Biol Chem 274:32108–32111
70. Kamiguti AS, Markland FS, Zhou Q, Laing GD, Theakston RD, Zuzel M (1997) Proteolytic cleavage of the beta1 subunit of platelet alpha2beta1 integrin by the metalloproteinase jararhagin compromises collagen-stimulated phosphorylation of pp72. J Biol Chem 272:32599–32605
71. Kang L, Ayala JE, Lee-Young RS, Zhang Z, James FD, Neufer PD, Pozzi A, Zutter MM, Wasserman DH (2011) Diet-induced muscle insulin resistance is associated with extracellular matrix remodeling and interaction with integrin alpha2beta1 in mice. Diabetes 60:416–426
72. Kapp TG, Rechenmacher F, Sobahi TR, Kessler H (2013) Integrin modulators: a patent review. Expert Opin Ther Pat 23:1273–1295
73. Karjalainen M, Kakkonen E, Upla P, Paloranta H, Kankaanpaa P, Liberali P, Renkema GH, Hyypia T, Heino J, Marjomaki V (2008) A Raft-derived, Pak1-regulated entry participates in alpha2beta1 integrin-dependent sorting to caveosomes. Mol Biol Cell 19:2857–2869
74. Kassiotis G, Gray D, Kiafard Z, Zwirner J, Stockinger B (2006) Functional specialization of memory Th cells revealed by expression of integrin CD49b. J Immunol 177:968–975

75. Keely PJ, Parise LV (1996) The alpha2beta1 integrin is a necessary co-receptor for collagen-induced activation of Syk and the subsequent phosphorylation of phospholipase Cgamma2 in platelets. J Biol Chem 271:26668–26676
76. Kehrel B (1995) Platelet receptors for collagens. Platelets 6:11–16
77. Keizer RJ, Funahashi Y, Semba T, Wanders J, Beijnen JH, Schellens JHM, Huitema ADR (2011) Evaluation of α2-integrin expression as a biomarker for tumor growth inhibition for the investigational integrin inhibitor E7820 in preclinical and clinical studies. AAPS J 13:230–239
78. Kern A, Eble J, Golbik R, Kuhn K (1993) Interaction of type IV collagen with the isolated integrins alpha 1 beta 1 and alpha 2 beta 1. Eur J Biochem 215:151–159
79. Kirchhofer D, Languino LR, Ruoslahti E, Pierschbacher MD (1990) Alpha 2 beta 1 integrins from different cell types show different binding specificities. J Biol Chem 265:615–618
80. Klekotka PA, Santoro SA, Zutter MM (2001) alpha 2 integrin subunit cytoplasmic domain-dependent cellular migration requires p38 MAPK. J Biol Chem 276:9503–9511
81. Klekotka PA, Santoro SA, Wang H, Zutter MM (2001) Specific residues within the alpha 2 integrin subunit cytoplasmic domain regulate migration and cell cycle progression via distinct MAPK pathways. J Biol Chem 276:32353–32361
82. Knight CG, Morton LF, Onley DJ, Peachey AR, Messent AJ, Smethurst PA, Tuckwell DS, Farndale RW, Barnes MJ (1998) Identification in collagen type I of an integrin alpha2 beta1-binding site containing an essential GER sequence. J Biol Chem 273:33287–33294
83. Knight CG, Morton LF, Peachey AR, Tuckwell DS, Farndale RW, Barnes MJ (2000) The collagen-binding A-domains of integrins alpha(1)beta(1) and alpha(2)beta(1) recognize the same specific amino acid sequence, GFOGER, in native (triple-helical) collagens. J Biol Chem 275:35–40
84. Kritzik M, Savage B, Nugent DJ, Santoso S, Ruggeri ZM, Kunicki TJ (1998) Nucleotide polymorphisms in the alpha2 gene define multiple alleles that are associated with differences in platelet alpha2 beta1 density. Blood 92:2382–2388
85. Kunicki TJ, Orchekowski R, Annis D, Honda Y (1993) Variability of integrin alpha 2 beta 1 activity on human platelets. Blood 82:2693–2703
86. Kunicki TJ, Kritzik M, Annis DS, Nugent DJ (1997) Hereditary variation in platelet integrin alpha 2 beta 1 density is associated with two silent polymorphisms in the alpha 2 gene coding sequence. Blood 89:1939–1943
87. Kunicki TJ, Williams SA, Diaz D, Farndale RW, Nugent DJ (2012) Platelet adhesion to decorin but not collagen I correlates with the integrin alpha2 dimorphism E534 K, the basis of the human platelet alloantigen (HPA)-5 system. Haematologica 97:692–695
88. Kunicki TJ, Williams SA, Nugent DJ, Yeager M (2012) Mean platelet volume and integrin alleles correlate with levels of integrins alpha(IIb)beta(3) and alpha(2)beta(1) in acute coronary syndrome patients and normal subjects. Arterioscler Thromb Vasc Biol 32:147–152
89. Lahti M, Bligt E, Niskanen H, Parkash V, Brandt AM, Jokinen J, Patrikainen P, Kapyla J, Heino J, Salminen TA (2011) Structure of collagen receptor integrin alpha(1)I domain carrying the activating mutation E317A. J Biol Chem 286:43343–43351
90. Langsenlehner U, Renner W, Yazdani-Biuki B, Eder T, Wascher TC, Paulweber B, Clar H, Hofmann G, Samonigg H, Krippl P (2006) Integrin alpha-2 and beta-3 gene polymorphisms and breast cancer risk. Breast Cancer Res Treat 97:67–72
91. Lee SH, Hatakeyama S, Yu S-Y, Bao X, Ohyama C, Khoo K-H, Fukuda MN, Fukuda M (2009) Core3 O-glycan synthase suppresses tumor formation and metastasis of prostate carcinoma PC3 and LNCaP cells through down-regulation of alpha2beta1 integrin complex. J Biol Chem 284:17157–17169
92. Maaser K, Wolf K, Klein CE, Niggemann B, Zänker KS, Bröcker EB, Friedl P (1999) Functional hierarchy of simultaneously expressed adhesion receptors: integrin alpha2beta1 but not CD44 mediates MV3 melanoma cell migration and matrix reorganization within three-dimensional hyaluronan-containing collagen matrices. Mol Biol Cell 10:3067–3079
93. Marjomaki V, Pietiainen V, Matilainen H, Upla P, Ivaska J, Nissinen L, Reunanen H, Huttunen P, Hyypia T, Heino J (2002) Internalization of echovirus 1 in caveolae. J Virol 76:1856–1865
94. Marjoram RJ, Voss B, Pan Y, Dickeson SK, Zutter MM, Hamm HE, Santoro SA (2009) Suboptimal activation of protease-activated receptors enhances alpha2beta1 integrin-mediated platelet adhesion to collagen. J Biol Chem 284:34640–34647
95. Matsubara Y, Murata M, Maruyama T, Handa M, Yamagata N, Watanabe G, Saruta T, Ikeda Y (2000) Association between diabetic retinopathy and genetic variations in alpha2beta1 integrin, a platelet receptor for collagen. Blood 95:1560–1564
96. Matsuoka T, Yashiro M, Nishimura S, Inoue T, Fujihara T, Sawada T, Kato Y, Seki S, Hirakawa-Ys Chung K (2000) Increased expression of alpha2beta1-integrin in the peritoneal dissemination of human gastric carcinoma. Int J Mol Med 5:21–25
97. Mazzucato M, Cozzi MR, Battiston M, Jandrot-Perrus M, Mongiat M, Marchese P, Kunicki TJ, Ruggeri ZM, De Marco L (2009) Distinct spatio-temporal $Ca2+$ signaling elicited by integrin alpha2beta1 and glycoprotein VI under flow. Blood 114:2793–2801

98. McCall-Culbreath KD, Li Z, Zutter MM (2008) Crosstalk between the alpha2beta1 integrin and c-met/HGF-R regulates innate immunity. Blood 111:3562–3570
99. Messent AJ, Tuckwell DS, Knauper V, Humphries MJ, Murphy G, Gavrilovic J (1998) Effects of collagenase-cleavage of type I collagen on alpha2beta1 integrin-mediated cell adhesion. J Cell Sci 111(Pt 8):1127–1135
100. Miller MW, Basra S, Kulp DW, Billings PC, Choi S, Beavers MP, McCarty OJ, Zou Z, Kahn ML, Bennett JS et al (2009) Small-molecule inhibitors of integrin alpha2beta1 that prevent pathological thrombus formation via an allosteric mechanism. Proc Natl Acad Sci USA 106:719–724
101. Mita M, Kelly KR, Mita A, Ricart AD, Romero O, Tolcher A, Hook L, Okereke C, Krivelevich I, Rossignol DP et al (2011) Phase I study of E7820, an oral inhibitor of integrin alpha-2 expression with antiangiogenic properties, in patients with advanced malignancies. Clin Cancer Res Off J Am Assoc Cancer Res 17:193–200
102. Miura Y, Ohnuma M, Jung SM, Moroi M (2000) Cloning and expression of the platelet-specific collagen receptor glycoprotein VI. Thromb Res 98:301–309
103. Miura Y, Takahashi T, Jung SM, Moroi M (2002) Analysis of the interaction of platelet collagen receptor glycoprotein VI (GPVI) with collagen. A dimeric form of GPVI, but not the monomeric form, shows affinity to fibrous collagen. J Biol Chem 277:46197–46204
104. Moroi M, Jung SM (1997) Platelet receptors for collagen. Thromb Haemost 78:439–444
105. Moroi M, Jung SM (2004) Platelet glycoprotein VI: its structure and function. Thromb Res 114:221–233
106. Nieuwenhuis HK, Akkerman JW, Houdijk WP, Sixma JJ (1985) Human blood platelets showing no response to collagen fail to express surface glycoprotein Ia. Nature 318:470–472
107. Nieuwenhuis HK, Sakariassen KS, Houdijk WP, Nievelstein PF, Sixma JJ (1986) Deficiency of platelet membrane glycoprotein Ia associated with a decreased platelet adhesion to subendothelium: a defect in platelet spreading. Blood 68:692–695
108. Nissinen L, Pentikäinen OT, Jouppila A, Käpylä J, Ojala M, Nieminen J, Lipsanen A, Lappalainen H, Eckes B, Johnson MS et al (2010) A small-molecule inhibitor of integrin alpha2 beta1 introduces a new strategy for antithrombotic therapy. Thromb Haemost 103:387–397
109. Phng LK, Potente M, Leslie JD, Babbage J, Nyqvist D, Lobov I, Ondr JK, Rao S, Lang RA, Thurston G et al (2009) Nrarp coordinates endothelial Notch and Wnt signaling to control vessel density in angiogenesis. Dev Cell 16:70–82
110. Pilcher BK, Dumin JA, Sudbeck BD, Krane SM, Welgus HG, Parks WC (1997) The activity of collagenase-1 is required for keratinocyte migration on a type I collagen matrix. J Cell Biol 137:1445–1457
111. Ramirez NE, Zhang Z, Madamanchi A, Boyd KL, O'Rear LD, Nashabi A, Li Z, Dupont WD, Zijlstra A, Zutter MM (2011) The $\alpha_2\beta_1$ integrin is a metastasis suppressor in mouse models and human cancer. J Clin Invest 121:226–237
112. Raynal N, Hamaia SW, Siljander PR, Maddox B, Peachey AR, Fernandez R, Foley LJ, Slatter DA, Jarvis GE, Farndale RW (2006) Use of synthetic peptides to locate novel integrin alpha2beta1-binding motifs in human collagen III. J Biol Chem 281:3821–3831
113. Richter M, Ray SJ, Chapman TJ, Austin SJ, Rebhahn J, Mosmann TR, Gardner H, Kotelianski V, deFougerolles AR, Topham DJ (2007) Collagen distribution and expression of collagen-binding alpha1beta1 (VLA-1) and alpha2beta1 (VLA-2) integrins on CD4 and CD8 T cells during influenza infection. J Immunol 178:4506–4516
114. Saarialho-Kere UK, Kovacs SO, Pentland AP, Olerud JE, Welgus HG, Parks WC (1993) Cell-matrix interactions modulate interstitial collagenase expression by human keratinocytes actively involved in wound healing. J Clin Invest 92:2858–2866
115. San Antonio JD, Zoeller JJ, Habursky K, Turner K, Pimtong W, Burrows M, Choi S, Basra S, Bennett JS, DeGrado WF et al (2009) A key role for the integrin alpha2beta1 in experimental and developmental angiogenesis. Am J Pathol 175:1338–1347
116. Santoro SA (1986) Identification of a 160,000 dalton platelet membrane protein that mediates the initial divalent cation-dependent adhesion of platelets to collagen. Cell 46:913–920
117. Santoro SA, Zutter MM (1995) The alpha 2 beta 1 integrin: a collagen receptor on platelets and other cells. Thromb Haemost 74:813–821
118. Santoro SA, Walsh JJ, Staatz WD, Baranski KJ (1991) Distinct determinants on collagen support alpha 2 beta 1 integrin-mediated platelet adhesion and platelet activation. Cell Regul 2:905–913
119. Santoso S, Kunicki TJ, Kroll H, Haberbosch W, Gardemann A (1999) Association of the platelet glycoprotein Ia C807T gene polymorphism with nonfatal myocardial infarction in younger patients. Blood 93:2449–2453
120. Sasaki K, Tsuji T, Jinushi T, Matsuzaki J, Sato T, Chamoto K, Togashi Y, Koda T, Nishimura T (2003) Differential regulation of VLA-2 expression on Th1 and Th2 cells: a novel marker for the classification of Th subsets. Int Immunol 15:701–710
121. Sato Y, Morimoto K, Kubo T, Yanagihara K, Seyama T (2012) High mannose-binding antiviral lectin PFL from Pseudomonas fluorescens Pf0-1 promotes cell death of gastric cancer cell MKN28 via interaction with alpha2-integrin. PLoS One 7:e45922
122. Senger DR, Claffey KP, Benes JE, Perruzzi CA, Sergiou AP, Detmar M (1997) Angiogenesis

promoted by vascular endothelial growth factor: regulation through alpha1beta1 and alpha2beta1 integrins. Proc Natl Acad Sci USA 94:13612–13617

123. Shimaoka M, Lu C, Palframan RT, von Andrian UH, McCormack A, Takagi J, Springer TA (2001) Reversibly locking a protein fold in an active conformation with a disulfide bond: integrin alphaL I domains with high affinity and antagonist activity in vivo. Proc Natl Acad Sci USA 98:6009–6014
124. Siljander PR, Hamaia S, Peachey AR, Slatter DA, Smethurst PA, Ouwehand WH, Knight CG, Farndale RW (2004) Integrin activation state determines selectivity for novel recognition sites in fibrillar collagens. J Biol Chem 279:47763–47772
125. Sixma JJ, van Zanten GH, Huizinga EG, van der Plas RM, Verkley M, Wu YP, Gros P, de Groot PG (1997) Platelet adhesion to collagen: an update. Thromb Haemost 78:434–438
126. Slavka G, Perkmann T, Haslacher H, Greisenegger S, Marsik C, Wagner OF, Endler G (2011) Mean platelet volume may represent a predictive parameter for overall vascular mortality and ischemic heart disease. Arterioscler Thromb Vasc Biol 31:1215–1218
127. Sottnik JL, Daignault-Newton S, Zhang X, Morrissey C, Hussain MH, Keller ET, Hall CL (2013) Integrin alpha2beta 1 (α2β1) promotes prostate cancer skeletal metastasis. Clin Exp Metastasis 30:569–578
128. Staatz WD, Rajpara SM, Wayner EA, Carter WG, Santoro SA (1989) The membrane glycoprotein Ia-IIa (VLA-2) complex mediates the Mg ++-dependent adhesion of platelets to collagen. J Cell Biol 108:1917–1924
129. Stenzel D, Franco CA, Estrach S, Mettouchi A, Sauvaget D, Rosewell I, Schertel A, Armer H, Domogatskaya A, Rodin S et al (2011) Endothelial basement membrane limits tip cell formation by inducing Dll4/Notch signalling in vivo. EMBO Rep 12:1135–1143
130. Sun H, Santoro SA, Zutter MM (1998) Downstream events in mammary gland morphogenesis mediated by reexpression of the alpha2beta1 integrin: the role of the alpha6 and beta4 integrin subunits. Cancer Res 58:2224–2233
131. Suzuki-Inoue K, Ozaki Y, Kainoh M, Shin Y, Wu Y, Yatomi Y, Ohmori T, Tanaka T, Satoh K, Morita T (2001) Rhodocytin induces platelet aggregation by interacting with glycoprotein Ia/IIa (GPIa/IIa, Integrin alpha 2beta 1). Involvement of GPIa/IIa-associated src and protein tyrosine phosphorylation. J Biol Chem 276:1643–1652
132. Sweeney SM, DiLullo G, Slater SJ, Martinez J, Iozzo RV, Lauer-Fields JL, Fields GB, San Antonio JD (2003). Angiogenesis in collagen I requires alpha2beta1 ligation of a GFP*GER sequence and possibly p38 MAPK activation and focal adhesion disassembly. J Biol Chem 278:30516–30524
133. Takahashi K, Nakamura T, Koyanagi M, Kato K, Hashimoto Y, Yagita H, Okumura K (1990) A murine very late activation antigen-like extracellular matrix receptor involved in CD2- and lymphocyte function-associated antigen-1-independent killer-target cell interaction. J Immunol Baltim Md 1950 145:4371–4379
134. Tsuji M, Ezumi Y, Arai M, Takayama H (1997) A novel association of Fc receptor gamma-chain with glycoprotein VI and their co-expression as a collagen receptor in human platelets. J Biol Chem 272:23528–23531
135. Tuckwell DS, Smith L, Korda M, Askari JA, Santoso S, Barnes MJ, Farndale RW, Humphries MJ (2000) Monoclonal antibodies identify residues 199-216 of the integrin alpha2 vWFA domain as a functionally important region within alpha2beta1. Biochem J 350(Pt 2):485–493
136. Tulla M, Pentikäinen OT, Viitasalo T, Käpylä J, Impola U, Nykvist P, Nissinen L, Johnson MS, Heino J (2001) Selective binding of collagen subtypes by integrin alpha 1I, alpha 2I, and alpha 10I domains. J Biol Chem 276:48206–48212
137. Tulla M, Lahti M, Puranen JS, Brandt AM, Kapyla J, Domogatskaya A, Salminen TA, Tryggvason K, Johnson MS, Heino J (2008) Effects of conformational activation of integrin alpha 1I and alpha 2I domains on selective recognition of laminin and collagen subtypes. Exp Cell Res 314:1734–1743
138. Upla P, Marjomaki V, Kankaanpaa P, Ivaska J, Hyypia T, Van Der Goot FG, Heino J (2004) Clustering induces a lateral redistribution of alpha 2 beta 1 integrin from membrane rafts to caveolae and subsequent protein kinase C-dependent internalization. Mol Biol Cell 15:625–636
139. Ura H, Denno R, Hirata K, Yamaguchi K, Yasoshima T (1998) Separate functions of alpha2beta1 and alpha3beta1 integrins in the metastatic process of human gastric carcinoma. Surg Today 28:1001–1006
140. Van de Walle GR, Vanhoorelbeke K, Majer Z, Illyes E, Baert J, Pareyn I, Deckmyn H (2005) Two functional active conformations of the integrin {alpha}2{beta}1, depending on activation condition and cell type. J Biol Chem 280:36873–36882
141. Veit G, Zwolanek D, Eckes B, Niland S, Käpylä J, Zweers MC, Ishada-Yamamoto A, Krieg T, Heino J, Eble JA et al (2011) Collagen XXIII, novel ligand for integrin alpha2beta1 in the epidermis. J Biol Chem 286:27804–27813
142. Vogel KG, Trotter JA (1987) The effect of proteoglycans on the morphology of collagen fibrils formed in vitro. Coll Relat Res 7:105–114
143. Vogel KG, Paulsson M, Heinegard D (1984) Specific inhibition of type I and type II collagen fibrillogenesis by the small proteoglycan of tendon. Biochem J 223:587–597
144. Werr J, Johansson J, Eriksson EE, Hedqvist P, Ruoslahti E, Lindbom L (2000) Integrin alpha(2)beta(1) (VLA-2) is a principal receptor

used by neutrophils for locomotion in extravascular tissue. Blood 95:1804–1809
145. Woodall BP, Nystrom A, Iozzo RA, Eble JA, Niland S, Krieg T, Eckes B, Pozzi A, Iozzo RV (2008) Integrin alpha2beta1 is the required receptor for endorepellin angiostatic activity. J Biol Chem 283:2335–2343
146. Wu JE, Santoro SA (1994) Complex patterns of expression suggest extensive roles for the alpha 2 beta 1 integrin in murine development. Dev Dyn 199:292–314
147. Xia H, Seeman J, Hong J, Hergert P, Bodem V, Jessurun J, Smith K, Nho R, Kahm J, Gaillard P et al (2012) Low alpha(2)beta(1) integrin function enhances the proliferation of fibroblasts from patients with idiopathic pulmonary fibrosis by activation of the beta-catenin pathway. Am J Pathol 181:222–233
148. Yebra M, Diaferia GR, Montgomery AM, Kaido T, Brunken WJ, Koch M, Hardiman G, Crisa L, Cirulli V (2011) Endothelium-derived Netrin-4 supports pancreatic epithelial cell adhesion and differentiation through integrins alpha2beta1 and alpha3beta1. PLoS One 6:e22750
149. Zhang Z, Ramirez NE, Yankeelov TE, Li Z, Ford LE, Qi Y, Pozzi A, Zutter MM (2008) alpha2beta1 integrin expression in the tumor microenvironment enhances tumor angiogenesis in a tumor cell-specific manner. Blood 111:1980–1988
150. Zutter MM, Sun H, Santoro SA (1998) Altered integrin expression and the malignant phenotype: the contribution of multiple integrated integrin receptors. J Mammary Gland Biol Neoplasia 3:191–200
151. Zutter MM, Santoro SA, Wu JE, Wakatsuki T, Dickeson SK, Elson EL (1999) Collagen receptor control of epithelial morphogenesis and cell cycle progression. Am J Pathol 155:927–940
152. Zweers MC, Davidson JM, Pozzi A, Hallinger R, Janz K, Quondamatteo F, Leutgeb B, Krieg T, Eckes B (2007) Integrin alpha2beta1 is required for regulation of murine wound angiogenesis but is dispensable for reepithelialization. J Invest Dermatol 127:467–478

Integrin $\alpha 10\beta 1$: A Collagen Receptor Critical in Skeletal Development

4

Evy Lundgren-Åkerlund and Attila Aszòdi

Abstract

Integrin $\alpha 10\beta 1$ is the most abundant collagen-binding integrin in cartilaginous tissues and its expression pattern is distinct from that of other collagen-binding integrins. In vitro and in vivo studies have identified integrin $\alpha 10\beta 1$ as a unique phenotypic marker for chondrocyte differentiation and a crucial mediator of cell-matrix interactions required for proper cartilage development. This chapter describes the structure of the integrin subunit $\alpha 10$, the tissue distribution of the integrin $10\beta 1$ and updates available information regarding its regulation and ligand binding. We also summarize current information on the functional roles of $\alpha 10\beta 1$ in chondrogenesis of mesenchymal stem cells and in skeletal growth.

Keywords

Integrin $\alpha 10\beta 1$ · Collagen · Chondrocyte · Mesenchymal stem cell · Chondrodysplasia

4.1 Introduction

The integrin $\alpha 10\beta 1$ was identified as a collagen type II binding receptor on chondrocytes in 1998 by Camper et al. [9]. Earlier studies in our laboratory had indicated that an unknown α subunit in the $\beta 1$ integrin family with a molecular weight of approximately 160 kDa was present on chondrocytes and chondrosarcoma cells [16]. To identify this integrin subunit, large quantities of chondrocytes (2.5 billion cells) were collected from bovine articular cartilage and integrin $\alpha 10$ was isolated by affinity purification of the chondrocyte lysate on a collagen II-Sepharose column. The human ortholog of the $\alpha 10$ subunit was subsequently characterized using a human chondrocyte library [9]. With an antibody raised against the cytoplasmic domain of $\alpha 10$ we could show that integrin $\alpha 10\beta 1$ is a major collagen-binding integrin on chondrocytes and that it is highly expressed in cartilage, both during development and in adult tissues [9, 10]. We have also

E. Lundgren-Åkerlund (✉)
Xintela AB, Medicon Village, SE-223 81 Lund, Sweden
e-mail: Evy@xintela.se

A. Aszòdi
Department of Surgery, Ludwig-Maximilians-University, Nussbaumstr. 20, 80336, Munich, Germany

D. Gullberg (ed.), *I Domain Integrins*, Advances in Experimental Medicine and Biology 819,
DOI: 10.1007/978-94-017-9153-3_4,

shown that fibroblast growth factor -2 (FGF-2) upregulates expression of α10 and improves chondrogenic potential of mesenchymal stem cells (MSCs) [30]. In 2005 we demonstrated that mice lacking the integrin α10β1 have defects in the cartilaginous growth plate and, as a consequence, develop growth retardation of the long bones [4]. A recent study revealed that a naturally occurring mutation in the canine α10 integrin gene is responsible for chondrodysplasia in hunting dog breeds [19], supporting a critical role for α10β1 in skeletal development.

This chapter summarizes current knowledge on structure, distribution and function of the integrin α10β1.

4.2 The α10 Structure

The α10 subunit has a M_r of around 160 kDa under reducing conditions as determined by SDS-PAGE. When compared to the other collagen-binding integrins on cultured chondrocytes, the integrin subunit α10 appears distinctly smaller than α1 in SDS-PAGE, slightly smaller than α11 and similar in size to the α2 subunit under reducing conditions [9].

Sequence analysis of the four collagen-binding integrins shows that α10 displays 43 % sequence identity to α11, 33 % to α1 and 31 % identity to α2 at the amino acid level. Similar to the other collagen-binding integrin subunits α1, α2 and α11, the I-domain in α10 is encoded by exons 6-9. At the protein level, α10 I domain extends from C^{140}-G^{337} and consists of 198 amino acids which is inserted in the *N*-terminal region of the extracellular domain between cation binding sites two and three. Three cysteine residues are present within the α10 I-domain, compared to one cysteine in α11, two in α1 and three in α2. Like the other I-domain-containing collagen-binding integrins, the α10 I-domain contains a MIDAS (metal ion-dependent adhesion site) motif [9]. The overall identity between α10 I-domain and the I-domains of the other collagen-binding integrins is high with the highest identity to the α11 I-domain (60 %) [14].

The conserved sequence in the transmembrane/cytoplasmic region of α10 is GFFAH and not GFFR/KR as in most other integrins [9]. Analysis by an in vitro glycosylation method has shown that the length of the α10 transmembrane (TM) domain is 29 amino acids and extends to the Ala in the sequence GFFAH, 1-2 amino acids further at the C terminus compared with the TM domain of other α subunits [26]. Thus, the cytoplasmic domain of α10 consists of 16 amino acids.

4.3 The α10 Gene

The human α10 gene, (*ITGA10*), which consists of 30 translated exons distributed over a region of 19kb has been located to chromosome 1, locus q21 [5, 20]. It is the only integrin located on this chromosome [5].

The mouse α10 gene (*Itga10*), located on chromosome 3, F2.2, has a homology of 90 % with *ITGA10*. The highest homology is found in the I-domain (97 %) [5]. *Itga10* consists of 30 translated exons spanning a region of about 18 kb genomic DNA. Primer extension analysis determined that a major transcription start site is located 38 nucleotides (nt) upstream from the translation initiation site ATG. The 5′-flanking region of the transcription site at –38 nt lacks a TATA box, as is the case for most other integrin subunits [5].

We have previously demonstrated that the human α10 subunit exists as two splice variants due to alternative splicing of exon 25 (114 nt). The spliced domain is located extracellularly, close to the transmembrane region. Both forms are expressed at the mRNA level in human chondrocytes but it is not clear if the smaller form, lacking exon 25, is present on the cell surface of chondrocytes [5].

In contrast to human *ITGA10*, mouse exon 26 and not exon 25 is alternatively spliced [5]. In the spliced variant, exon 26 containing 144 nt is extended into the intron by 62 nt and results in a shift in the reading frame and a premature stop codon [5].

Extracellular splice variants have been reported for other integrin α-subunits, although the functions of the variants are not understood [11]. Similar to α10, α7 and αIIb have alternatively spliced extracellular regions close to the transmembrane spanning domain [7, 21]. In addition, α11 has an inserted region of 22 amino acids in exon 20 [32].

Sequence analysis of a promotor region of human *ITGA10*, together with functionality tests, have identified the transcription factors AP-2ε and Ets-1 as regulators of the integrin α10 gene in chondrocytes [34]. These transcription factors are also known to control expression of integrin α10 in melanoma cells [33].

Interestingly, a truncating mutation of the *ITGA10* gene on canine choromosome 17 was recently shown to cause chondrodysplasia, short stature dwarfism, in dogs (see below) [19].

4.4 Distribution of Integrin α10β1

Expression analysis of mouse and human tissues has shown that distribution of integrin α10β1 is quite restricted [10, 14]. It is strongly expressed in the cartilage of joints and in other cartilage-containing tissues such as vertebrae, ribs, trachea and bronchi [10]. Integrin α10β1 is found where collagen type II is expressed and is an excellent cellular marker of cartilage tissue. Integrin α10β1 is also expressed on chondrocytes in the growth plate and on bone lining cells in the trabecular bone (Fig. 4.1).

We have also found α10β1 on cells in some fibrous tissues such as perichondrium, periosteum and endosteum, and in the fascia lining skeletal muscle fibers [10, 30]. These tissues are known to house mesenchymal progenitor cells with the potential to develop into different cell types [3, 15, 22]. In addition, integrin α10 is present in the junctions between cartilage/bone and ligaments [10].

In the vertebral column, α10 was detected by immunohistochemistry in the cartilage of the vertebral body and in the inner annulus fibrosus of the intervertebral discs [10]. In both tissues, α10 expression was co-localized with collagen II. The outer annulus fibrosus, which contains collagen I, was negative for α10. Interestingly, α11 has been detected in the outer annulus fibrosus [28].

An mRNA array of human tissues has also suggested expression of α10 outside of the musculoskeletal system, e.g. in heart and aorta [14]. When we analyzed α10 protein in the mouse heart, we found immunodetectable expression in the aortic and atrioventricular valves but not in the heart muscle [10]. Interestingly, collagen II is expressed in the heart valves during early development but the role for this collagen during heart development is not known [27]. On the other hand, we have not been able to confirm expression of α10 on the protein level in aorta. However, we have detected α10 in atherosclerotic plaque both in human and in a mouse model (Lundgren-Åkerlund and Hultgård-Nilsson, unpublished results). The plaques may represent the strong aorta signal in the mRNA tissue array [14]. In unpublished studies we found that α10 colocalized with collagen II in atherosclerotic plaques.

The expression of integrin α10β1 in tissues that are exposed to high mechanical load such as articular cartilage, vertebral column and heart valves implicates a role for α10β1 integrin in mechanical integrity and/or in mechanical signaling of these tissues.

We have earlier published that malignant melanoma express integrin α10 and that antibodies blocking α10 reduce migration of the melanoma cells in vitro [33]. It is not known if α10 is present in other tumors.

4.5 Integrin α10β1 in Mouse Limb Development

The major part of the mammalian skeleton is laid down by a process called endochondral bone formation. Cartilage moulds form during embryogenesis via the sequential steps of mesenchymal cell condensation, chondroprogenitor commitment and chondrocyte differentiation under the control of multiple mechanisms, including cell-cell and cell-matrix interactions,

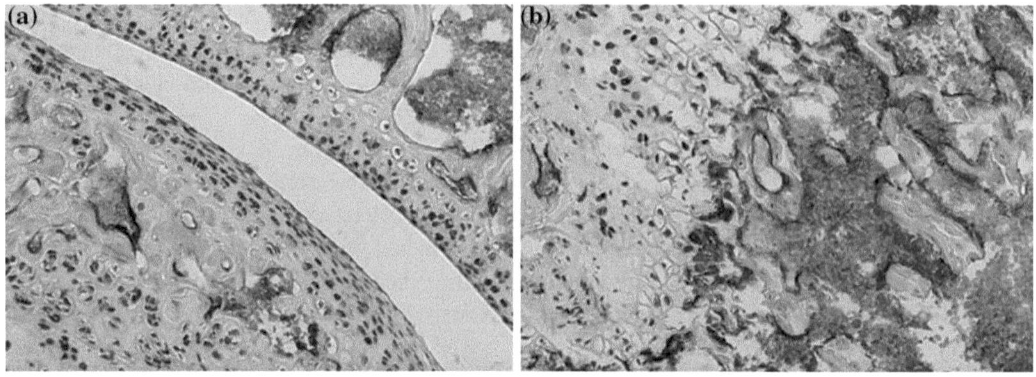

Fig. 4.1 Immunohistochemical localization of the integrin $\alpha10\beta1$. Mouse hind limbs from 8-week-old mice were cryosectioned and stained with an affinity purified polyclonal antibody recognizing the cytoplasmic domain of $\alpha10$. The integrin $\alpha10\beta1$ was expressed by all chondrocytes in the epiphyseal cartilage and in the growth plate and also by cells in the trabecular bone

cellular signaling and transcriptional and translational regulation. Most fetal cartilages are transient and will gradually be replaced by trabecular bone during a series of events including chondrocyte proliferation, hypertrophy and apoptosis followed by cartilage matrix mineralization, vascularization and matrix degradation. These events take place within the growth plate. Proliferation, matrix production and hypertrophy of chondrocytes in the growth plate are essential to achieve longitudinal elongation of endochondral bones. Oppositely, in articular cartilage, the chondrocytes acquire a stable phenotype that resists hypertrophy and vascular invasion, thus maintain a mechanically adequate ECM throughout the life [1].

We have demonstrated that $\alpha10\beta1$ is the major collagen-binding integrin in the cartilage during development of the skeleton in mouse [10]. Expression analysis of the hind limb revealed that $\alpha10$ appears at embryonic day 11.5 (E11.5) at the onset of chondrogenesis as determined by collagen type II expression. At E13.5, $\alpha10$ is present throughout the anlage as well as in the perichondrium and in the mesenchyme just outside of the perichondrium. In newborn mice, $\alpha10\beta1$ is expressed by all chondrocytes in the growing epiphyseal and growth plate cartilage. 4 weeks after birth, $\alpha10$ is prominent both at the articular surface and in the growth plate [10]. $\alpha10$ is also detected in the inner cartilaginous region of the meniscus where collagen type II is expressed.

Interestingly, $\alpha10$ is expressed by cells in the ossification groove of Ranvier [10]. The ossification groove contains precursors for both chondrocytes and osteoblasts and has been suggested to be involved in growth of the bone [23, 24].

4.6 Ligands Specificity

The integrin subunit $\alpha10$ was originally isolated by affinity purification on a collagen type II-column [9]. Experiments using $\alpha10\beta1$-expressing C2C12 cells and an $\alpha10$ blocking monoclonal antibody have confirmed that $\alpha10\beta1$ is a receptor for fibril-forming collagen types II and XI and for the FACIT (fibril-associated collagen with interruptions in triple helix) collagen type IX in vitro (Fig. 4.2). The $\alpha10$-expressing cells also bound to the beaded filament-forming collagen VI but, in contrast to the other collagens, we could not block the adhesion of $\alpha10$ expressing C2C12 cells to collagen VI with the monoclonal antibody. This might be explained by a different binding mechanism of $\alpha10\beta1$ to collagen VI or, alternatively, that other collagen interacting receptors are involved in the attachment of C2C12 cells to collagen VI. We have also found that collagen type X, expressed by

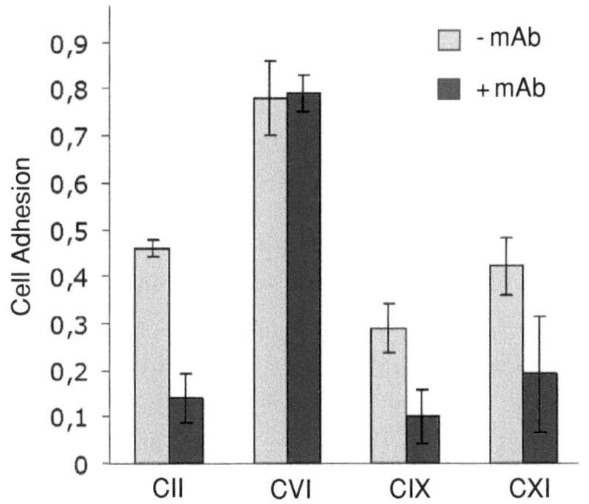

Fig. 4.2 Adhesion of α10β1 expressing cells to the cartilage collagen types II, VI, IX and XI. C2C12 cells, expressing α10β1 as the only collagen-binding integrin, were allowed to adhere for 1 h to collagen-coated (10 μg/ml) culture dishes in the absence (−mAb) or in the presence (+mAb) of a blocking monoclonal antibody directed to the I-domain of α10 (5 μg/ml). The numbers of adhered cells were compared to the total number of cells added to the wells (1 = 100 %). The numbers represent the mean ± S.D. of triplicate experiments. The antibody reduced adhesion of α10β1 cells co collagen II, IX and XI but not to collagen VI

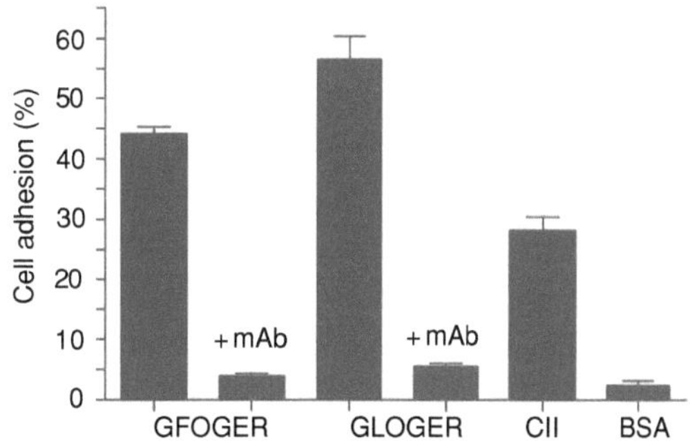

Fig. 4.3 Adhesion of α10β1 expressing cells to the collagen peptides GFOGER and GLOGER. C2C12 cells, expressing α10β1 as the only collagen-binding integrin, were allowed to adhere for 1 h to culture dishes coated with 10 μg/ml of the collagen peptides GFOGER and GLOGER and collagen type II (CII) and blocked with 1 % bovine serum albumin (BSA). The adhesion experiments were performed in the absence or in the presence (+mAb) of an α10 blocking monoclonal antibody (5 μg/ml). The number of adhered cells is shown as a percentage of the total number of cells added to the wells. The numbers represent the mean ± S.D. of triplicate experiments. From Bryngelson Ohlsson [8]

hypertrophic chondrocytes, can mediate adhesion of α10β1-expressing cells (Lundgren-Åkerlund, unpublished results).

Our findings that α10β1 is present in non-cartilaginous tissues implicates that α10β1 in vivo interacts with other ligands than cartilage collagens. Indeed, we have found that α10β1-expressing cells also interact with collagen types I and IV (Lundgren-Åkerlund, unpublished results).

Using recombinant I-domains, the study by Tulla et al. [29] suggested that the α10 I-domain, similar to the α1 I-domain, has a preferred affinity for collagen IV (a basement membrane collagen) and collagen VI over the fibrillar collagen types I-III.

It has previously been reported that I domains of the collagen-binding integrin subunits α1, α2 and α11 interact with the trippel helical collagen I peptides GFOGER and GLOGER [25, 35, 37]. We have found that α10β1-expressing cells adhere to the peptides GFOGER and GLOGER and that the adhesion can be inhibited by an α10-blocking monoclonal antibody recognizing the I-domain of α10 (Fig. 4.3) [8]. Recently, Käpylä et al. has demonstrated that all collagen integrin receptors, including α10β1, bind collagen IX via a novel, GFOGER-independent mechanism which does not resemble interactions with other collagen types [17].

4.7 Integrin α10β1: A Chondrogenic Differentiation and Potency Marker

Primary chondrocytes, with a differentiated phenotype, express α10β1 on the cell surface and synthesize the cartilage specific molecules, collagen type II and aggrecan. During monolayer cultures chondrocytes are known to dedifferentiate as characterized by a decrease in expression of collagen II and increased synthesis of fibrous matrix molecules, such as collagen type I [6]. We have found that expression of integrin α10β1 is gradually downregulated from the cell surface during dedifferentiation of chondrocytes in monolayer cultures (Fig. 4.4).

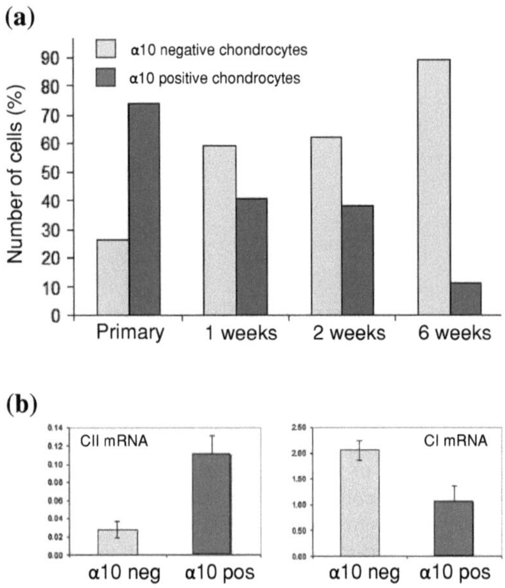

Fig. 4.4 Expression of α10β1 and collagen types I and II in cultured chondrocytes. **a** Chondrocytes were isolated from human articular cartilage and the expression of integrin α10β1 was analyzed on primary chondrocytes and on chondrocytes cultured in monolayer for 1, 2 or 6 weeks by flow cytometry. The cells were passaged once a week. The bars show percentage of α10 negative and α10 positive chondrocytes at each time point. Expression of α10 gradually decreased with time in culture and was approximately 10% after 6 weeks. **b** After 6 weeks of culture, mRNA levels of collagens I and II were analyzed in α10 positive and α10 negative sorted chondrocytes. Expression of collagen II was higher on α10 positive chondrocytes while expression of collagen type I was higher on α10 negative chondrocytes

After 6 weeks of culture, only about 10% of the cells expressed α10. However, after FACS sorting, we found that the α10 positive cells had a higher collagen II/collagen I mRNA ratio compared to the α10 negative cells. Furthermore, dedifferentiated chondrocytes lacking α10 were able to restore expression of integrin α10β1 and redifferentiate when the cells were transferred to three-dimensional alginate cultures (Fig. 4.5). This observation implicates that integrin α10β1 is a unique cellular marker for the differentiation state of the chondrocytes. In agreement, Gouttenoire et al. [13] showed that α10 together with the chondrogenic collagen type IIB isoform are expressed by differentiated mouse chondrocytes after treatment with BMP2, while TGFβ1 stimulated the expression of the non-chondrogenic procollagen type IIA and α11 integrin. In another study it was demonstrated that aggregation of phenotypically stable chondrocytes is mediated by integrin α10β1 and collagen type II interaction [12].

We have previously reported that integrin α10β1 is present on human mesenchymal stem cells (MSCs) and that its expression increases during *in vitro* chondrogenesis in aggregate cultures [30]. We have also reported that extended monolayer culturing of MSCs down-regulates integrin α10, while treatment of the cultured MSCs with fibroblast growth factor-2 (FGF-2) increases expression of α10. In contrast, FGF-2 treatment of the MSCs decreases expression of α11 [30] (Fig. 4.6). Transforming growth factor-β3 (TGF-β3), on the other hand, was found to decrease expression of α10 and increase expression of α11 on MSCs [30]. The effects of FGF-2 and TGF-β3 on α10 and α11 expression observed in MSCs appears to extend to cultured human and bovine chondrocytes (Lundgren-Åkerlund and Aszòdi, unpublished results).

We have also reported that FGF-2-induced upregulation of α10 in MSCs enhances chondrogenesis and synthesis of cartilage molecules such collagen type II and aggrecan in pellet cultures [30]. This demonstrates that α10β1 is a unique cell surface biomarker and potency marker of MSCs with chondrogenic potential and will serve as a valuable tool in the quality assurance of chondrocytes and chondrogenic MSCs used in cartilage repair.

4.8 Loss of Integrin α10β1 Leads to Chondrodysplasia

The essential role of integrin-mediated attachment and signaling in endochondral bone formation was first demonstrated by conditional inactivation of the floxed β1 integrin gene in the entire cartilaginous skeleton using a transgene which drives the expression of the cre recombinase under the control of the collagen II

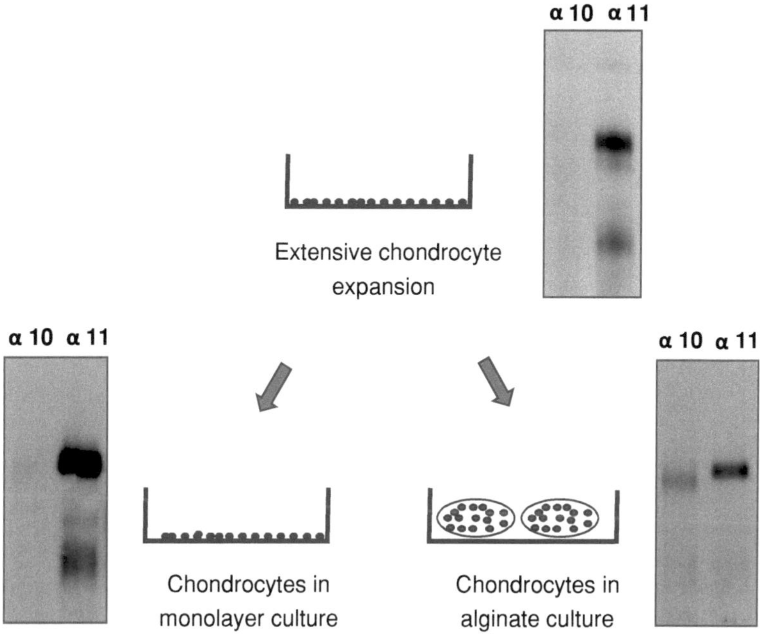

Fig. 4.5 Integrin α10β1 expression is restored when dedifferentiated chondrocytes are transferred to three-dimensional alginate culture. Human chondrocytes were dedifferentiated in monolayer cultures until expression of the α10 protein was lost (8 weeks of culture, five passages, in this experiment). One part of the chondrocytes was then transferred to alginate beads for redifferentiation in three-dimensional culture. The other part was kept in monolayer culture. The chondrocytes were cultured for additional 2 weeks and then cell surface biotinylated, lysed and immunoprecipitated with antibodies directed to α10 and α11 followed by separation by SDS-PAGE and western blot analysis. The results demonstrate that expression of α10 was restored in chondrocytes in alginate culture while expression of α11 appeared to decrease in alginate compared to monolayer cultures. This demonstrates that integrin α10β1 is a cellular marker for staging the differentiation status of chondrocytes

promoter (*Col2a1cre*) [2]. We have shown that mutant mice ($\beta 1^{fl/fl}$-*Col2a1cre*$^{+}$) develop perinatal lethal chondrodysplasia, characterized by the lack of columnar growth plate, reduced chondrocyte proliferation, abnormal cell shape and distorted collagen fibrillar network in the ECM. These observations have demonstrated that β1 integrin-mediated cell-matrix interactions are mandatory for chondrocyte geometry, motility and cytokinesis, essential mechanisms necessary for the proper formation and function of the growth plate.

Among mice lacking an α integrin subunit, only the knockout of the α10 integrin gene [4] results in skeletal abnormalities which partially recapitulate the phenotype of the $\beta 1^{fl/fl}$-*Col2a1cre*$^{+}$ mice (Fig. 4.7). We have shown that α10-deficiency is dispensable for life but causes chondrocyte shape change and mild disorganization of columnar arrangement resulting in moderate growth retardation. Some other abnormalities such as the shorter hypertrophic zone, increased apoptosis and reduced chondrocyte proliferation also contribute for growth plate dysfunction in both mouse models. It is particularly interesting that the chondrocyte cell cycle apparently is modulated by α10β1 integrin. Both α10- and β1-deficient chondrocytes display delayed G1/S transition accompanied by increased nuclear translocation of Stat1 and Stat5a, two members of the family of signal transducers and activators of transcription, inducing the upregulation of cell cycle inhibitors p16 and/or p21 which in turn decreases the proliferation rate [2, 4]. Another striking phenotype present in the two knockout strains is the

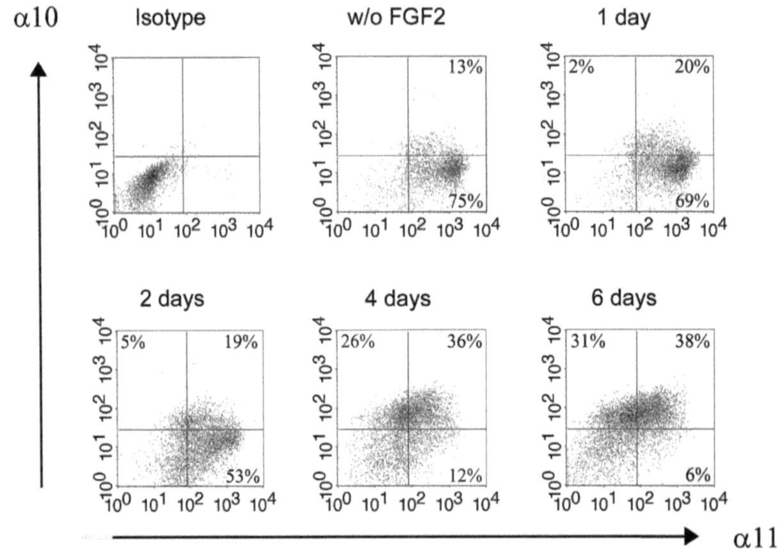

Fig. 4.6 FGF-2 increase expression of α10 and decrease expression of α11 on mesenchymal stem cells. Human bone marrow derived mesenchymal stem cells (MSCs) were isolated by plastic adherence and cultured for 4 weeks. The MSCs were then transferred to 6-well dishes and stimulated with 10 ng/ml of fibroblast growth factor-2 (FGF-2,) for 1, 2, 4 and 6 days and subsequently analyzed by flow cytometry using antibodies directed to α10 and α11. The *upper left panel* represent the background staining using an isotype antibody. The results show that treatment with FGF-2 gradually increases expression of α10 (y-axis) and decreases expression of α11 (x-axis). From Varas et al. [30]

reduced density of the collagen matrix implying a role of integrin α10β1 for matrix assembly. Since the β1-deficient chondrocytes express normal levels of collagen II, this matrix defect suggest the direct involvement of β1 integrins in collagen fibril polymerization and/or incorporation of the fibrils into the collagen meshwork.

Despite the aforementioned similarities, the skeletal phenotype is more severe in *$\beta1^{fl/fl}$-$Col2a1cre^{+}$* mice and some abnormalities such as the defective cytokinesis or the disrupted actin network are only observed in β1 null chondrocytes suggesting partial redundancy among β1 integrins. In vitro assays have indeed demonstrated comparable adhesion and spreading of primary wild type and α10 null chondrocytes on fibrillar collagens [4]. The integrin α1β1, which is expressed on chondrocytes in the articular cartilage, is a strong candidate for compensating the α10-deficiency. The integrins α2β1 and α11β1, on the other hand, appeared to be absent or only weakly expressed in the cartilage and on isolated chondrocytes [4, 10]. However, mice lacking α1 integrin show no growth plate phenotype but develop early onset osteoarthritis [36], an ageing-dependent degeneration of the articular cartilage. We have shown in 4-week-old mice that in contrast to the epiphyseal cartilage, α1 is only weakly expressed in the growth plate [10]. This may explain why mice lacking α1 integrin have no growth plate abnormalities. Integrin α10, on the other hand, was strongly expressed both in the epiphyseal cartilage and in the growth plate. Taken together, the studies with genetically modified mice support a hypothesis for both overlapping and distinct role of collagen-binding integrins in skeletal development and function where α10β1 may play a specific role in growth plate morphogenesis and skeletal growth.

Inherited chondrodysplasias are caused by mutations in a range of gene families encoding e.g. ECM proteins, transcription factors, growth factor receptors, enzymes, or signaling molecules, but so far no integrin mutations have been associated with these skeletal disorders in

(a)

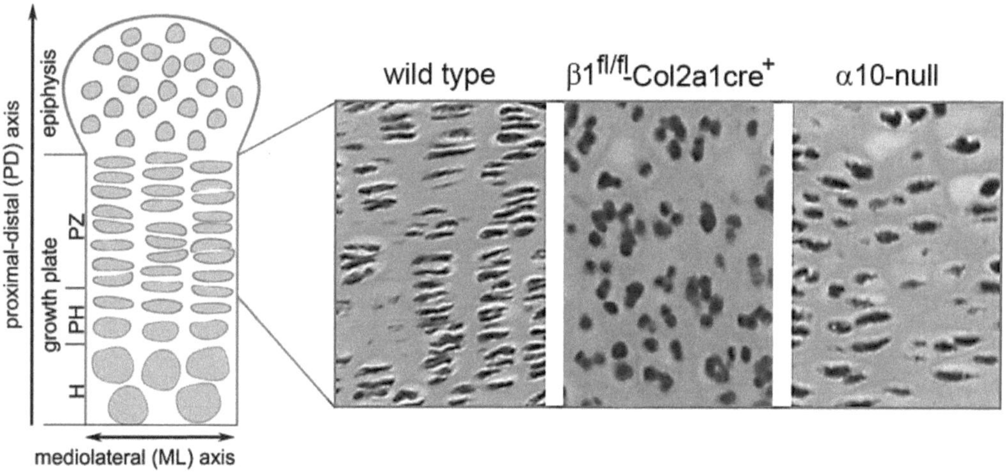

(b)

phenotype	β1 fl/fl - Col2a1cre +	α10-null
chondrodysplasia	severe, lethal	mild, vital
shape of proliferative chondrocytes	round	moderatelly rounded
growth plate disorganization	complete lack of columns	moderate disorganization
hypertrophic zone	reduced	reduced
chondrocyte proliferation	decreased	decreased
apoptosis	increased	increased
cytokinesis defect	severe	none
adhesion to collagen II	diminished	normal
actin cytoskeleton	abnormal	normal
collagen network	reduced density	slightly reduced density
collagen fibril diameter	thickened	thickened

Fig. 4.7 The lack of α10β1 integrin results in moderate chondrodysplasia and growth plate dysfunction. **a** In normal growth plate, chondrocytes form horizontal zones reflecting their differentiation stage (proliferative, pre-hypertrophic and hypertrophic). The flattened proliferative cells are oriented with their long axes perpendicular to the direction of the longitudinal growth and arranged into vertical columns. Proliferative chondrocytes are gradually rounding up and lose their columnar organization in mice lacking α10β1 integrin (*α10-null*) or all β1 integrin-containing heterodimers (*β1 fl/fl Col2a1cre*+) on chondrocytes. **b** Comparison of the cartilage phenotype in *α10-null* and *β1 fl/fl Col2a1cre*+mice

human [18, 31]. Recently, genetic analysis of two dog breeds has revealed that a naturally occurring mutation in the α10 integrin gene is responsible for canine chondrodysplasia [19]. The Nordic hunting dogs Norwegian Elkhound and Karelian Bear Dog display disproportionate short-limbed dwarfism characterized by growth plate abnormalities resembling the α10 knockout mouse phenotype. Using a genome-wide approach, a recessive mutation in *ITGA10* was shown to be segregated with the disease in both breeds. The nonsense mutation p.Arg695* in exon 16 was predicted to produce a truncated protein lacking the cytosolic tail, the transmembrane domain and part of the extracellular domain. As judged by western blot, the truncated protein was undetectable in tracheal cartilage suggesting a loss of function of $\alpha10\beta1$ [19] in these dogs.

The current nosology and classification of human genetic skeletal disorders show that the causative gene for many human genetic skeletal disorders are still unknown [31]. Our results from the $\alpha10$ knockout mouse model together with the results from the natural mutation in the canine ITGA10, showing that loss of function of the integrin $\alpha10\beta1$ gene leads to disproportionate chondrodysplasia, suggest that ITGA10 is a likely candidate gene responsible also for human disproportionate chondrodysplasias.

4.9 Perspectives

Integrin $\alpha10\beta1$ has a specific role during skeletal development that does not overlap with other collagen-binding integrins. However, very little is known about the molecular mechanisms behind the specific function of $\alpha10$ in skeletal development and its involvement in different pathological conditions such as chondrodysplasias and osteoarthritis. The fact that integrin $\alpha10$ is expressed in tissues that are exposed to high mechanical load implicates a role for $\alpha10$ in mechanical integrity and/or in mechanical signaling of these tissues.

Integrin $\alpha10\beta1$ is a prominent collagen receptor on chondrocytes and its expression correlates with expression of cartilage matrix molecules such as collagen type II. This makes $\alpha10$ a unique and very useful differentiation quality/potency marker of chondrocytes as well as MSCs in tissue engineering of cartilage.

References

1. Aszodi A, Bateman JF, Gustafsson E, Boot-Handford R, Fassler R (2000) Mammalian skeletogenesis and extracellular matrix: what can we learn from knockout mice? Cell Struct Funct 25:73–84
2. Aszodi A, Hunziker EB, Brakebusch C, Fassler R (2003) Beta1 integrins regulate chondrocyte rotation, G1 progression, and cytokinesis. Genes Dev 17:2465–2479
3. Balduino A, Hurtado SP, Frazao P, Takiya CM, Alves LM, Nasciutti LE et al (2005) Bone marrow subendosteal microenvironment harbours functionally distinct haemosupportive stromal cell populations. Cell Tissue Res 319:255–266
4. Bengtsson T, Aszodi A, Nicolae C, Hunziker EB, Lundgren-Akerlund E, Fassler R (2005) Loss of alpha10beta1 integrin expression leads to moderate dysfunction of growth plate chondrocytes. J Cell Sci 118:929–936
5. Bengtsson T, Camper L, Schneller M, Lundgren-Akerlund E (2001) Characterization of the mouse integrin subunit alpha10 gene and comparison with its human homologue. Genomic structure, chromosomal localization and identification of splice variants. Matrix Biol 20:565–576
6. Benya PD, Padilla SR (1986) Modulation of the rabbit chondrocyte phenotype by retinoic acid terminates type II collagen synthesis without inducing type I collagen: the modulated phenotype differs from that produced by subculture. Dev Biol 118:296–305
7. Bray PF, Leung CS, Shuman MA (1990) Human platelets and megakaryocytes contain alternately spliced glycoprotein IIb mRNAs. J Biol Chem 265:9587–9590
8. Bryngelson Ohlsson L (2005) Mesenchymal progenitor cells and tumor growth, Integrins and matrix mettalloproteinases. Thesis, Lund University, Sweden
9. Camper L, Hellman U, Lundgren-Akerlund E (1998) Isolation, cloning, and sequence analysis of the integrin subunit alpha10, a beta1-associated collagen binding integrin expressed on chondrocytes. J Biol Chem 273:20383–20389
10. Camper L, Holmvall K, Wangnerud C, Aszodi A, Lundgren-Akerlund E (2001) Distribution of the collagen-binding integrin alpha10beta1 during mouse development. Cell Tissue Res 306:107–116
11. de Melker AA, Sonnenberg A (1999) Integrins: alternative splicing as a mechanism to regulate ligand binding and integrin signaling events. BioEssays 21:499–509
12. Gigout A, Jolicoeur M, Nelea M, Raynal N, Farndale R, Buschmann MD (2008) Chondrocyte aggregation in suspension culture is GFOGER-GPP- and beta1 integrin-dependent. J Biol Chem 283:31522–31530
13. Gouttenoire J, Bougault C, Aubert-Foucher E, Perrier E, Ronziere MC, Sandell L et al (2010) BMP-2 and TGF-beta1 differentially control expression of type II procollagen and alpha 10 and alpha 11 integrins in mouse chondrocytes. Eur J Cell Biol 89:307–314
14. Gullberg DE, Lundgren-Akerlund E (2002) Collagen-binding I domain integrins–what do they do? Prog Histochem Cytochem 37:3–54

15. Hanada K, Solchaga LA, Caplan AI, Hering TM, Goldberg VM, Yoo JU et al (2001) BMP-2 induction and TGF-beta 1 modulation of rat periosteal cell chondrogenesis. J Cell Biochem 81:284–294
16. Holmvall K, Camper L, Johansson S, Kimura JH, Lundgren-Akerlund E (1995) Chondrocyte and chondrosarcoma cell integrins with affinity for collagen type II and their response to mechanical stress. Exp Cell Res 221:496–503
17. Kapyla J, Jaalinoja J, Tulla M, Ylostalo J, Nissinen L, Viitasalo T et al (2004) The fibril-associated collagen IX provides a novel mechanism for cell adhesion to cartilaginous matrix. J Biol Chem 279:51677–51687
18. Krakow D, Rimoin DL (2010) The skeletal dysplasias. Genet Med 12:327–341
19. Kyostila K, Lappalainen AK, Lohi H (2013) Canine chondrodysplasia caused by a truncating mutation in collagen-binding integrin alpha subunit 10. PLoS ONE 8:e75621
20. Lehnert K, Ni J, Leung E, Gough S, Morris CM, Liu D et al (1999) The integrin alpha10 subunit: expression pattern, partial gene structure, and chromosomal localization. Cytogenet Cell Genet 87:238–244
21. Leung E, Lim SP, Berg R, Yang Y, Ni J, Wang SX et al (1998) A novel extracellular domain variant of the human integrin alpha 7 subunit generated by alternative intron splicing. Biochem Biophys Res Commun 243:317–325
22. Li G, Zheng B, Meszaros LB, Vella JB, Usas A, Matsumoto T et al (2011) Identification and characterization of chondrogenic progenitor cells in the fascia of postnatal skeletal muscle. J Mol Cell Biol 3:369–377
23. Schollmeier G, Uhthoff HK, Lewandrowski KU, Fukuhara K (1999) Role of bone bark during growth in width of tubular bones. A study in human fetuses. Clin Orthop Relat Res 367:291–299
24. Shapiro F, Holtrop ME, Glimcher MJ (1977) Organization and cellular biology of the perichondrial ossification groove of ranvier: a morphological study in rabbits. J Bone Joint Surg Am 59:703–723
25. Siljander PR, Hamaia S, Peachey AR, Slatter DA, Smethurst PA, Ouwehand WH et al (2004) Integrin activation state determines selectivity for novel recognition sites in fibrillar collagens. J Biol Chem 279:47763–47772
26. Stefansson A, Armulik A, Nilsson I, von Heijne G, Johansson S (2004) Determination of N- and C-terminal borders of the transmembrane domain of integrin subunits. J Biol Chem 279:21200–21205
27. Swiderski RE, Daniels KJ, Jensen KL, Solursh M (1994) Type II collagen is transiently expressed during avian cardiac valve morphogenesis. Dev Dyn 200:294–304
28. Tiger CF, Fougerousse F, Grundstrom G, Velling T, Gullberg D (2001) alpha11beta1 integrin is a receptor for interstitial collagens involved in cell migration and collagen reorganization on mesenchymal nonmuscle cells. Dev Biol 237:116–129
29. Tulla M, Pentikainen OT, Viitasalo T, Kapyla J, Impola U, Nykvist P et al (2001) Selective binding of collagen subtypes by integrin alpha 1I, alpha 2I, and alpha 10I domains. J Biol Chem 276:48206–48212
30. Varas L, Ohlsson LB, Honeth G, Olsson A, Bengtsson T, Wiberg C et al (2007) Alpha10 integrin expression is up-regulated on fibroblast growth factor-2-treated mesenchymal stem cells with improved chondrogenic differentiation potential. Stem Cells Dev 16:965–978
31. Warman ML, Cormier-Daire V, Hall C, Krakow D, Lachman R, LeMerrer M et al (2011) Nosology and classification of genetic skeletal disorders: 2010 revision. Am J Med Genet A 155A:943–968
32. Velling T, Kusche-Gullberg M, Sejersen T, Gullberg D (1999) cDNA cloning and chromosomal localization of human alpha(11) integrin. A collagen-binding, I domain-containing, beta(1)-associated integrin alpha-chain present in muscle tissues. J Biol Chem 274:25735–25742
33. Wenke AK, Kjellman C, Lundgren-Akerlund E, Uhlmann C, Haass NK, Herlyn M et al (2007) Expression of integrin alpha10 is induced in malignant melanoma. Cell Oncol 29:373–386
34. Wenke AK, Rothhammer T, Moser M, Bosserhoff AK (2006) Regulation of integrin alpha10 expression in chondrocytes by the transcription factors AP-2epsilon and Ets-1. Biochem Biophys Res Commun 345:495–501
35. Xu Y, Gurusiddappa S, Rich RL, Owens RT, Keene DR, Mayne R et al (2000) Multiple binding sites in collagen type I for the integrins alpha1beta1 and alpha2beta1. J Biol Chem 275:38981–38989
36. Zemmyo M, Meharra EJ, Kuhn K, Creighton-Achermann L, Lotz M (2003) Accelerated, aging-dependent development of osteoarthritis in alpha1 integrin-deficient mice. Arthritis Rheum 48:2873–2880
37. Zhang WM, Kapyla J, Puranen JS, Knight CG, Tiger CF, Pentikainen OT et al (2003) alpha 11beta 1 integrin recognizes the GFOGER sequence in interstitial collagens. J Biol Chem 278:7270–7277

Integrin $\alpha 11\beta 1$: A Major Collagen Receptor on Fibroblastic Cells

5

Cédric Zeltz, Ning Lu, and Donald Gullberg

Abstract

Integrin $\alpha 11$ is the last addition to the vertebrate integrin family. In this chapter we will summarize some basic facts about this integrin and update with information that has been gained in the last decade. Integrin $\alpha 11\beta 1$ is a major collagen receptor on a subset of fibroblasts. Extensive characterization of the expression pattern in developing mouse embryos has demonstrated expression restricted to subsets of fibroblasts and a transient expression in odontoblasts, but comprehensive characterization of corresponding expression in adult tissues is still lacking. Mice lacking integrin $\alpha 11$ are dwarfed, primarily due to defective incisor eruption defect, which can be traced back to need for $\alpha 11$ on periodontal ligament fibroblasts during incisor eruption. Separate studies have suggested reduced levels of IGF-1 in mice lacking $\alpha 11$. Analysis of lung cancer has identified $\alpha 11\beta 1$ as a functional important collagen receptor on carcinoma associated fibroblasts (CAFs) and a number of disease models are awaiting analysis to see the importance of this collagen receptor in pathological models.

Keywords

Integrin $\alpha 11$ • Collagen • Fibroblasts • Carcinoma-associated fibroblasts • Myofibroblast

5.1 Introduction

Integrins are evolutionary old molecules that have been around for millions of years [11, 14], but we first identified integrin $\alpha 11$ in 1995 [12]. Integrin $\alpha 11$ is the last member of the integrin family to be discovered. This integrin subunit was initially named α_{mt}, since it was first identified on cultured human fetal myotubes. In parenthesis, this discovery was indeed serendipitous since the

C. Zeltz · N. Lu · D. Gullberg (✉)
Department of Biomedicine and Centre for Cancer Biomarkers, Norwegian Centre of Excellence, University of Bergen, Jonas Lies vei 91, 5009, Bergen, Norway
e-mail: donald.gullberg@biomed.uib.no

D. Gullberg (ed.), *I Domain Integrins*, Advances in Experimental Medicine and Biology 819,
DOI: 10.1007/978-94-017-9153-3_5,

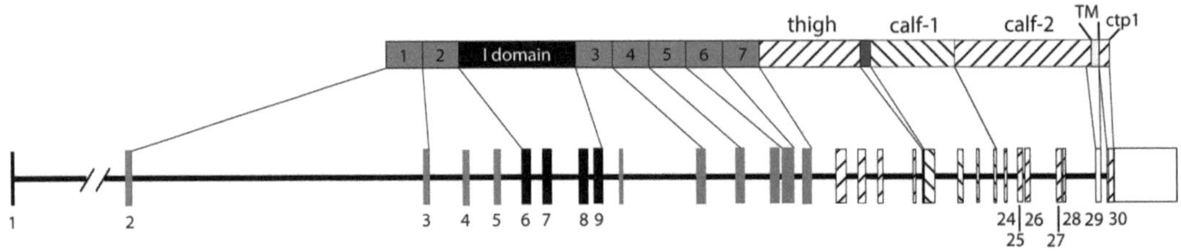

Fig. 5.1 Schematic representation of *ITGA11* gene. Top part shows a schematic representation of human α11 protein, and lower part shows a schematic overview of the organization of the *ITGA11* gene with its 30 exons. For the protein, the I domain, the 7 FG-GAP repeats (1–7), transmembrane part (TM) and cytoplasmic tail (ctpl) are marked. In the gene, exonic sequences representing untranslated regions are marked with open boxes

expression of α11 on myogenic cells is only seen upon tissue culture (in muscle tissue in vivo α11 is present on muscle fibroblasts). In 1999, we characterized this integrin and identified it as the β1-associated α11 subunit with properties of a collagen-binding integrin chain [26]. Two years later, we functionally described the α11β1 integrin as a collagen receptor involved in cell migration and collagen reorganization [24] and in 2004 we described the mouse variant of integrin α11 [21]. The generation of α11 integrin-deficient mice was a major advance in our efforts to elucidate α11 integrin function in health and disease [20].

5.2 The *ITGA11* Gene

The human α11 integrin gene (*ITGA11*) is localized on chromosome 15q23 and spans 130 kb, whereas the mouse α11 integrin gene (*Itga11*; length of 106 kb), has been mapped to chromosome 9. Both the human and mouse genes contain 30 exons and 29 introns (Fig. 5.1).

5.2.1 Promoter and Transcription Start

The *ITGA11* promoter lacks both TATA and CCAAG boxes. Promoters of other integrin α-chains contain features of a conserved initiator element often associated with an upstream Sp1 site found close to the transcription start [10, 32].

We used oligo-capping to identify a transcription start site (TSS) 30 nucleotides upstream of ATG in ITGA11 [30]. While the consensus sequence for initiator sequences is pypyANT/Apypy, the experimentally found sequence ACACC in *ITGA11* functions as an abbreviated initiator sequence. Furthermore an Sp1 binding site is located upstream of the putative initiator sequence, supporting the view that it is functional [18].

Using a panel of 15 serially-deleted promoter constructs, the α11 integrin proximal promoter spanning nt −176 and +25 nt has been characterized in the 3 kb *ITGA11* promoter region and found to convey high level of transcription activity. The presence of two Sp1 sites and an Ets-1 site in the proximal promoter is essential for its promoter activity [18] (Fig. 5.2).

Cytokines are able to regulate the α11 integrin expression by inducing signaling molecules, which regulate transcription factor binding to promoters. Thus, TGF-β1 was shown to up-regulate α11 expression in HT1080 fibrosarcoma cell line, as well as in human dermal fibroblasts and MRC-5 fibroblasts [13, 17]. The responsiveness to TGF-β1 is dependent on Smad2/3 and Sp1-regulated transcription. The Smad-binding element SBE2 and the Sp1-binding site SBS1 are closely located in the proximal promoter (nt –182/–176 and –140/–134, respectively). This proximity could promote a possible interaction between the Smad and Sp1 transcription factors. Activin A, which belongs to the TGF-β family, is involved in the up-regulation of α11 in mouse embryonic fibroblasts (MEFs), in a mechanosensitive manner [8]. This induction of α11 expression requires the Smad3 transcription factor. In transgenic reporter mice the human 3 kb *ITGA11* promoter drives a

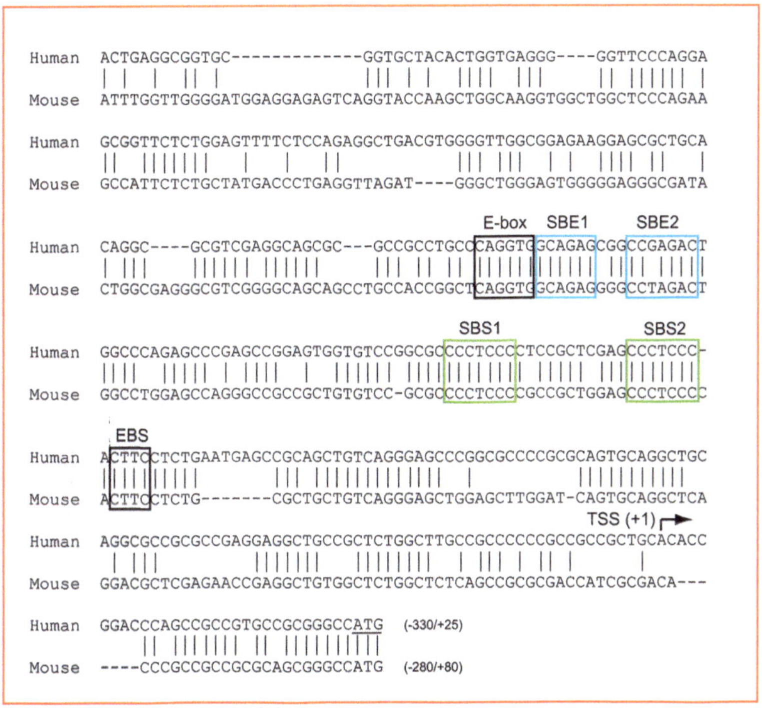

Fig. 5.2 Conserved promoter elements in human and mouse integrin α11 promoters. Alignment of the human integrin α11 promoter region −330 to +25 with the corresponding mouse promoter sequence. An E-box, two putative Smad binding elements (SBE1 and SBS2), tandem Sp1 binding sites (SBS1 and SBS2) and an Ets binding site (EBS) are boxed in the sequences. The human α11 transcription start site (TSS) is indicated by an *arrow* and the translation start codon ATG is *underlined*. From [17]

fibroblast-specific expression in developing mouse embryos (Fig. 5.3).

Type I interferons, including IFN-α and IFN-β, have also been described to regulate α11 expression. IFNs are able to stimulate α11 mRNA and protein expression in the glioblastoma derived cell line T98G [16].

Down-regulation of α11 has been reported to occur in mesenchymal stem cells and mouse embryonic fibroblasts treated with FGF-2 [8, 25]. However, the responsive elements involved in this down-regulation have not yet been determined in the α11 promoter.

5.2.2 Exon Structure

As in other integrin genes, the signal peptide is split by two exons, exon 1 and 2. Unlike some other integrin genes, exon 1 in *ITGA11* does not solely contain an UTR sequence, but also encodes the major part of the signal peptide, similar to the organization observed in the α10 integrin gene [4].

Previous analysis of the exon structure of leukocyte integrins has been very informative in identifying I domain borders. In the case of α11, the I domain is encoded by 4 exons (exons 6-9) and flanking introns are in phase 1.

We have previously identified an inserted region of 22 amino acids that distinguishes α11 from other integrin α chains [26]. It is likely that the inserted region has arisen by reshuffling of the existing sequence, which has then been inserted at the beginning of exon 20 during evolution. Comparison with *ITGA2* exon 20 (GenBank:AC016619.5) shows that the homologous sequence starts in a position immediately after the insert observed in exon 20 of *ITGA11*. Comparison of the α11 sequence with other integrin α-

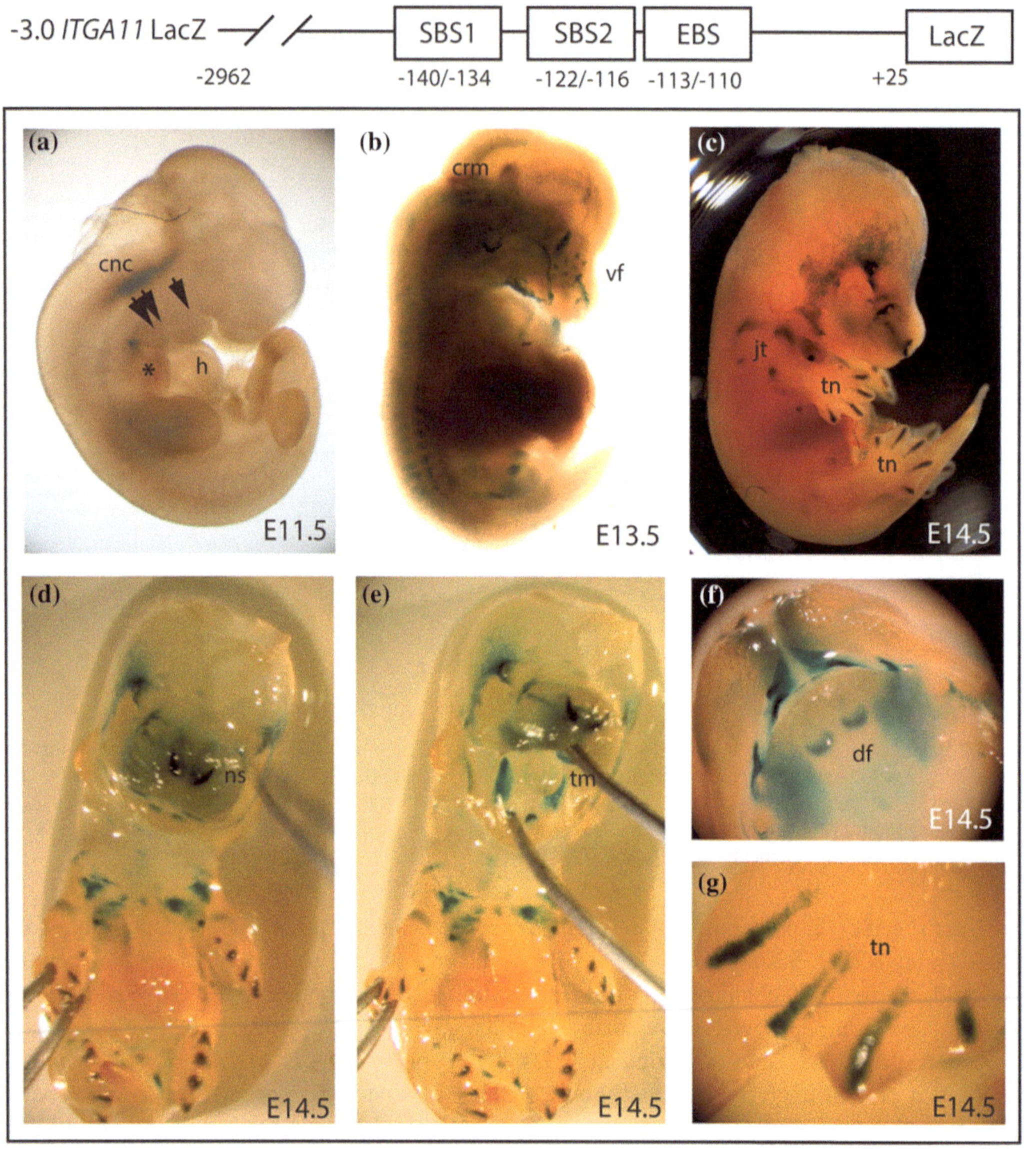

Fig. 5.3 A 3 kb ITGA11 promoter drives LacZ reporter expression in fibroblast precursors. The *upper* panel shows a schematic map of the −3.0 ITGA11promoter-LacZ reporter transgene used for generating transgenic mice. The previously identified Sp1 binding sites (SBS1 and SBS2) and an Ets family binding site (EBS) are indicated. The lower panel shows the whole-mount X-Gal staining of transgenic embryos at different embryonic days. LacZ expression is indicated by the blue staining. At E11.5 **a** LacZ expression was observed in the subpopulations of cranial neural crest cells cnc-cranial neural crest cells; in the first branchial arch which give rise to maxillary (*single arrow*) and mandibular (*double arrow*) prominences, and in the second branchial arch (*asterisk*) which will give rise to hyoid; h—heart; at E13.5 **b** LacZ expression was detected in *crm* cranial mesenchyme around calvarial bone, future location of the intervertebral disc and in *vf* vibrissae follicle; at E14.5 **c** LacZ expression was shown in *jt* joints, *tn* tendons of limbs, **d** in *ns* nasal septum, **e** in *tm* tongue mesenchyme, **f** in *df* dental follicles and **g** close up for forelimb showing positive staining in *tn* tendon (**g**). From [17]

chains places the insertion in a region of the stalk region called the calf-1 domain. Future studies will reveal the possible importance of this region. Interestingly, comparison with other integrins α-chains has identified this region as being involved in α/β chain interactions with the ability to influence integrin activation [28].

Careful examination of intron sequences using the program Genscan resulted in the identification of three potential in-frame exon sequences (with the tentative names 10B, 21L and 22B). We, however, failed to detect RNA messages for these variants, and currently no experimental data is thus available to support their existence. In *ITGA10* alternative splicing in a region corresponding to exon 25 has been described [4]. PCR amplifications of RNAs from different tissues covering this region in α11 transcripts suggest that corresponding splicing does not occur in α11.

The cytoplasmic tail in integrin α chains has a conserved sequence GFFXX, which in human α11 corresponds to the sequence GFFRS. It is interesting to note that for those integrin α chains that undergo alternative splicing in the cytoplasmic tail, the alternative exons also encode GFFKR, supporting the view that GFFKR together with the cytoplasmic tail denotes a functional unit. Biochemical analysis of the exact border of the transmembrane domain has recently suggested that GFFK residues are part of the transmembrane domain [1]. Comparison with the gene structure for α11 shows that if the homologous sequence GFFR is considered to be part of the transmembrane domain, the majority of this domain is encoded by exon 29, with the final four residues being encoded by the terminating exon 30, which also encodes the cytoplasmic tail.

5.2.3 Comparison with Other αI-Encoding Integrin Genes

Comparison of *ITGA10* and *ITGA11* genes lends further support to the view that α10 and α11 have arisen by a gene duplication event. Unlike *ITGA1* and *ITGA2*, which are both located on chromosome 5, *ITGA10* and *ITGA11* have been mapped to human chromosome 1 and 15, respectively. Unlike the *ITGA11* gene, which spans at more than 130 kb, *ITGA10* is much more compact and spans less than 19 kb. Comparison of *ITGA11* with the αX exon structure reveals a striking conservation of exon borders, underlining the close evolutionary relationship of integrin αI domain encoding genes.

The closely related *ITGA2* has been shown to display polymorphisms in the promoter region, also identified as risk factors for thrombotic disease [6, 7]. Based on the high expression of integrin α11 in the periodontal ligament (PDL) we hypothesized that single nucleotide polymorphisms (SNPs) in *ITGA11* might predispose to periodontitis. However, analyses of patients with juvenile periodontitis failed to identify polymorphism in the *ITGA11* basal promoter [3]. Further studies of the promoter will be instructive in determining: the regions in the upstream region that direct the fibroblast and carcinoma- associated fibroblast (CAF) -specific expression of α11 observed in vivo, the underlying mechanism for its mechanosensitivity and finally the regions mediating responsiveness to fibrogenic growth factors.

5.3 The Integrin α11 Subunit

ITGA11 encodes a mature protein of 1166 amino acids with a predicted integrin alpha chain structure. In SDS-PAGE, it runs as a 155kD band in a position above integrin α2 and α10 chains, indicating a higher degree of glycosylation than these two related integrins (Fig. 5.4). The extracellular domain contains seven FG-GAP repeats and a 195-amino acid-long I domain inserted between the repeats 2 and 3. The I domain presents a metal ion-dependent adhesion site (MIDAS) motif and three potential divalent cation binding motifs. As already mentioned the short cytoplasmic tail of 24 amino acids contains the motif GFFRS instead of the conserved GFFKR sequence most commonly found in integrin α subunits. A 23-amino acid-

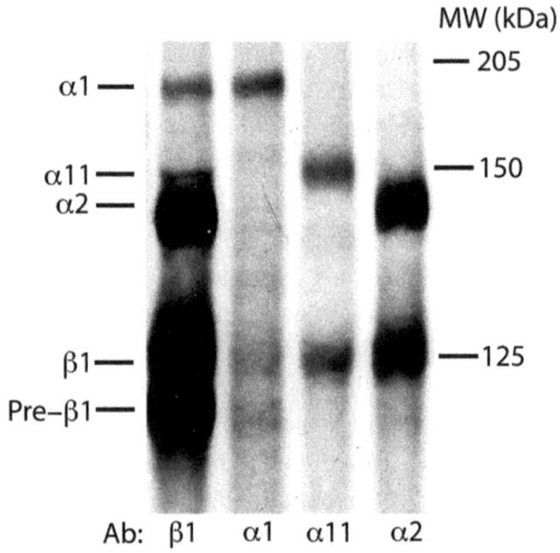

Fig. 5.4 Expression of α11 in cultured human fibroblasts. Cultures of subconfluent 1518 human foreskin fibroblasts were metabolically labeled, proteins were immunoprecipitated with antibodies, separated on a 6 % SDS–PAGE gel under nonreducing conditions, and visualized by fluorography. The antibodies used were directed to integrin subunits *β*1, α1, α11 and α2. Positions of different integrin chains are marked. From [24]

long transmembrane domain links the extracellular and cytoplasmic domains [26]. The mouse α11 integrin chain shows an 89 % identity with human α11 at the protein level and 97 % identity in the I domain [21].

5.3.1 Expression of the α11 Integrin Chain In Vivo

The expression of integrin α11 was first described in cultured human fetal muscle cells in vitro [12]. In human adult tissue, α11 mRNA was expressed in high levels in uterus and heart and in intermediate levels in skeletal muscle. However, in human and mouse embryos, no expression of α11 was detected in muscle cells [21, 24]. Later it was shown that in muscle tissue α11 is expressed in fibroblasts, hence explaining the detection of α11 RNA in muscle tissues in Northern blotting. α11 is present in fibroblasts around ribs, vertebrae, in intervertebral discs and in keratocytes of the cornea of 8-week human embryos [24]. In the mouse embryo, α11 has been localized to the ectomesenchyme in the head including the PDL, in tendons and intestinal villi fibroblasts [21]. The α11 chain expression appears to be specific to mesenchymal nonmuscle cells in vivo (Fig. 5.5), but a complete characterization in adult tissues has not yet been performed. α11 expression has also been reported in tumor tissue from melanoma and lung carcinoma [27, 31]. The high levels of α11 integrin expression in lung carcinoma in situ are derived from the CAFs and is thus in the lung not contributed by the cancer cells. Recent data indicates that α11 RNA is regulated during epithelial mesenchymal transition [15].

5.4 Integrin α11β1 Functions

5.4.1 In Vivo Functions

The in vivo function of the α11 integrin has been partially elucidated using the knockout mouse model. The α11-deficient mice are smaller and display an increased mortality compared to heterozygous and wild-type mice [20]. Dwarfism observed in these α11-deficient mice appears not to be due to structural defects in forming cartilage or bone. Instead the smaller size and malnutrition of weaned α11-deficient mice appear to correlate with delayed incisor eruption and altered tooth shape (Fig. 5.6). The incisor PDL, which plays a central role during rodent incisor eruption, showed increased thickness due to increased amount of collagen. In this mutant tissue, a decrease of MT1-MMP and MMP-13 mRNA levels were also noted. A reproducible result was obtained in vitro, where MEFs isolated from α11-deficient embryos showed reduced MT1-MMP and MMP-13 mRNA expression, whereas MMP-2 and MMP-9 activities were not affected. These observations suggest that α11 could be involved in the regulation of metalloproteinases as MMP-13 and -14, thus controlling the collagen turnover in PDL. Later studies of PDL fibroblasts isolated from mouse incisors confirmed a role for integrin α11 *β*1 in regulating MMP-13 expression, but failed to show regulation of MMP-14 at the

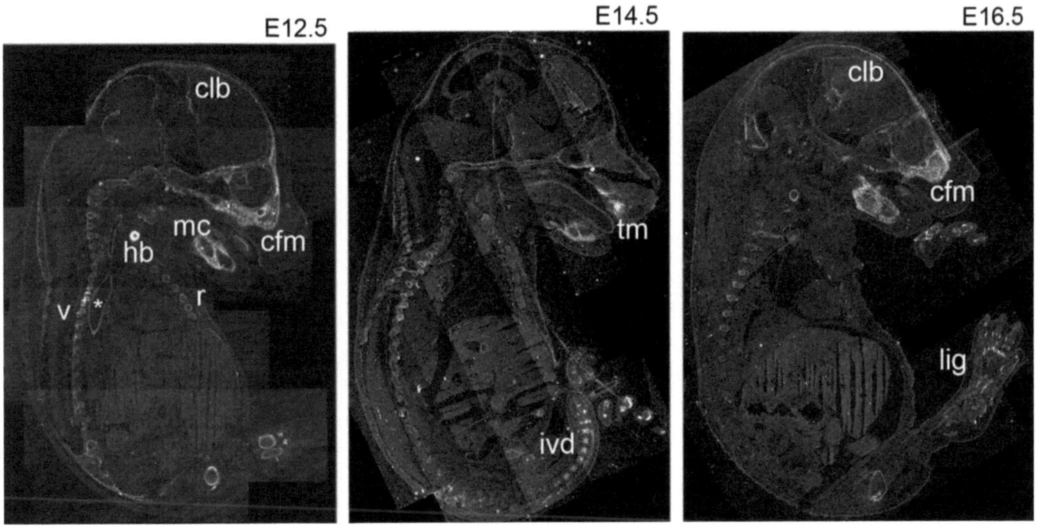

Fig. 5.5 Localization of α11 mRNA during mouse embryogenesis. Sagittal sections of mouse embryos from *E* embryonic days E12.5–E16.5 were subjected to in situ hybridization using an antisense RNA probe specific for mouse α11. Darkfield images are shown. At E12.5 α11 mRNA can be detected around *clb* calvarian bone, in *cfm* craniofacial mesenchyme, around *mc* Meckel's cartilage, around *hb* hyoid bone, around *v* vertebrae, and around the *r* ribs. * denotes signal in descending aorta, not confirmed by immunohistochemistry. At E14.5 α11 can also be detected in *ivd* intervertebral discs in the tail region and in *tm* tongue mesenchyme. At E16.5 tendons and *lig* ligaments in the hind limb express high levels of α11 RNA. From [21]

protein level under the conditions used [2]. Intriguingly, later data has shown that α11-deficient mice display reduced serum levels of IGF-1 [5]. The mechanism for how a fibroblast specific protein would affect the pituitary axis responsible for IGF-1 secretion is unclear at this stage. The study is however important since it stresses that α11-deficient mice are smaller already at birth, before the incisor eruption effect on body weight has come into play and further stresses that more detailed studies are needed to sort out the underlying molecular mechanism for reduced body weight observed in α11-/- mice.

As described above, α11β1 has been reported to be up-regulated in the CAFs in non-small cell lung cancer [31] (Fig. 5.7) and was also in a xenograft model shown to enhance tumorigenicity by regulation of the IGF-2 expression. However, the exact role of α11 in the tumor stroma during TGF-β1-dependent myofibroblast differentiation, tumor growth and tumor metastasis remains to be determined and it will be important to determine if α11 upregulation occurs on CAFs also in other types of tumors.

5.4.2 In Vitro Functions

The α11 integrin chain is exclusively associated with the β1 subunit at the cell surface, to form the α11β1 integrin. The first study that demonstrated that α11β1 promoted cell attachment to collagen I was performed in 2001, when we started using the system of transfecting cDNAs encoding collagen-binding integrins into the mouse satellite cell line C2C12 [24]. Using these cells we could show that integrin α11β1 displays certain collagen specificity, since it binds preferentially type I collagen, whereas it interacts with collagen IV with a low affinity. The α11 I domain recognizes the triple-helical GFOGER sequence present in collagen I as well as the GLOGER motif [22, 29].

Another study has identified the GLPGER motif of the recombinant Scl1 protein, a prokaryotic collagen, as an α11β1 binding sequence [9]. The interaction between the cell surface streptococcal Scl1 and the human α11β1 integrin might increase host colonization by pathogenic bacteria, but this process remains to be determined.

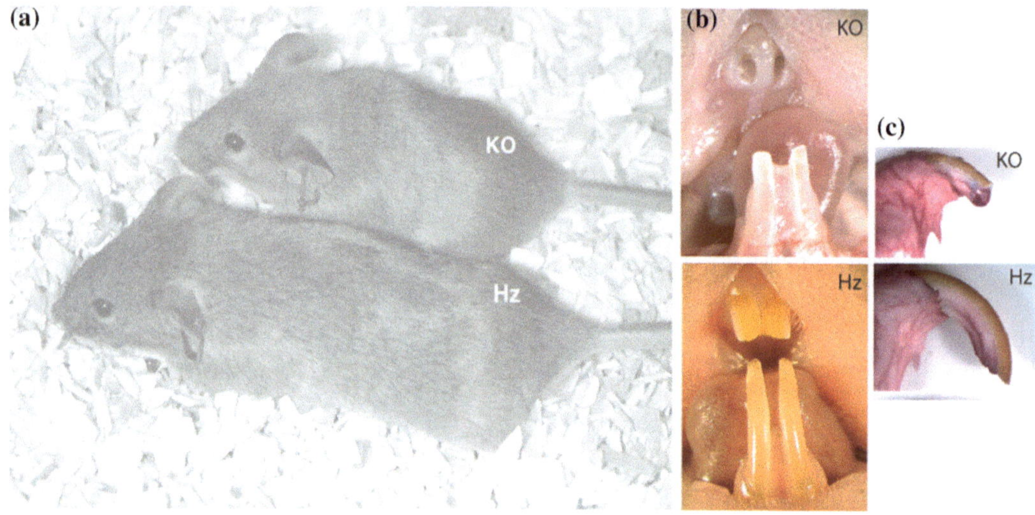

Fig. 5.6 Phenotype of integrin α11-deficient mice. **a** 10-week-old male mutant mice showed a reduction in size. Hz, heterozygous ($\alpha 11^{+/-}$); KO, homozygous ($\alpha 11^{-/-}$). **b** A lack of the outer portion of the upper incisors was observed in the α11 null mice at 1 year of age. **c** The incisors of 4-month-old mutant control (*lower panel*) and mutant (*upper panel*) mice were excised from their sockets, and the soft tissue was digested away. Note the altered size and shape of the KO incisors. From [20]

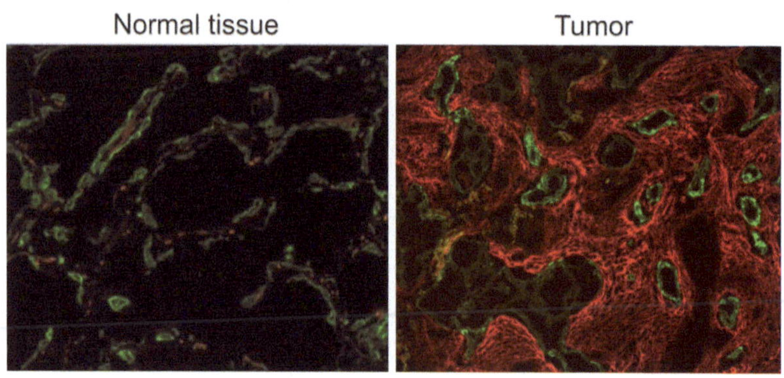

Fig. 5.7 Expression of integrin α11 chain in tumor stroma. Immunofluorescence images of human Normal tissue (lung) and Tumor tissue (non small cell lung adenocarcinoma) that were double stained with antibodies to α11 (*red*) and epithelial cell marker cytokeratin (*green*). The α11 staining was negligible in non-neoplastic lung tissue and was mainly confined to the stroma in the tumor sample. From [31]

The role of α11β1 in PDGF-stimulated cell migration on collagen I coating seems to be cell type dependent. The C2C12 mouse cells, stably transfected with human α11 integrin cDNA, showed a stronger chemotactic response to PDGF-BB, compared to C2C12 wild-type cells, lacking endogenous collagen receptors [24]. In contrast, MEFs depleted in α11β1 migrated more on collagen I in comparison to wild-type embryonic fibroblasts [21]. However, in this last case, a compensatory mechanism, involving other collagen receptors, cannot be excluded.

In several studies we have shown that α11β1 mediates the contraction of collagen lattices, an important function, which contributes to the regulation of the reorganization of collagen matrices [2, 3, 20, 24]. Interestingly, when we isolated PDL fibroblasts from α11 deficient mice, these fibroblasts displayed reduced levels of MMP-13 and cathepsin K, which in α11β1

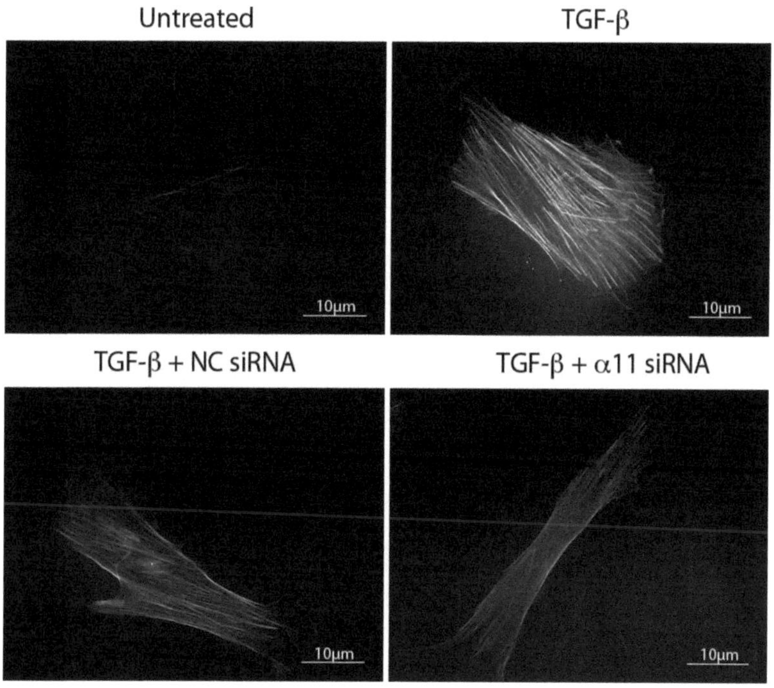

Fig. 5.8 $\alpha 11$ influences myofibroblast differentiation in human corneal fibroblasts. Immunolocalization of α-SMA in corneal fibroblasts that remained untreated (*upper left*), were stimulated with 5 ng/ml TGF-$\beta 1$ only (*upper right*), or were treated with 5 ng/ml TGF-$\beta 1$ and $\alpha 11$-specific siRNA (100 nM) (*lower right*), or an off-target siRNA (*NC* negative control; *lower left*). The exposure time when acquiring pictures was identical in all conditions. From [8]

expressing fibroblasts seemed to facilitate the collagen remodeling process during collagen contraction [2].

Under certain conditions fibroblasts become activated and differentiate into so-called myofibroblasts. Myofibroblasts are characterized by α-smooth muscle actin (α-SMA) incorporated into stress fibers. Corneal fibroblasts, under action of TGF-$\beta 1$, overexpress α-SMA. Since siRNA directed against the $\alpha 11$ integrin completely abrogated α-SMA up-regulation, these data demonstrate that $\alpha 11\beta 1$ also plays a role in myofibroblast differentiation [8] (Fig. 5.8). The regulation of myofibroblast differentiation by $\alpha 11\beta 1$ could be relevant in pathological processes such as tumor-stroma interactions and fibrosis, where myofibroblasts are involved (Fig. 5.9).

More recently $\alpha 11$- and TGF-$\beta 2$-dependent myofibroblast differentiation in cardiac fibroblasts has been observed, suggesting a potential role for $\alpha 11$ in cardiac fibrosis [23].

Integrin turnover is an essential process involved in cell adhesion and migration. Generally, integrins present on the cell surface are either released and used in new adhesion sites or internalized by endocytosis. Rab proteins, including Rab21, regulate the traffic of endocytotic vesicles via interaction with the cytoplasmic tail of α integrin subunit, as shown for integrin $\alpha 2\beta 1$ [19]. The C-terminal part of Rab21 was also able to bind to the cytoplasmic domain of $\alpha 11$ integrin, thus suggesting that $\alpha 11\beta 1$ could be regulated by endocytosis. Since Rab21 activity has been shown to regulate the motility of breast and prostate cancer cells, it could be interesting to examine if an association between this small GTPase and $\alpha 11\beta 1$ occurs in cancer-associated fibroblasts and if it might have an impact on the tumor progression.

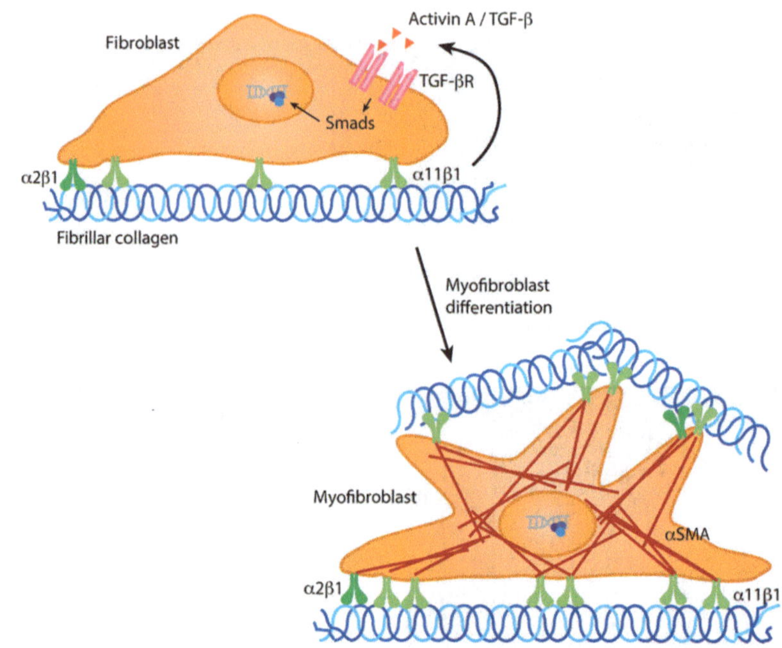

Fig. 5.9 Schematic illustration of role of integrin α11 in myofibroblast differentiation. Under mechanical stress, on a stiff substrate, fibroblasts secrete members of TGF-β, influenced by α11β1 integrin. TGF-β signaling, via Smad proteins, induces fibroblast differentiation into myofibroblast by up-regulation of α-SMA and α11β1 expression. Myofibroblasts synthesize collagen and are able to reorganize collagen matrices in an α11β1-dependent manner. We hypothesize that the regulation of myofibroblast differentiation by α11β1 is relevant in fibrosis and tumor-stroma interactions

5.5 Perspectives

Integrin α11β1 is expressed in mesenchymal non-muscle cells in vivo at sites where collagens are organized in a highly ordered manner. It appears as a multifunctional integrin in different contexts. However, little is known about the detailed molecular mechanisms involved in α11β1 functions including the major signaling pathways utilized by α11β1 and its involvement in various pathological conditions, and thus much still remain to be learned about this collagen receptor.

References

1. Armulik A, Nilsson I, von Heijne G, Johansson S (1999) Determination of the border between the transmembrane and cytoplasmic domains of human integrin subunits. J Biol Chem 274:37030–37034
2. Barczyk MM, Lu N, Popova SN, Bolstad AI, Gullberg D (2013) Alpha11beta1 integrin-mediated MMP-13-dependent collagen lattice contraction by fibroblasts: Evidence for integrin-coordinated collagen proteolysis. J Cell Physiol 228:1108–1119
3. Barczyk MM, Olsen LH, da Franca P, Loos BG, Mustafa K, Gullberg D et al (2009) A role for alpha11beta1 integrin in the human periodontal ligament. J Dent Res 88:621–626
4. Bengtsson T, Camper L, Schneller M, Lundgren-Akerlund E (2001) Characterization of the mouse integrin subunit a10 gene and comparison with its human homologue. Genomic structure, chromosomal localization and identification of splice variants. Matrix Biol 20:565–576
5. Blumbach K, Niehoff A, Belgardt BF, Ehlen HW, Schmitz M, Hallinger R et al (2012) Dwarfism in mice lacking collagen-binding integrins alpha2beta1 and alpha11beta1 is caused by severely diminished IGF-1 levels. J Biol Chem 287:6431–6440
6. Bray PF (1999) Integrin polymorphisms as risk factors for thrombosis. Thromb Haemost 82:337–344
7. Bray PF (2000) Platelet glycoprotein polymorphisms as risk factors for thrombosis. Curr Opin Hematol 7:284–289
8. Carracedo S, Lu N, Popova SN, Jonsson R, Eckes B, Gullberg D (2010) The fibroblast integrin alpha11beta1

is induced in a mechanosensitive manner involving activin A and regulates myofibroblast differentiation. J Biol Chem 285:10434–10445
9. Caswell CC, Barczyk M, Keene DR, Lukomska E, Gullberg DE, Lukomski S (2008) Identification of the first prokaryotic collagen sequence motif that mediates binding to human collagen receptors, integrins alpha2beta1 and alpha11beta1. J Biol Chem 283:36168–36175
10. Corbi AL, Garcia-Aguilar J, Springer TA (1990) Genomic structure of an integrin α subunit, the leukocyte p150,95 molecule. J Biol Chem 265: 2782–2788
11. Gullberg D, Fessler LI, Fessler JH (1994) Differentiation, extracellular matrix synthesis, and integrin assembly by *Drosophila* embryo cells cultured on vitronectin and laminin substrates. Dev Dyn 199:116–128
12. Gullberg D, Velling T, Sjoberg G, Sejersen T (1995) Up-regulation of a novel integrin alpha-chain (alpha mt) on human fetal myotubes. Dev Dyn 204:57–65
13. Honda E, Yoshida K, Munakata H (2010) Transforming growth factor-beta upregulates the expression of integrin and related proteins in MRC-5 human myofibroblasts. Tohoku J Exp Med 220:319–327
14. Johnson MS, Lu N, Denessiouk K, Heino J, Gullberg D (2009) Integrins during evolution: evolutionary trees and model organisms. Biochim Biophys Acta 1788:779–789
15. Ke X, Qu Y, Goldfinger N, Rostad K, Hovland R, Akslen L et al (2008) Epithelial to mesenchymal transition of a primary prostate cell line with switches of cell adhesion modules but without malignant transformation. PLoS One 3:e3368
16. Leomil Coelho LF, Mota BE, Sales PC, Marques JT, de Oliveira JG, Bonjardim CA et al (2006) Integrin alpha 11 is a novel type I interferon stimulated gene. Cytokine 33:352–361
17. Lu N, Carracedo S, Ranta J, Heuchel R, Soininen R, Gullberg D (2010) The human alpha11 integrin promoter drives fibroblast-restricted expression in vivo and is regulated by TGF-beta1 in a Smad- and Sp1-dependent manner. Matrix Biol 29:166–176
18. Lu N, Heuchel R, Barczyk M, Zhang WM, Gullberg D (2006) Tandem Sp1/Sp3 sites together with an Ets-1 site cooperate to mediate alpha11 integrin chain expression in mesenchymal cells. Matrix Biol 25:118–129
19. Pellinen T, Arjonen A, Vuoriluoto K, Kallio K, Fransen JA, Ivaska J (2006) Small GTPase Rab21 regulates cell adhesion and controls endosomal traffic of beta1-integrins. J Cell Biol 173:767–780
20. Popova SN, Barczyk M, Tiger CF, Beertsen W, Zigrino P, Aszodi A et al (2007) Alpha11 beta1 integrin-dependent regulation of periodontal ligament function in the erupting mouse incisor. Mol Cell Biol 27:4306–4316
21. Popova SN, Rodriguez-Sanchez B, Liden A, Betsholtz C, Van Den Bos T, Gullberg D (2004) The mesenchymal alpha11beta1 integrin attenuates PDGF-BB-stimulated chemotaxis of embryonic fibroblasts on collagens. Dev Biol 270:427–442
22. Siljander PR, Hamaia S, Peachey AR, Slatter DA, Smethurst PA, Ouwehand WH et al (2004) Integrin activation state determines selectivity for novel recognition sites in fibrillar collagens. J Biol Chem 279:47763–47772
23. Talior-Volodarsky I, Arora PD, Wang Y, Zeltz C, Gullberg D, McCulloch CA (2014) Glycated collagen induces $\alpha 11$ integrin expression through TGF-$\beta 2$ and Smads (submitted)
24. Tiger CF, Fougerousse F, Grundstrom G, Velling T, Gullberg D (2001) Alpha11beta1 integrin is a receptor for interstitial collagens involved in cell migration and collagen reorganization on mesenchymal nonmuscle cells. Dev Biol 237:116–129
25. Varas L, Ohlsson LB, Honeth G, Olsson A, Bengtsson T, Wiberg C et al (2007) Alpha10 integrin expression is up-regulated on fibroblast growth factor-2-treated mesenchymal stem cells with improved chondrogenic differentiation potential. Stem Cells Dev 16:965–978
26. Velling T, Kusche-Gullberg M, Sejersen T, Gullberg D (1999) cDNA cloning and chromosomal localization of human alpha11 integrin. A collagen-binding, I domain-containing, beta1-associated integrin alpha-chain present in muscle tissues. J Biol Chem 274:25735–25742
27. Vuoristo M, Vihinen P, Vlaykova T, Nylund C, Heino J, Pyrhonen S (2007) Increased gene expression levels of collagen receptor integrins are associated with decreased survival parameters in patients with advanced melanoma. Melanoma Res 17:215–223
28. Xie C, Shimaoka M, Xiao T, Schwab P, Klickstein LB, Springer TA (2004) The integrin alpha-subunit leg extends at a Ca2+-dependent epitope in the thigh/genu interface upon activation. Proc Natl Acad Sci USA 101:15422–15427
29. Zhang WM, Kapyla J, Puranen JS, Knight CG, Tiger CF, Pentikainen OT et al (2003) Alpha11beta1 integrin recognizes the GFOGER sequence in interstitial collagens. J Biol Chem 278:7270–7277
30. Zhang WM, Popova SN, Bergman C, Velling T, Gullberg MK, Gullberg D (2002) Analysis of the human integrin alpha11 gene (ITGA11) and its promoter. Matrix Biol 21:513–523
31. Zhu CQ, Popova SN, Brown ER, Barsyte-Lovejoy D, Navab R, Shih W et al (2007) Integrin alpha 11 regulates IGF2 expression in fibroblasts to enhance tumorigenicity of human non-small-cell lung cancer cells. Proc Natl Acad Sci USA 104:11754–11759
32. Ziober BL, Kramer RH (1996) Identification and characterization of the cell type-specific and developmentally regulated $\alpha 7$ integrin gene promoter. J Biol Chem 271:22915–22922

Regulation of Integrin Activity by Phosphorylation

6

Carl G. Gahmberg, Mikaela Grönholm, and Liisa M. Uotila

Abstract

Integrins are heterodimeric complex type I membrane proteins involved in cellular adhesion and signaling. They exist as inactive molecules in resting cells, and need activation to become adhesive. Although much is known about their structure, and a large number of interacting molecules have been described, we still only partially understand how their activities are regulated. In this review we focus on the leukocyte-specific $\beta 2$—integrins and, specifically, on the role of integrin phosphorylation in the regulation of activity. Phosphorylation reactions can be fast and reversible, thus enabling strictly directed regulatory activities both time-wise and locally in specific regions of the plasma membrane in different leukocytes.

Keywords

Integrin · Phosphorylation · Leukocyte · Adhesion · Signaling

6.1 Introduction

About 30 years ago leukocyte cell surface proteins were described to be involved in various functions such as antibody production, cytotoxicity, phagocytosis and chemotaxis [1, 2]. They were given names such as leukocyte function associated antigen (LFA-1) and macrophage antigen-1 (Mac-1). Later work used phorbol esters to induce cell adhesion. In the presence of an array of monoclonal antibodies reacting with the leukocyte surface we looked for antibodies, which could inhibit the induced adhesion. One antibody, called 60.3, was efficient, and immune precipitation resulted in the identification of protein dimers [3, 4]. Subsequently, these proteins were shown make up a subfamily of adhesion proteins named integrins [5, 6]. Because of their functional importance, they drew large interest and the literature on integrins is currently impressively large [7–11].

Integrins are present in the animal kingdom from nematodes and fruit flies to humans, and their amino acid sequences are remarkably well conserved [10, 12]. The ligand binding domains

C. G. Gahmberg (✉) · M. Grönholm · L. M. Uotila
Division of Biochemistry and Biotechnology, Department of Biosciences, University of Helsinki, Viikinkaari 5, 00014 Helsinki, Finland
e-mail: carl.gahmberg@helsinki.fi

D. Gullberg (ed.), *I Domain Integrins*, Advances in Experimental Medicine and Biology 819,
DOI: 10.1007/978-94-017-9153-3_6, © Springer Science+Business Media Dordrecht 2014

on the outside of cells are large, and some integrins, including the $\beta2$-integrins, contain an I/A-domain. When present, the I-domain first characterized in von Willebrand factor, forms the ligand binding site and is located in the α-chains of the integrins. In humans four $\beta2$-integrins exist. The lymphocyte enriched LFA-1 ($\alpha L\beta2$, CD11a/CD18) binds to the intercellular adhesion molecules (ICAM) −1 to −5 [13–15], but the other family members Mac-1 ($\alpha M\beta2$, CD11b/CD18), $\alpha X\beta2$ (CD11c/CD18) and $\alpha D\beta2$ (CD11d/CD18) bind to several additional types of ligands including plasma proteins, extracellular matrix components and even carbohydrates. In integrins lacking the I-domain, the ligand binding site is formed by the β-propeller domain on the α-chain and the βI-domain of the β-subunit.

The $\beta2$ integrins are expressed on white blood cells, but the expression profile on different leukocytes is unique for each member of the family. LFA-1 is expressed on all leukocytes, whereas $\alpha M\beta2$ is found on monocytes, macrophages, NK cells, neutrophils and on $\gamma\delta$ subsets of T cells. $\alpha X\beta2$ is expressed on monocytes, macrophages, dendritic cells, NK cells and some subsets of T and B cells, and $\alpha D\beta2$ on macrophages and eosinophils [8, 10] The functions of leukocyte integrins are vital for a functional immune system, but, reflecting the variance in the expression pattern, also the functions of the family members are somewhat different. Leukocyte integrins and especially $\alpha L\beta2$ are essential for the extravasation of the immune cells from the circulation to inflamed tissues. $\alpha L\beta2$ is also necessary for the proper formation of the immunological synapse that forms between an antigen presenting cell and a T cell [16]. The complement receptors $\alpha X\beta2$ and especially $\alpha M\beta2$ have been shown to be important in phagocytosis. Other functions of $\alpha M\beta2$ include remodeling of ECM, survival of neutrophils and development of immune tolerance [17, 18]. $\alpha X\beta2$ on the dendritic cells has been found to be an efficient target for antigen uptake in eliciting T cell immune responses [19]. Other studies have demonstrated the role of $\alpha X\beta2$ on hypercholesterolemic mouse monocytes in the development of atherosclerosis [20]. $\alpha D\beta2$ on macrophages possibly takes part in phagocytosis and migration [18].

The integrins are remarkable in the sense that they can signal across the plasma membrane in two directions: inside–out and outside–in. In the former case, a ligand or activating agent is bound to a non-integrin receptor and eventually the signal is transmitted to the integrins. In outside–in signaling the ligand binds directly to the integrin on the outside of the cell and induces signaling. Increasing evidence shows that inside–out signaling results in partial activation of the integrins where the head piece remains closed and the integrin weakly binds to the ligand. When the ligand binds, the binding region in the integrin opens up followed by outside–in signaling. Thus inside–out and outside–in signaling may be coupled and occur subsequently in an individual integrin.

6.2 Integrin Structural Features

Figure 6.1 shows schematic figures of leukocyte $\beta2$-integrins [9]. The different domains are schematically drawn in Fig. 6.1a. The external domains are complex and interact non-covalently on the outside of the cell. A major advance in understanding integrins came when the I-domain from αM was crystallized and the structure determined [21]. Later the structures of the external parts of the $\alpha V\beta3$ and the $\alpha X\beta2$ integrins were determined [22, 23]. Surprisingly, the binding site formed by the α-chain propeller and the β-chain I-like domain was turned towards the lipid membrane (Fig. 6.1b). Later work showed that upon activation the molecules straighten out and now show intermediate binding affinity (Fig. 6.1c). In the fully active protein, the binding site has opened up, the hybrid domain has moved out and presumably the "legs" including the cytoplasmic tails have moved apart (Fig. 6.1d). All results indicate that the integrin α- and β-subunits play very different

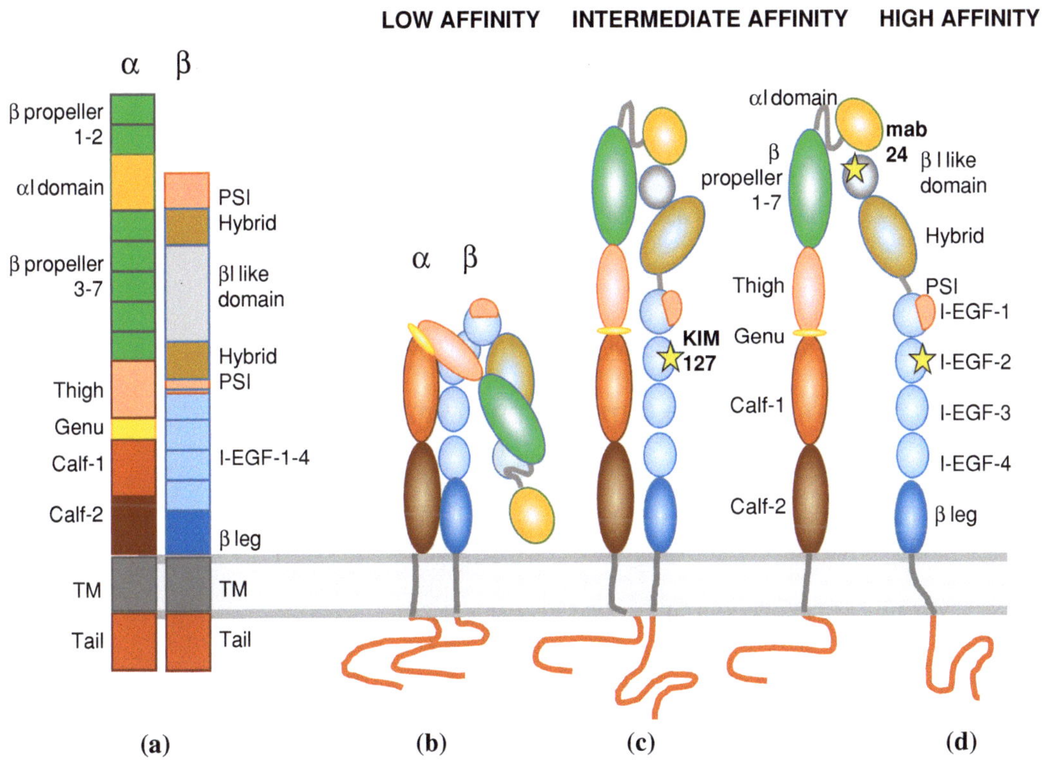

Fig. 6.1 Schematic figure of the αLβ2 integrin. **a** The domains are depicted. **b** In the resting state the integrin head piece containing the binding site is turned towards the membrane. **c** The integrin of intermediate affinity is straightened out, but the head piece remains closed, and the cytoplasmic domains are more closely packed. The binding site of the KIM127 antibody, which recognizes an integrin of intermediate affinity is shown. **d** The fully activated integrin. Note that the legs, the transmembrane domain and the cytoplasmic parts have moved apart, and the binding site in the I-domain is open. The location of the binding site of antibody mab24, which only binds to fully activated integrins is shown

roles in adhesion. The α-chains have an important structural role, including ligand specificity, but not necessarily a regulatory role. The *β* subunit cytoplasmic domains, on the other hand, are largely conserved and are able to regulate integrin activity. It has been proposed that in the resting state, a salt bridge exists between an aspartate in the *β*-subunit (SDLR in *β*2) close to the membrane, and the arginine in the conserved α-chain GFFKR sequence [24, 25]. Upon activation, this salt bridge may be broken enabling a switch from a relatively parallel heterodimer to a more X-like structure, which reaches out and modifies the structure also on the outside of the membrane. How this could take place is discussed below.

6.3 The Integrin Cytoplasmic Domains Bind Several Intracellular Regulatory Proteins

Much work has focused on the cytoplasmic domains of the integrins (Fig. 6.2). We now know that the cytoplasmic regions of the integrins are pivotal in the regulation of activity. The structure of the cytoplasmic tails of αL*β*2, αM*β*2 and αX*β*2 [26–28] as well as those of αIIb [24] and α4 [29] have been solved by NMR. The α-chains of *β*2-integrins show quite striking differences, even though the membrane proximal structures, forming a conserved helical structure,

1140

αL IVLYKVGFFKRNLKEKMEAGRGVPNGIPAEDSEQLASGQEAGDPGCLKPLHEKDSESGGGKD

1126

αM AALYKLGFFKRQYKDMMSEGGPPGAEPQ

1158

αX AVLYKVGFFKRQYKEMMEEANGQIAPENGTQTPSPPSEK

αD ATLYKLGFFKRHYKEMLEDKPEDTATFSGDDFSCVAPNVPLS

735 745 756 758

β2 LVI WKALIHLSDLREYRRFEKEKLKSQWNNDNPLFKSATTTVMNPKFAES

α4 YVMWKAGFFKRQYKSILQEENRRDSWSYINSKSNDD

788

β1A LLIWKLLMIIHDRREFAKFEKEKMNAKWDTGENPIYKSAVTTVVNPKYEGK

αIIb LAMWKVGFFKRNRPPLEEDDEEGE

β3 LLIWKLLITIHDRKEFAKFEEERARAKWDTANNPLYKEATSTFTNITYRGT

αPS1 YVLWKVGFFKRIRPTDPTLSGNLEKMNEEKPFLAPSKNTHHVF

βPS LLLWKLLTTIHDRREFARFEKERMNAKWDTGENPIYKQATSTFKNPMYAGK

αPAT2 LLLWRCGFFKRNRPPTEHAELRADRQPNAQYADSQSRYTSQDQYNQGRHGQML

βPAT3 LLLWKLLTVLHDRSEYATFNNERLMAKWDTNENPIYKQATTTFKNPVYAGKAN

Fig. 6.2 The cytoplasmic sequences of the β2-integrins and those of some other integrin subunits dealt with in the text are shown. The functionally important α-chain regions, the potential phosphorylation sites, the NPXY sequences and other possibly important amino acids are *shaded*. Confirmed phosphorylation sites are numbered

remain similar. All leukocyte integrin α and β cytoplasmic tails studied appear to interact with each other through multiple ionic and hydrogen bonds in the membrane proximal helical area, but the rest of the cytoplasmic α-chains form different structures. αL, which has the longest cytoplasmic part, forms three alpha helices that are sustained by salt bridges and/or hydrogen bonds. This structure forms a large negatively charged surface that is able to bind metal ions [26]. The membrane distal residues of two other leukocyte integrins, αM and αX, form loops that can adopt more conformational variations. These findings suggest that the different members of the leukocyte integrin family may have different cytoplasmic binding partners, which can lead to different signaling events and outcomes. The reported phosphorylation sites of αL, αM and αX cytoplasmic tails as well as the β2 tail are situated outside the membrane-proximal helices and are thus more easily available for kinases and phosphatases involved in phosphorylation as well as for other cytoplasmic molecules for interactions. There are several recent articles and reviews dealing with the binding of cytoplasmic proteins to integrin intracellular domains [12, 30–37] and we will not try here to cover all aspects of this fascinating, but large subject. Instead, we focus on how integrin activities can be regulated rapidly and specifically. We want to put forward the proposition that integrin regulation primarily takes place through specific phosphorylation reactions, which in turn affect the binding of specific adaptor proteins. Further downstream binding components then relay various integrin related cellular functions.

Few cytoplasmic proteins have been found to specifically bind to the integrin α-chain cytoplasmic tails. These include paxillin, calreticulin, CD45, RapL and the adhesion inhibitory protein SHARPIN [38]. In contrast, more than 40 proteins have been claimed to bind to the β-subunits [35]. The integrin β-chain cytoplasmic regions contain three important "hot spots" for binding proteins. These are the two NPL(I)Y(F) and N(XX)Y(F)(or NPXY for short) sequences and the Ser/Thr enriched sequence between them. Depending on the integrin, all three sequences are also potential phosphorylation sites. It should, however, be pointed out that several integrin cytoplasmic domains have been used for studies on interactions with cytoplasmic proteins, and certainly different integrins have different binding preferences. Therefore, results obtained with one cytoplasmic sequence must not necessarily be true for another one. Furthermore, the cytoplasmic parts of the integrins are relatively short with a limited binding capacity. Obviously, the interactions are competitive to a large extent and binding of a given cytoplasmic protein often excludes the binding of another one.

The best studied cytoplasmic proteins interacting with integrin β-chain COOH-terminal regions are talin, filamin, kindlins and 14-3-3 (Fig. 6.3). The talin molecule is large with a "head" and a long "tail" (rod). The head binds to integrins whereas the tail may interact with the actin cytoskeleton, but also with integrins [39]. The properties of talin and its interactions with integrins have been described in several recent reviews [12, 30, 32, 36, 39]. Importantly, the integrin β-subunits have two binding sites for talin, an acidic sequence close to the membrane [40], and the proximal NPXY(F) sequence. Filamin binds to a region extending from the first to the end of the second NPXY(F) sequence whereas kindlins bind to the region beginning from the Ser-Thr enriched region (Thr-Thr-Thr in β2) and covering the second NPXY(F) sequence. The 14-3-3 proteins bind to Thr-758 in the Thr-Thr-Thr sequence in β2, but only after phosphorylation [41, 42]. A weaker binding may also occur, which is not dependent on phosphorylation [12].

6.4 Integrin Phosphorylations are of Pivotal Importance for Activity

An accumulating amount of recent results indicate that specific integrin phosphorylations are important in the regulation of integrin associated activities [43]. The leukocyte integrins have turned out to be especially useful models for studies on phosphorylation mediated regulation. In contrast to other cells, blood cells such as lymphocytes, are normally completely resting, but can be activated by various agents including chemokines and lectins, but also through the T cell receptor. In addition, immunologists have obtained a vast knowledge of leukocyte functions. Certainly, there exists a large clinical interest in leukocytes and in several diseases in which leukocytes play an important role. These facts have further stimulated work on leukocyte integrins.

The early finding that integrins can be activated by phorbol esters, which were known to activate protein kinase C isoenzymes, led to studies on integrin phosphorylation. It soon turned out that the α-chains are constitutively

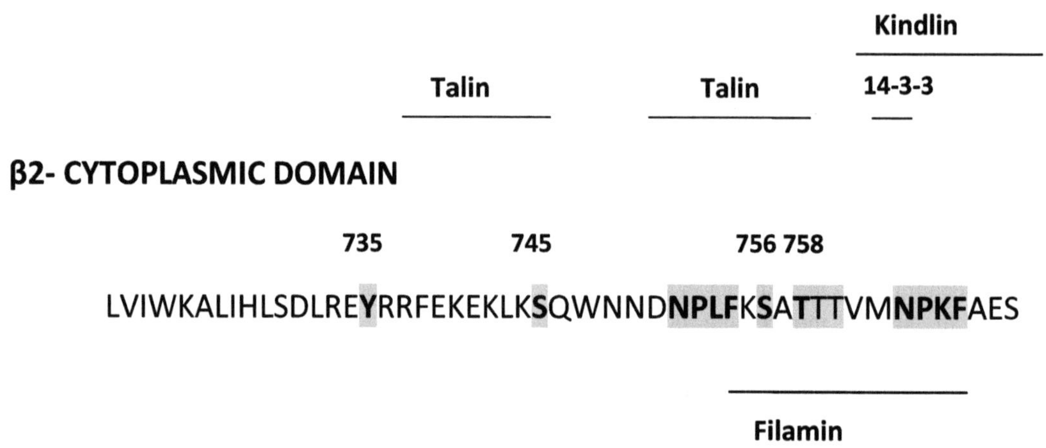

Fig. 6.3 The cytoplasmic domain of $\beta 2$. The phosphorylation sites are numbered, and the functionally important amino acids shaded. The binding regions of some major cytoplasmic proteins binding to the $\beta 2$-integrin cytoplasmic domain are indicated. Note that talin has two binding sites in $\beta 2$

phosphorylated, whereas the β-chains only became phosphorylated upon activation [44, 45]. The leukocyte $\beta 2$-integrins and the platelet αIIb/$\beta 3$ integrin have been best studied in this respect. The possible and identified phosphorylation sites in the most studied integrin subunits are shown in Fig. 6.2. The αL, αM and αX subunits have single phosphorylation sites, all on serine residues. In contrast, the β-chains contain several phosphorylation sites. In $\beta 2$, phosphate has been found on Ser-745, Ser-756, Thr-758, Thr-759, Thr-760 and possibly on Tyr-735 [41, 46]. αL is phosphorylated on Ser-1140, αM on Ser-1126 and αX on Ser-1158 [47–49]. In human $\beta 1$ and $\beta 3$ integrin subunits, both NPXY sequences can be phosphorylated, but not in $\beta 2$ where the tyrosines are replaced by phenylalanines. $\beta 7$ is similar to $\beta 2$ in that the first threonine in the threonine triplet is phosphorylated [46].

In addition to the protein kinase C family enzymes known to be responsible for integrin $\beta 2$-chain phosphorylations [41, 50] other Ser/Thr kinases may be important. Ser-756 is strongly phosphorylated upon phorbol ester treatment, and it seems possible that it is phosphorylated by calcium/calmodulin kinase II, because the antagonist W7 inhibits the phosphorylation [51]. In addition, the integrin linked kinase (ILK), and the mammalian sterile20-like1 kinase (Mst1) have been implicated in integrin regulation [52, 53]. Also protein kinase A may participate in integrin related intracellular communication [54]. The p21-activated kinase 4 (Pak 4) phosphorylates Ser-759 and Ser-762 in the integrin $\beta 5$ subunit [55]. This is proximal to the Ser/Thr phosphorylation site present in several β-chains. Mutation of the serines in this SERS sequence reduced the migration of the αV$\beta 5$-containing cells .

When T lymphocytes are activated through the T cell receptor, the inside-out signaling results in phosphorylation of Thr-758 in the $\beta 2$ subunit of αL$\beta 2$ [46]. The phosphorylated $\beta 2$-chain now binds 14-3-3 proteins, which are dimers with two Ser/Thr-phosphate binding domains [56]. This is followed by binding of the adaptor protein Tiam1 [57], followed by activation of the small G protein Rac-1 by Tiam1 [47, 58], (Fig. 6.4). Importantly, the phosphorylation signaling from the integrin $\beta 2$-chain can be mimicked using a membrane permeable peptide containing phosphate at Thr-758. This peptide, when introduced into lymphocytes, was able to activate the pathway through the $\beta 2$-chain resulting in increased adhesion [47]. Filamin is bound to the $\beta 2$-chain in resting cells, but phosphorylation of Thr-758 results in a switch from filamin binding to 14-3-3 binding.

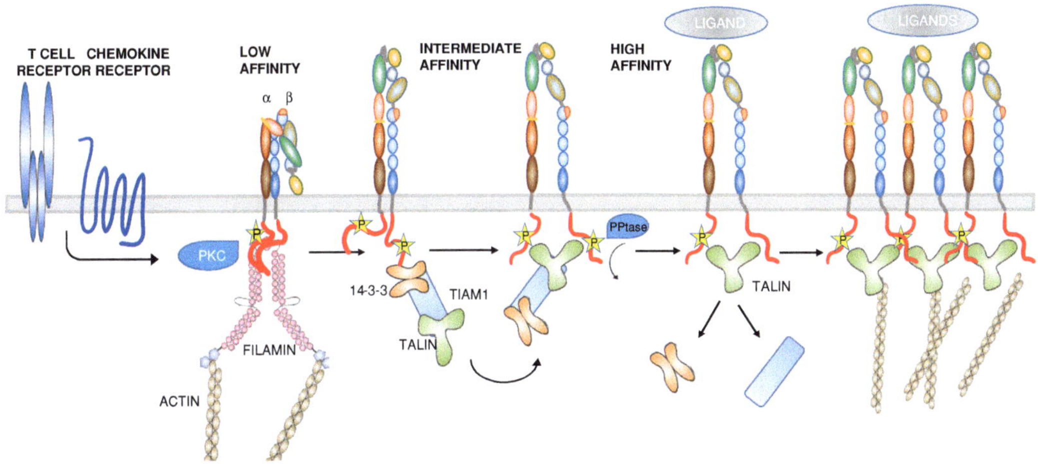

Fig. 6.4 Signaling from the T cell receptor/chemokine receptor resulting in active integrins. Ligand binding to the receptors results in activation of protein kinase C,s (PKC). This leads to phosphorylation of Thr-758 in the integrin $\beta2$ subunit. The cytoplasmic tail of the integrin is released from filamin and 14-3-3 proteins bind to the $\beta2$ chain, and the protein straightens out. A complex of the adaptor protein Tiam1 and talin is recruited to the integrin. The phosphate is subsequently released by phosphatase activity and talin can now directly bind to the integrin and activates it to high affinity. The interaction of talin with the actin cytoskeleton further increases the cellular binding to ligands by clustering of the integrin molecules resulting in increased avidity

Surface plasmon resonance experiments showed that the affinity of 14-3-3 proteins for phosphorylated $\beta2$ binding is high whereas filamin showed no binding [42]. Talin can in fact bind both to the unphosphorylated and phosphorylated chains, but its binding to the phosphorylated molecule is competed out when 14-3-3 is present. The structural explanation for 14-3-3 and filamin binding to the $\beta2$ cytoplasmic fragment has been determined. The phosphate on Thr-758 interacts electrostatically with Arg-56 and Arg-127 and by a hydrogen bond to Tyr-128 in 14-3-3. The filamin pocket in domain 21 can accomodate the unphosphorylated $\beta2$ peptide, but after phosphorylation there is no room for the peptide with the hydrophilic phosphate [42].

Less is known about the effect of phosphorylation on Ser-756. Mutation of it to methionine resulted in inhibition of phagocytosis of C3bi-coated erythrocytes [17, 59]. The small G protein Rap1 was shown to bind to the phosphorylation mimicking Ser-756/Asp, but not to the non-phosphorylatable mutant [59].

Relatively little is known the connection between outside-in signaling and phosphorylation. $\alpha L\beta2$ can be activated by integrin binding activating antibodies [60] soluble ligands such as ICAM-2 [61] and a peptide from the external part of ICAM-2 [62]. Ser-745 in the $\beta2$-chain was found to be phosphorylated by soluble ICAM-2. This phosphorylation resulted in the release of the transcription co-activator JAB-1 from the $\beta2$-chain enabling its downstream signaling to the nucleus [63].

The α-chain phosphorylations have been shown to be important. When αL, αM or αX, which are phosphorylated at positions Ser-1140, Ser-1126 and Ser-1158, respectively, are mutated with non-phosphorylatable alanines, adhesion was abrogated [47–49]. In $\alpha X\beta2$ the α-chain phosphorylatable residue could be replaced by aspartic acid regaining adhesion [49]. Whether this is true for the other $\beta2$-integrins is not known. A large proportion of the α-chains are constitutively phosphorylated, but there is still a continuous turnover of the phosphate [64]. Interestingly, outside-in activation by an integrin activating antibody resulted in activation of the Syk kinase both in wild type $\alpha X\beta2$ and αXSer-1158/Ala transfected cells, but adhesion was blocked in the Ser-1158/Ala transfected cells [49]. The fact that adhesion was normal in

Ser-1158/Asp transfected cells shows that the negative charge on residue-1158 is important, but it must not be phosphate. Furthermore, when cells were transfected with the constitutively active small G protein Rap1, wild type cells adhered but cells with the Ser-1158/Ala mutation did not [49].

The αM cytoplasmic domain is the shortest of the β2-integrin α-chain cytoplasmic domains and it is phosphorylated on Ser-1126 (Fig. 6.2). Interestingly, the corresponding amino acid in the other α-chain domains is glutamic acid. This could mean that the negative charge at this position is important for activity, but it can be either phosphate or amino acid based.

Tyrosine phosphorylation is important in signaling events such as those taking place downstream from the T cell receptor and it may involve a number of kinases and substrates [65]. Src family tyrosine kinases (Src, Lck, Fyn etc.) could be responsible for integrin tyrosine phosphorylation, but further details are poorly known [43]. Integrin tyrosine phosphorylation has been shown to take place upon ligand binding to the αIIb/β3 integrin [66]. This integrin contains tyrosines in the two NP(I)XY sequences and they become phosphorylated when platelets bind to ligands such as fibronectin, fibrinogen and von Willebrand factor [66–69]. In β1-integrins similar tyrosine phosphorylations occur and these may contribute to the phenotype of transformed cells. When the tyrosines were mutated to phenylalanines, cell movement was inhibited [70, 71]. Importantly, tyrosine phosphorylation of β1 has been shown to inhibit talin binding to the proximal NPXY sequence leaving room for other proteins to bind, for example filamin and tensin [72]. The integrin β2-subunit does not contain tyrosine in the NPXY sequences but phenylalanine. Evidently, the β1- and β2-integrins are regulated differently, in both cases by phosphorylation, but it can be either on Ser/Thr residues or on tyrosine. Interestingly, the β2 subunit has tyrosine at position 735 and mutation of this residue impairs integrin recycling [73]. In contrast, the β1- and β3-subunits contain phenylalanine in the corresponding position. Little is known about tyrosine phosphatases in this connection. The CD45 tyrosine phosphatase has been implicated in the regulation of T cell signaling probably by activating Src family kinases by removing the COOH-terminal phosphate. Interestingly, CD45 has been found to bind to integrin α-chains and it could be part of a signaling complex and the immunological synapse [16].

6.5 How Does Integrin Phosphorylation Affect Talin Binding?

How does talin and kindlins fit into the story on phosphorylations? Talin has been claimed to be the final activator of integrins. Using nanodiscs with a single lipid-embedded integrin Ye et al. [36] showed that purified talin could activate the αIIbβ3 integrin. Although several details are still incompletely understood, we propose the following model. Upon activation through the T cell receptor or through chemokines, phosphorylation takes place on Thr-758 in the β-subunit. This results in release of filamin from the β-chain and replacement by 14-3-3 proteins. The 14-3-3 proteins are homodimers and the free binding site binds in turn to Tiam1 and activates the small G protein Rac-1 resulting in remodeling of the cytoskeleton [57]. On the other hand, it is possible that Tiam1 binds to talin [32] and after dephosphorylation talin could bind to the integrin β-subunit. Kindlin-3 has been shown to bind to threonines-758-760 and the distal NPKF sequence [31, 74]. Probably, phosphorylation of Thr-758 inhibits the binding, which can be restored after dephosphorylation.

The negative charge on the integrin α-subunit due to phosphorylation, combined with recruiting of the β-chain-14-3-3-Tiam1-talin complex to the actin cytoskeleton after activation, could induce breaking of the bonds between the α- and β-subunits resulting in separation of the integrin tails. The talin head domain could then intercalate between the subunits. Serine/threonine phosphatase(s) like PP1 could remove the β2-chain phosphate enabling talin and kindlin binding to the β-chain. The talin and kindlin associations with the β-subunit would result in an allosteric change in the integrin resulting in activation of the molecule across the membrane.

6.6 Inhibition of Adhesion

Recently, natural inhibitors of integrins have been reported. The Del-1 protein is deposited on the surface of endothelial cells and binds to $\alpha L\beta 2$ and $\alpha M\beta 2$ integrins [75]. Evidently, it competes with the ICAM-molecules for integrin binding, and due to the fact that it is a soluble protein with relatively low affinity to integrins, it does not support leukocyte adhesion in vivo, but instead it inhibits binding by interfering with integrin/ICAM binding. Whereas Del-1 binds to the external surface of leukocytes, the intracellular protein SHARPIN binds to the cytoplasmic part of integrin α-chains and inhibits integrin activation [38, 76]. The SHARPIN/integrin interaction may compete out other α-chain cytoplasmic protein interactions, such as that of paxillin or inhibit the interaction of α-chains with the integrin β-chains.

6.7 Concluding Remarks

Protein phosphorylation is certainly extremely complex and because of its transient nature difficult to study. On the other hand it is fascinating to see how this relatively small protein modification can induce remarkable cellular changes enabling fast and precise adjustments in the highly variable environments of circulating and more stationary leukocytes. It is possible that the large number of protein kinase (>500) and phosphatase (>100) coding genes in higher organisms may forever preclude a detailed understanding of integrin function. But we optimistic and believe that a stepwise well defined approach can reveal the inner secrets of cellular adhesion complexity.

References

1. Arnaout MA, Pitt J, Cohen HJ, Melamed J, Rosen FS, Colten HR (1982) Deficiency of a granulocyte-membrane glycoprotein (gp150) in a boy with recurrent bacterial Infections. New Engl J Med 306:693–699
2. Kurzinger K, Reynolds T, Germain RN, Davignon D, Martz E, Springer TA (1981) A novel lymphocyte function-associated antigen (LFA-1): cellular distribution, quantitative expression, and structure. J Immunol 127:602
3. Patarroyo M, Beatty PG, Fabre JW, Gahmberg CG (1985) Identification of a cell surface protein complex mediating phorbol ester-induced adhesion (binding) among human mononuclear leukocytes. Scand J Immunol 22:171–182
4. Patarroyo M, Beatty PG, Serha CN, Gahmberg CG (1985) Identification of a cell surface glycoprotein mediating adhesion in human granulocytes. Scand J Immunol 22:619–631
5. Hynes RO (1987) Integrins: a family of cell surface receptors. Cell 48:549–554
6. Pytela R, Pierschbacher MD, Ruoslahti E (1985) Identification and isolation of a 140 kd cell surface glycoprotein with properties expected of a fibronectin receptor. Cell 40:191–198
7. Gahmberg CG, Tolvanen M, Kotovuori P (1997) Leukocyte adhesion. Structure and function of human leukocyte β_2-integrins and their cellular ligands. Eur J Biochem 245:215–232
8. Gahmberg CG, Fagerholm SC, Nurmi SM, Chavakis T, Marchesan S, Grönholm M (2009) Regulation of integrin activity and signaling. Biochim Biophys Acta 1790:431–444
9. Hogg N, Patzak I and Willenbrock F (2011) The insider's guide to leukocyte integrin signaling and function. Nat Rev Immunol 11:416–426
10. Hynes RO (2002) Integrins: bidirectional, allosteric signaling machines. Cell 110:673–687
11. Springer TA (1990) Adhesion receptors of the immune system. Nature 346:425–434
12. Calderwood DA (2004) Integrin activation. J Cell Sci 117:657–666
13. Gahmberg CG (1997) Leukocyte adhesion. CD11/CD18 integrins and intercellular adhesion molecules. Curr Opin Cell Biol 9:643–650
14. Patarroyo M, Clark EA, Prieto J, Kantor C, Gahmberg CG (1987) Identification of a novel adhesion molecule in human leukocytes by monoclonal antibody LB-2. FEBS Lett 210:127–131
15. Rothlein R, Dustin ML, Marlin SD, Springer TA (1986) A human intercellular adhesion molecule (ICAM-1) distinct from LFA-1. J Immunol 137:1270–1274
16. Grakoui A, Bromley SK, Sumen C, Davis MM, Shaw AS, Allen PM (1999) The immunological synapse: a molecular machine controlling T cell activation. Science 285:221–227

17. Lim J, Hotchin NA (2012) Signalling mechanisms of the leukocyte integrin $\alpha M\beta 2$: current and future perspectives. Biol Cell 104:631–640
18. Tan SM (2012) The leucocyte $\beta 2$ (CD18) integrins: the structure, functional regulation and signalling properties. Biosc rep 32:241–269
19. Castro FVV, Tutt AL, White AL, Teeling JL, James S, French RR, Glennie MJ (2008) CD11c provides an effective immunotarget for the generation of both CD4 and CD8 T cell responses. Eur J Immunol 38:2263–2273
20. Wu H, Gower RM, Wang H, Dai Perrard X-Y, Ma R, Bullar DC, Burns AR, Paul A, Smith WC, Simon SI, Ballantyne CM (2009) Functional role of CD11c + monocytes in atherogenesis associated with hypercholesterolemia. Circulation 119:2708–2717
21. Lee JO, Rieu P, Arnaout AM, Liddington R (1995) Crystal structure of the A domain from the α subunit of integrin CR3 (CD11b/CD18). Cell 80:631–638
22. Xiong JP, Stehle T, Diefenbach B, Zhang R, Dunker R, Scott DL, Joachimiak A, Goodman SL, Arnaout AM (2001) Crystal structure of the extracellular segment of integrin $\alpha V\beta 3$. Science 294:339–345
23. Xie C, Zhu J, Chen X, Mi L, Nishida N (2010) Structure of an integrin with an αI domain, complement receptor type 4. EMBO J 29:666–679
24. Vinogradova O, Haas T, Plow EF, Qin J (2000) A structural basis for integrin activation by the cytoplasmic tail of the aIIb-subunit. Proc Natl Acad Sci USA 97:1450–1455
25. Vinogradova O, Velyvis A, Velyviene A, Bin Hu TA, Haas EF, Plow EF, Qin J (2002) A structural mechanism of integrin $\alpha_{IIb}\beta_3$ "Inside-out" activation as regulated by its cytoplasmic face. Cell 110:587–597
26. Bhunia A, Tang XY, Mohanram H, Tan SM, Bhattacharjya S (2009) NMR solution conformations and interactions of integrin alpha Lbeta2 cytoplasmic tails. J Biol Chem 284:3873–3884
27. Chua GL, Tang XY, Amalraj M, Tan SM, Bhattacharjya S (2011) Structures and interaction analyses of integrin $\alpha M\beta 2$ cytoplasmic tails. J Biol Chem 286:43842–43854
28. Chua GL, Tang XY, Patra AT, Tan SM, Bhattacharjya S (2012) Structure and binding interface of the cytosolic tails of αX $\beta 2$ integrin. PLoS ONE 7:e41924. doi:10.1371/journal.pone.0041924
29. Chua GL, Patra AT, Tan SM, Bhattacharjya S (2013) NMR structure of integrin $\alpha 4$ cytosolic tail and its interactions with paxillin. PLoS One 8:e55184. doi:10.1371/journal.pone.0055184
30. Anthis NJ, Campbell ID (2011) The tail of integrin activation. Trends Biochem Sci 36:191–198
31. Bonet R, Vakonakis I, Campbell ID (2013) Characterization of 14-3-3-ζ interactions with integrin tails. J Mol Biol 425:3060–3072
32. Calderwood DA, Campbell ID, Critchley DR (2013) Talins and kindlins: partners in integrin-mediated adhesion. Nat Rev Mol Cell Biol 14:503–517
33. Das M, Ithychanda Qin J, Plow EF (2011) Migfilin and filamin as regulators of integrin activation in endothelial cells and neutrophils. PLoS ONE 6:e26355
34. Kinashi T (2005) Intercellular signaling controlling integrin activation in lymphocytes. Nat Rev Immunol 5:546–555
35. Legate KR, Fässler R (2009) Mechanisms that regulate adaptor binding to β-integrin cytoplasmic tails. J Cell Sci 122:187–198
36. Ye F, Hu G, Taylor D, Ratnikov B, Bobkov AA, McLean MA, Sligar SG, Taylor KA, Ginsberg MH (2010) Recreation of the terminal events in physiological integrin activation. J Cell Biol 188:157–173
37. Zahng Y, Wang H (2012) Integrin signalling and function in immune cells. Immunol 135:268–275
38. Rantala JK, Pouwels J, Pellinen T, Veltel S, Laasola P, Mattila E, Potter CS, Duffy T, Sundberg JP, Kallioniemi O, Askari JA, Humphries MJ, Parsons M, Salmi M, Ivaska J (2011) SHARPIN is an endogenous inhibitor of $\beta 1$-integrin activation. Nat Cell Biol 13:1315–1324
39. Shattil SJ, Kim C, Ginsberg MH (2010) The final steps of integrin activation: the end game. Nat Rev Mol Cell Biol 11:288–300
40. Tremuth L, Kreis S, Melchior C, Hoebeke J, Ronde P, Plancon S, Takeda K, Kieffer N (2004) A fluorescence cell biology approach to map the second integrin-binding site of talin to a 130-amino acid sequence within the rod domain. J Biol Chem 279:22258–22266
41. Fagerholm S, Morrice N, Gahmberg CG, Cohen P (2002) Phosphorylation of the cytoplasmic domain of the integrin CD18 chain by protein kinase C isoforms in leukocytes. J Biol Chem 277:1728–1738
42. Takala H, Nurminen E, Nurmi SM, Aatonen M, Strandin T, Takatalo TK, Gahmberg CG, Ylänne, Fagerholm SC (2008) $\beta 2$ integrin phosphorylation on Thr758 acts as a molecular switch to regulate 14-3-3 and filamin binding. Blood 112:1853–1862
43. Fagerholm S, Hilden TJ, Gahmberg CG (2004) P marks the spot: site-specific integrin phosphorylation regulates molecular interactions. Trends Biochem Sci 29:504–512
44. Chatila TA, Geha RS, Arnaout MA (1989) Constitutive and stimulus-induced phosphorylation of CD11/CD18 leukocyte adhesion molecules. J Cell Biol 109:3435–3444
45. Valmu L, Autero M, Siljander P, Patarroyo M, Gahmberg CG (1991) Phosphorylation of the ß-subunit of CD11/CD18 integrins by protein kinase C correlates with leukocyte adhesion. Eur J Immunol 21:2857–2862
46. Hilden TJ, Valmu L, Kärkkäinen S, Gahmberg CG (2003) Threonine phosphorylation sites in the β_2 and

β_7 leukocyte integrin polypeptides. J Immunol 170:4170–4177
47. Fagerholm SC, Hilden TJ, Nurmi SM, Gahmberg CG (2005) Specific integrin α and β chain phosphorylations regulate LFA-1 activation through affinity-dependent and –independent mechanisms. J Cell Biol 171:705–715
48. Fagerholm SC, Varis M, Stefanidakis M, Hilden TJ, Gahmberg CG (2006) α-chain phosphorylation of the human leukocyte CD11b/CD18 (Mac-1) integrin is pivotal for integrin activation to bind ICAMs and leukocyte extravasation in vivo. Blood 108:3379–3386
49. Uotila L, Aatonen M, Gahmberg CG (2013) Integrin CD11c/CD18 α-chain phosphorylation is functionally important. J Biol Chem 288:33494–33499
50. Valmu L, Hilden T, van Willigen G, Gahmberg CG (1999) Characterization of β_2 (CD18) integrin phosphorylation in phorbol ester activated T lymphocytes. Biochem J 339:119–125
51. Fagerholm S, Prescott A, Cohen P, Gahmberg CG (2001) An essential role for calmodulin in regulating human T cell aggregation. FEBS Lett 491:131–136
52. Hannigan GE, Leung-Hagesteijn C, Fitz-Gibbon L, Coppolino MG, Radeva G, Filmus J, Bell JC, Dedhar S (1996) Regulation of cell adhesion and anchorage-dependent growth by a new β1-integrin-linked protein kinase. Nature 379:91–96
53. Mou F, Praskova M, Xia F, Van Buren D, Hock H, Avruch J, Zhou D (2012) The Mst1 and Mst2 kinases control activation of rho family GTPases and thymic egress of mature thymocytes. J Exp Med 209:741–759
54. Gonzalez AM, Claiborne J, Jones JCR (2008) Integrin cross-talk in endothelial cells is regulated by protein kinase A and protein phosphatase 1. J Biol Chem 283:31849–31860
55. Li Z, Zhang H, Lundin L, Thullberg M, Liu Y, Wang Y, Claesson-Welsh L, Strömblad S (2010) p21-activated kinase 4 phosphorylation of integrin β5 Ser-759 and Ser-762 regulates cell migration. J Biol Chem 285:23699–23710
56. Fu H, Subramanian RR, Masters S (2000) 14-3-3 proteins: structure, function, and regulation. Ann Rev Pharmac Toxicol 40:617–647
57. Grönholm M, Jahan F, Marchesan S, Karvonen U, Aatonen M, Narumanchi S, Gahmberg CG (2011) TCR-induced activation of LFA-1 involves signaling through Tiam1. J Immunol 187:3613–3619
58. Nurmi SM, Autero M, Raunio AK, Gahmberg CG, Fagerholm SC (2007) Phosphorylation of the LFA-1 integrin β_2-chain on Thr-758 leads to adhesion, Rac-1/Cdc42 activation and stimulation of CD69 expression in human T cells. J Biol Chem 282:968–975
59. Lim J, Hotchin NA, Caron E (2011) Ser^{756} of β2 integrin controls Rap1 activity during inside-out activation of αMβ2. Biochem J 437:461–467
60. Petruzzelli L, Maduzia L, Springer TA (1995) Activation of lymphocyte function-associated molecule-1 (CD11a/CD18) and Mac-1 (CD11b/CD18) mimicked by an antibody directed against CD18. J Immunol 155:854–866
61. Kotovuori A, Pessa-Morikawa T, Kotovuori P, Nortamo P, Gahmberg CG (1999) ICAM-2 and a peptide from its binding domain are efficient activators of leukocyte adhesion and integrin affinity. J Immunol 162:6613–6620
62. Li R, Nortamo P, Kantor C, Kovanen P, Timonen T, Gahmberg CG (1993) A leukocyte integrin binding peptide from intercellular adhesion molecule-2 stimulates T cell adhesion and natural killer cell activity. J Biol Chem 268:21474–21477
63. Perez OD, Mitchell D, Jager GC, South S, Murriel C, McBride J, Herzenberg LA, Kinoshita S, Nolan GP (2003) Leukocyte functional antigen 1 lowers T cell activation thresholds and signaling through cytohesin-1 and Jun-activating binding protein 1. Nat Immunol 4:1083–1092
64. Valmu L, Gahmberg CG (1995) Treatment with okadaic acid reveals strong threonine phosphorylation of CD18 after activation of CD11/CD18 leukocyte integrin with phorbol esters or CD3 antibodies. J Immunol 155:1175–1183
65. Calderwood DA, Fujioka Y, de Pereda JM, Garcia-Alvarez B, Nakamoto T, Margolis B, McGlade J, Liddington RC, Ginsberg MH (2003) Integrin beta cytoplasmic domain interactions with phosphotyrosine-binding domains: A structural prototype for diversity in integrin signaling. Proc Natl Acad Sci USA 100:2272–2277
66. Law DA, Nannizzi-Alaimo L, Phillips DR (1996) Outside-in integrin signal transduction: αβ-(GPIIb-IIIa) tyrosine phosphorylation induced by platelet aggregation. J Biol Chem 271:10811–10815
67. Blystone SD, Lindberg FP, Williams MP, McHugh K, Brown EJ (1996) Inducible tyrosine phosphorylation of the β_3 integrin requires the α_v integrin cytoplasmic tail. J Biol Chem 271:31458–31462
68. Blystone SD (2002) Kinetic regulation of β_3 integrin tyrosine phosphorylation. J Biol Chem 277:46886–46890
69. Phillips DR, Prasad KSS, Manganello J, Bao M, Nannizzi-Alaimo L (2001) Integrin tyrosine phosphorylation in platelet signaling. Curr Opin Cell Biol 13:546–554
70. Sakai T, Jove R, Fässler R, Mosher DF (2001) Role of the cytoplasmic tyrosines of {beta}1A integrins in transformation by v-src. Proc Natl Acad Sci USA 98:3808–3813
71. Wennerberg K, Armulik A, Sakai T, Karlsson M, Fässler R, Schaefer EM, Mosher DF, Johansson S (2000) The cytoplasmic tyrosines of integrin subunit β1 are involved in focal adhesion kinase activation. Mol Cell Biol 20:5758–5768
72. Ling K, Doughman RL, Iyer VV, Firestone AJ, Bairstow SF, Mosher DF, Schaller MD, Anderson RA (2003) Tyrosine phosphorylation of type Iγ phosphatidylinositol phosphate kinase by Src

regulates an integrin–talin switch. J Cell Biol 163:1339–1349
73. Fabbri M, Fumagalli L, Bossi G, Bianchi E, Bender JR, Pardi R (1999) A tyrosine-based sorting signal in the $\beta 2$ integrin cytoplasmic domain mediates its recycling to the plasma membrane and is required for ligand-supported migration. EMBO J 18:4915–4925
74. Morrison VL, MacPherson M, Savinko T, Lek HS, Prescott A, Fagerholm SC (2013) The $\beta 2$ integrin–kindlin-3 interaction is essential for T-cell homing but dispensable for T-cell activation in vivo. Blood 122:1428–1436
75. Choi EY, Chavakis E, Czabanka MA, Langer HF, Fraemohs L, Economopoulou M, Kundu RK, Orlandi A, Zheng YY, Prieto RA, Ballantyne CM, Constant SL, Aird WC, Papayannopoulou T, Gahmberg CG, Udey MC, Vajkoczy P, Quertermous T, Dimmeler S, Weber C, Chavakis T (2008) Del-1, an endogenous leukocyte-endothelial adhesion inhibitor, limits inflammatory cell recruitment. Science 322:1101–1104
76. Pouwels J, de Franceschi N, Mattila E, Potter C, Sundberg JP, Hogg N, Gahmberg CG, Salmi M, Ivaska J (2013) SHARPIN regulates uropod detachment in migrating lymphocytes. Cell Rep 5:619–628

Integrin $\alpha E\beta 7$: Molecular Features and Functional Significance in the Immune System

7

Gregg A. Hadley and Jonathan M. G. Higgins

Abstract

Alpha E beta 7 ($\alpha E\beta 7$) is an α-I domain-containing integrin that is highly expressed by a variety of leukocyte populations at mucosal sites including intraepithelial T cells, dendritic cells, mast cells, and T regulatory cells (Treg). Expression depends largely or solely on transforming growth factor beta (TGF-β) isoforms. The best characterized ligand for $\alpha E\beta 7$ is E-cadherin on epithelial cells, though there is evidence of a second ligand in the human system. An exposed acidic residue on the distal aspect of E-cadherin domain 1 interacts with the MIDAS site in the αE α-I domain. By binding to E-cadherin, $\alpha E\beta 7$ contributes to mucosal specific retention of leukocytes within epithelia. Studies on αE knockout mice have identified an additional important function for this integrin in allograft rejection and have also indicated that it may have a role in immunoregulation. Recent studies point to a multifaceted role for $\alpha E\beta 7$ in regulating both innate and acquired immune responses to foreign antigen.

Keywords

Integrins · Intraepithelial T cells · Regulatory T cells · Dendritic cells · Mast cells · TGFβ

7.1 Introduction

Integrin $\alpha E\beta 7$ is, in many respects, an unusual integrin. The αE subunit (CD103) has unique structural features (Fig. 7.1) and is the only α-I domain-containing integrin chain that pairs with $\beta 7$. Beta 7, however, can pair with $\alpha 4$ as well as αE. Both heterodimers are expressed exclusively by leukocytes and have special significance for the mucosal immune system. Alpha 4 beta 7 is the principal mucosal homing receptor for

G. A. Hadley (✉)
Department of Microbial Infection and Immunity, The Ohio State University Wexner Medical Center, Columbus, OH 43210, USA
e-mail: Gregg.Hadley@osumc.edu

J. M. G. Higgins
Division of Rheumatology, Immunology and Allergy, Brigham and Women's Hospital, Harvard Medical School, Boston, MA 02115, USA

D. Gullberg (ed.), *I Domain Integrins*, Advances in Experimental Medicine and Biology 819,
DOI: 10.1007/978-94-017-9153-3_7,

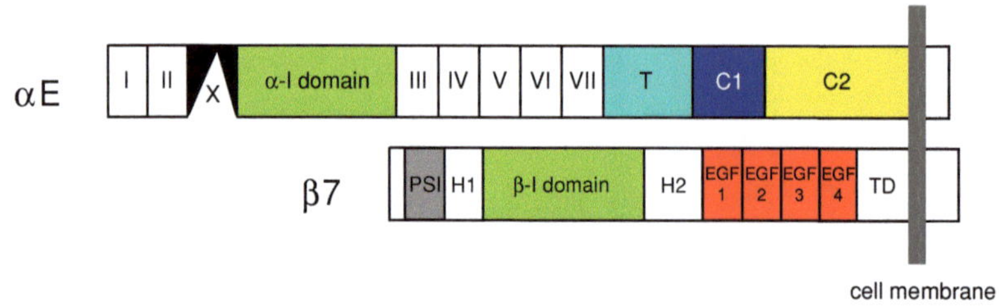

Fig. 7.1 Domain structure of integrin αEβ7. The seven blades of the β-propeller domain of αE are labeled I to VII; the thigh domain, T; the calf-1 and –2 domains, C1 and C2. The extra X-domain, not found in any other integrin chain and containing a post-translational cleavage site, is marked X. In β7, the plexin/semaphorin/integrin domain is labeled PSI; the two components of the hybrid domain, H1 and H2; and the β-tail domain, TD. Note that this figure does not illustrate the relative orientation of the different domains

leukocytes [11] whereas αEβ7 appears to play a role in retention of these cells within or near epithelia. In this review we shall discuss the tissue distribution and induction of αE and present a molecular perspective on the interaction between αEβ7 and its principal ligand, E-cadherin. The complex organization of the αE gene locus will be described. Finally, we aim to present current views of αEβ7 function.

Recent studies indicate that αEβ7 plays an important role in determining the localization of dendritic cell subsets, and therefore indirectly impacts all immune responses, both innate and adaptive. In fact, studies of the αE expressing subset of dendritic cells now dominate the literature on this subject. Of 94 αE integrin references published in 2013, 61 were about αEβ7 expressing dendritic cells. By contrast, in calendar year 2003, there were only 25 references to αE integrin, and none of these were about dendritic cells. The field initially focused on the role of αEβ7 in promoting the functional activities of mucosal T cells, then shifted to a focus to the functional relevance of αE on Tregs then to the current focus on the relevance of αE expression by dendritic cell subsets. The emphasis on the αE expressing subset of dendritic cells is warranted as it is now clear that dendritic cells initiate essentially all immune responses—both innate and adaptive—and thereby play a critical role in defining the nature and character of the immune response. Thus, αEβ7 likely controls these critical processes.

7.2 Tissue and Cellular Distribution

Integrin αEβ7 was originally discovered in the rat, human and mouse by screening panels of monoclonal antibodies for cell surface features that were distinctive for intestinal intraepithelial lymphocytes (IEL) [15, 16, 47]. The original mAbs to αEβ7, RGL-1, HML-1, and M290 (reactive in rat, man and mouse respectively) were subsequently shown to identify a novel integrin alpha chain now known as αE (CD103) [14, 49, 50, 54, 70, 76, 78, 89, 92, 115]. A fourth antibody, MRC-OX62, raised against rat lymphatic dendritic cells was later shown also to recognize αEβ7 [7, 8]. Thus, a distinguishing feature of αEβ7 is that it is expressed most prominently and abundantly in the gut, particularly on T cells in the epithelium [8, 15, 16, 25, 47, 49, 50]. At first, it seemed a foregone conclusion that αEβ7 functioned to retain T cells at mucosal sites, but recent studies reveal a more complex situation. In other compartments of the immune system and among other lymphoid/myeloid cell lineages expression is found on sub-populations which express αEβ7 at lower levels that are, nevertheless, functionally important. In particular, while αEβ7 is expressed by diverse leukocyte subsets, it is now clear that it defines a subset of dendritic cells, and thereby can have a global impact on immune responses.

$\alpha E\beta 7^+$ T cells are usually found in locations where active TGF-β isoforms are abundant. Expression of $\alpha E\beta 7$ on T cells is usually skewed towards the CD8 subset [16, 25, 47], a phenomenon that is readily seen in mixed T cell cultures stimulated with mitogen in the presence of TGF-βisoforms [9, 80, 87]. In the gut, almost all IEL and about half the T lymphocyte population in the lamina propria express $\alpha E\beta 7$ [16, 25, 47]. Similarly, the integrin is present on T cells in or near other epithelial surfaces, including those of the lung [80] and genital tract [22, 77]. In lymphoid tissues, including Peyer's patches and mesenteric lymph nodes and in peripheral blood the percentage of $\alpha E\beta 7^+$ T cells and their level of expression of $\alpha E\beta 7$ is generally low [2, 16, 50].

Although $\alpha E\beta 7$ was formerly considered to be a mucosal T cell marker, the molecule is also found on other cell lineages. Most studies on the distribution of $\alpha E\beta 7$ have failed to detect the molecule on tissue macrophages, but there is an exception in which a proportion of macrophages in lung, liver and lymph node sinuses is reported to have stained positively with mAb HML-1 [100]. An interesting observation was also made that mucosal-type mast cells generated in vitro from bone marrow precursors by culturing in the presence of stem cell factor, IL-3, IL-9 and TGF-β expressed $\alpha E\beta 7$ strongly [92, 111]. The presence of the integrin on mucosal mast cells in vivo is strongly supported by circumstantial evidence, but the functional significance and in vivo relevance of such expression remains to be demonstrated.

Significant subsets of dendritic antigen-presenting cells (DC) in the gut mucosa, the mesenteric lymph nodes and the epithelium of the airways of rats and mice are $\alpha E\beta 7^+$ [7, 8, 46, 65, 73] but in lymph nodes which have no mucosal involvement the proportion is considerably smaller [46] and in the spleen $\alpha E\beta 7$ expression is confined to the small subset of $CD8^+$ DC [69]. In man, expression of $\alpha E\beta 7$ by mucosal dendritic cells has been less extensively documented but $\alpha E\beta 7^+$ DC are present in the dome epithelium of Peyer's patches and in the lamina propria [25, 108]. In contrast, Langerhans-type DC generated in vitro from hematopoietic stem cells in the presence of TGF-β and other cytokines do not express the integrin [79].

Detailed scrutiny of B cell subsets for expression of $\alpha E\beta 7$ has revealed a complex picture. Early studies showed that the integrin was expressed by few if any B cells in the gut mucosa or elsewhere. However, Csencsits et al. [23] identified a population of $\alpha E\beta 7^+$ $B220^+$ cells in the intestinal mucosa following intranasal immunisation of mice with cholera toxin. That cells of B lymphocyte lineage can, in certain circumstances, express $\alpha E\beta 7$ is supported by the detection of a small population $CD19^+$ αE^+ B cells in peripheral blood [40] and also by much earlier observations that $\alpha E\beta 7$ expression is a diagnostic marker for hairy cell leukemias [70–72].

Studies of $\alpha E\beta 7$ expression during thymic ontogeny in the mouse have shown that 3–5 % of cells express the integrin and that it is represented in both TCR$\alpha\beta$ and $\gamma\delta$ lineages, particularly in the late developmental stages [2, 59]. The integrin is present on about half the population of thymic precursors of dendritic epidermal T cells (DETC) and on all mature cells of this subset [59]. In humans, $\alpha E\beta 7$ was found to be expressed by a major subpopulation of single positive $CD8^+$ human thymocytes and a smaller proportion of less mature double negative cells [56, 67]. Recent studies implicate Runx 3 in controlling $\alpha E\beta 7$ expression during thymocyte development [33, 112], and indicate that CXCR3 and $\alpha E\beta 7$ both are expressed by the $CD8^+$ single positive thymocyte subset [4], and that Treg likely derive from $Foxp3^+$ double positive ($CD8^+CD4^+$) cells that lack $\alpha E\beta 7$ expression [75]. It has also been reported that most, if not all, naïve $CD8^+$ that have recently emigrated from the thymus into the circulation express $\alpha E\beta 7$ [67]. Thus, the frequency of $\alpha E\beta 7^+$ CD8 T cells in the blood with a naïve phenotype appears to be a useful indicator of thymopoiesis. Maintenance of $\alpha E\beta 7$ expression by this cell population, and also by splenic and blood $CD8^+$ T cells, has been reported to depend on lymphotoxin alpha (LTα) [31]. However, the possibility was not excluded that LTα

induces expression of the αE subunit by an indirect effect on TGF-β processing. Single positive thymocytes expressing αEβ7 may migrate to the small intestine via a sphingosine 1-phosphate (S1P) dependent process [55].

The study of T regulatory (Treg) cells (formerly known as suppressor T cells) has undergone a renaissance and their importance in immune homeostasis and in the prevention of autoimmune diseases and allograft rejection is clear. In vivo models of suppression of autoimmunity involving adoptive cell transfer and in vitro studies on suppression of lymphocyte proliferation by spleen or lymph node T cell subpopulations have shown that Treg cells reside within a population that is $CD4^+$ $CD25^+$ $CD45RB^{low}$ [17, 64]. Four studies have shown that αEβ7 is expressed by 20–30 % of this T cell subset [32, 60, 68, 117]. Similarly, regulatory $CD8^+$ T cells generated by co-culture of intestinal epithelial cells and peripheral T cells were shown to express αEβ7 [1].

7.3 Induction of αEβ7

It is long been recognized that transcription of the αE subunit is regulated by transforming growth factor beta (TGF-β) [49, 50, 76, 85, 92]. Such induction is commonly attributed to the TGF-β1 isoform but all isoforms of TGF-β (also mouse TGF-β2, and -β3 for example) exhibit this property (GAH unpublished data); it has not yet been established which of the TGF-β isoforms contribute to αEβ7 induction in vivo. Recent studies point to a key role for membrane bound TGF-β in this process [113, 114], but a complete understanding of this important interaction is muddled by our poor understanding of how TGF-β isoforms are processed to their active forms in the particular cells used in these experiments. It has been reported that ligation of β1 integrins can act synergistically with TGF-β in αE induction [80], and that activation of naïve human $CD8^+$ T cells with anti-CD3 in the presence of IL-4 can also increase αEβ7 expression [99], though it is unclear if these apparent inducers operate through the indirect action of TGF-β isoforms.

It is widely held that αEβ7 expressed by T cells located in the vicinity of epithelia is induced locally by TGF-β isoforms produced mainly by epithelial cells. This view is supported by the observation that T cells stimulated in vitro by co-culture with allogeneic kidney epithelial cells, or T cells that migrate into epithelial monolayers, are induced to express αEβ7 and that expression is blocked by anti-TGF-β antibody [34, 90]. The results of a study of mucosal T cell memory by Kim et al. [51] are also consistent with the idea that αEβ7 is upregulated locally. Ovalbumin-specific transgenic $CD8^+$ T cells were adoptively transferred to recipients that were then infected with recombinant vesicular stomatitis virus expressing ovalbumin (VSV-OVA). Analysis of donor-type memory cells in various lymphoid compartments indicated that αEβ7 was strongly upregulated on IEL over the 5 week study period.

The notion that αEβ7 expression is mainly, if not solely, TGF-β-dependent is supported by a study showing that in transgenic mice which express the negative regulator of TGF-β isoform signalling, Smad7, under an *Lck* promoter, 50 % of intraepithelial T cells in the gut no longer express αE [96]. Expression of the integrin by the remaining cells probably reflects insufficient expression of the transgene in this population but leaves open the possibility that an alternative signaling pathway could be responsible for αE expression in these circumstances. Using a T cell line, Robinson et al. showed that TGF-β induces αE transcription de novo within 30 min [85]. The speed of induction suggests that synthesis of signaling intermediaries or new transcription factors was probably not required. These authors also looked for transcription control elements in the promoter region of the human αE gene using deletion analysis to examine 4 kb of genomic sequence upstream of the transcription start site. Although the promoter functioned well in reporter assays, it bestowed neither cell lineage specificity nor TGF-β responsiveness. Thus, transcription control mechanisms for αE are likely to be considerably more complex than those of most other integrin α-chain genes, whereas lineage specificity is determined by the proximal promoter in other integrins.

7.4 Gene Structure

Past studies established the complexity at the locus of the integrin αE gene, *Itgae*. Schön et al. [88] generated a partial map of murine *Itgae*, and subsequently the human genome sequencing project provided more complete information on human *Itgae* [37]. Human *Itgae* contains 31 exons spanning approximately 85 kb (Fig. 7.2). Comparison with the genes encoding the closely related αM and αX integrin proteins [21, 27, 74] reveals a highly conserved gene structure. All the introns are located in similar positions and have the same phase in the three genes, although *Itgae* contains an extra exon (exon 6) that encodes the X domain not present in other integrins (see Fig. 7.1). The α-I domain is encoded by exons 7–10. Human *Itgae* is found at chromosome 17p13.3 rather than in the αL/αM/αX/αD integrin cluster at chromosome 16p11 [21, 110], and is syntenic with that of murine *Itgae* on chromosome 11 [88]. Robinson et al. [85] analyzed the transcription start site of human *Itgae*, and identified two start sites 51 and 44 bp upstream of the start codon, and a third possible initiation site around position ~90 bp. Interestingly, another gene, *Gsg2*, that encodes the mitotic protein kinase Haspin is found on the opposite strand within an intron of *Itgae* Fig. 7.2. The Haspin promoter appears also to drive expression of a truncated and alternatively spliced *Itgae* transcript that is widely expressed and could function as a non-coding RNA [37].

The human β7 gene, *Itgb7*, comprises 14 exons spanning approximately 10 kb and maps to chromosome 12q13.13 [5, 42], syntenic with murine β7 on chromosome 15 [116]. The intron-exon structure of *Itgb7* is more similar to that of the β1 and β2 genes than the β3, β5 and β6 genes, consistent with a similar sub-grouping derived from analyses of sequence homology [42].

7.5 Ligand Binding

Expression of αEβ7 by T cells closely juxtaposed to epithelial surfaces suggested that this integrin might bind a counter-receptor on the surface of epithelial cells. In 1993 three groups reported that a ligand for αEβ7 was present on epithelial cell lines [12, 81, 82]. The epithelial ligand was later identified as the homophilic adhesion molecule E-cadherin [13, 39, 44], and mutagenesis studies combined with crystal structure determination and molecular modeling led to a detailed model for αEβ7 binding to E-cadherin in which the MIDAS motif within the α-I domain of αE makes direct contact with an acidic residue at the tip of domain 1 in E-cadherin (Fig. 7.3) [38, 41, 45, 98]. These findings strengthened the concept that αEβ7 retains leukocytes in epithelial tissues by binding to E-cadherin on epithelial cells.

E-cadherin expression is found on most epithelial cells, but is not limited to this population. Recent studies suggest that E-cadherin can act at the level of dendritic cells to impact immune responses. For example, Siddequi et al. [91] observed that monocyte-derived inflammatory DCs express E-cadherin, and that these promote intestinal inflammation. Similarly, Uchida et al. [102] reported that E-cadherin and αEβ7 on DETC regulate their activation threshold through binding to E-cadherin on keratinocytes. Van den Bossch et al. [104, 105] detailed the regulation and function of the E-cadherin/catenin complex in cells of the monocyte-macrophage lineage and DCs, and found that E-cadherin is expressed by alternatively activated macrophages. Thus, αEβ7 expressing cells potentially interact with and regulate diverse leukocyte populations, but the extent to which this occurs in vivo has yet to be established.

E-cadherin is the only well-defined counter-receptor of αEβ7, but there is preliminary evidence for at least one further ligand on keratinocyte cell lines and intestinal lamina propria endothelial cells that lack E-cadherin expression [10, 41, 93].

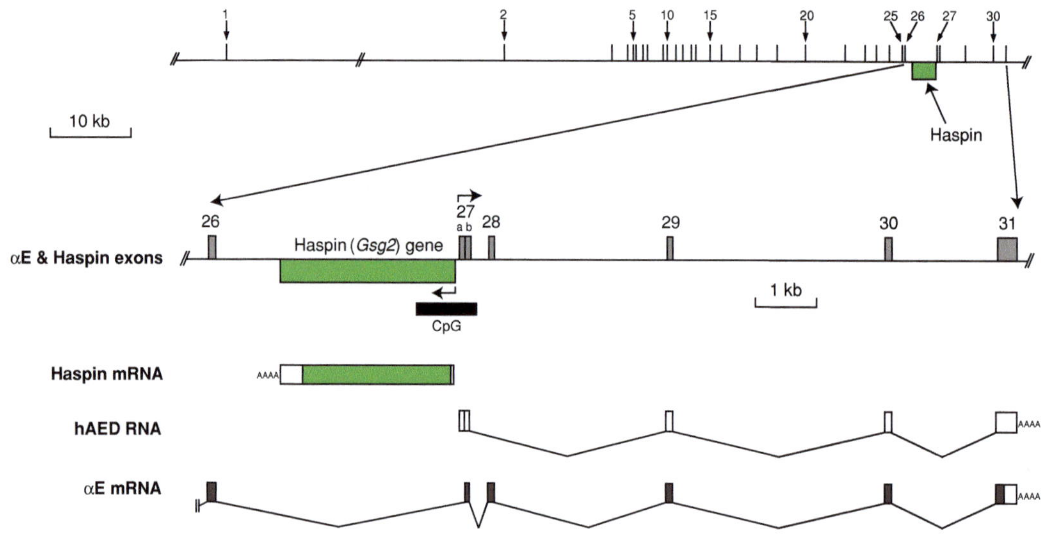

Fig. 7.2 Outline structure of the αE genomic locus. The *top line* shows the intron-exon structure of the integrin αE (*Itgae*) and Haspin (*Gsg2*) genes. On the *second line*, the 3′ region of the αE gene containing the Haspin gene is shown in more detail. Intronic regions are shown as *horizontal lines* and exons as *boxes*. *Bent arrows* represent transcription start sites, and the *thick black line* indicates the location of a CpG island. The *three lower lines* show the three transcribed products of this genomic region, including the Alpha-E derived mRNA, hAED. In each case, *boxes* represent exonic portions of RNA, *thin lines* indicate introns, *dark shading* indicates protein coding regions, and AAAA indicates a poly(A) tail

7.6 Function

7.6.1 Effector and Memory T Cells

It is now clear that αE controls the accumulation of effector and memory T cells (resident memory T cells, T_{rm}) in non-lymphoid tissues and thereby may promote their capacity to eliminate invading pathogens. Alpha-E expression marks T_{rm} cells in a variety of tissues [63, 107] (25), and there is good evidence that such expression promotes their local persistence, particularly for intraepithelial $CD8^+$ T cells in the intestinal and vaginal mucosa, where binding to E-cadherin may be critical [88, 107]. Moreover, it is not clear that αEβ7 expressing T cells present at all sites are exclusively memory T cells, in that many are present in naïve mice prior to specific antigen exposure. The underlying mechanisms regulating αEβ7 expression by T_{rm} remain poorly defined but are likely similar to those described above for other αEβ7 expressing cells. These include induction of αEβ7 and downregulation of the chemokine receptor CCR7 with a dominant role for local TGF-β activity in the process. Suvas et al. have shown that systemic and mucosal infection both are effective in generating mucosal $\alpha E\beta 7^+$ T_{rm} responses [95]. Yu et al. [113] have also reported that human $CD1c^+$ DCs express cell surface TGF-β and thereby drive the generation of αEβ7 expressing cytotoxic lymphocytes (CTL).

7.6.2 Allografts

A number of studies have examined whether αEβ7 on T cells could play a role in allograft rejection. Hadley et al. [35, 36] reported that up to 63 % of T cells infiltrating renal allografts undergoing a late rejection crisis expressed αEβ7 and that the cells were localized mainly in the tubular epithelium. Similar findings were reported by Robertson et al. [83, 84] who observed a correlation between the prevalence of

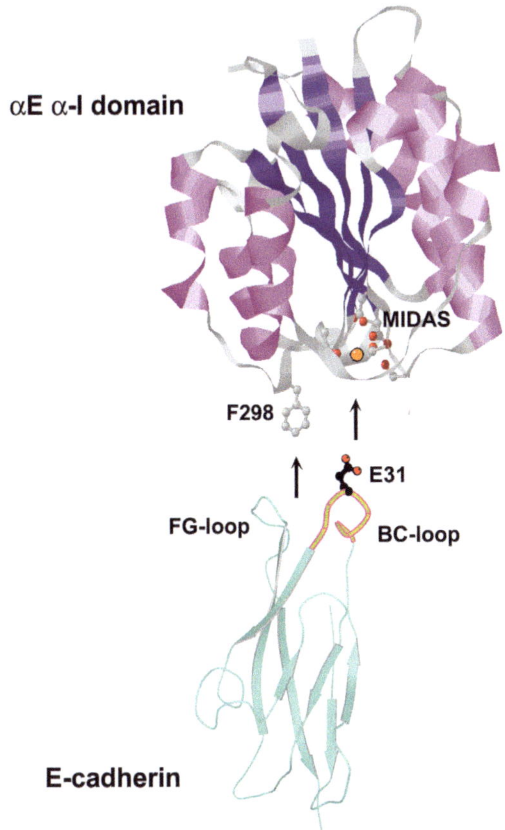

Fig. 7.3 A model of the $\alpha E\beta 7$ integrin α-I domain docked onto E-cadherin domain 1. Residue E31 in the BC-loop of E-cadherin is predicted to coordinate the MIDAS magnesium ion in the α-I domain and F298 of αE is predicted to project into a hydrophobic pocket between the BC and FG loops of E-cadherin. For details see ref [38]. Reproduced with permission from Agace WW, Higgins JMG, Sadasivan B, Brenner MB, Parker CM. Curr. Opin. Cell Biol. 2000; 12:563–568 (Copyright 2000, Elsevier Science)

$\alpha E\beta 7^+$ cells in the tubular epithelium, the severity of tubulitis and the levels of TGF-β in the epithelium. Earlier studies established that $\alpha E\beta 7$ was induced on $CD8^+$ T cells co-cultured with renal epithelial cells [34] and that $\alpha E\beta 7$ provided accessory function for cytotoxic lysis of target epithelial cells [86]. This evidence supports the view that $\alpha E\beta 7$ is induced on infiltrating $CD8^+$ cells by TGF-β produced locally in the allograft and causes the cells to accumulate in the graft epithelium by adhesion to E-cadherin expressed by tubular epithelial cells. The integrin/ligand interaction would then provide accessory function for cytotoxic lysis or cytokine production. This interaction may be especially important in rejection when other integrin/ligand interactions, principally $\alpha L\beta 2$ (LFA-1)/ICAM-1, are unavailable. This view is strongly supported by the observation that αE null/null mice are unable to reject pancreatic islet allografts [26, 48]. Although $CD8^+$ cells accumulate around the graft they do not come into intimate contact with islet cells, which are known to express E-cadherin but not ICAM-1. This view is further supported by the observation that T cells from αE null/null mice do not elicit gut graft-versus-host disease (GVHD) on transfer to wildtype allogenic recipients [24]. Zhou et al. [118] confirmed these findings in a rat GVHD model, and further observed that the skin epidermis in rats during GVHD is infiltrated by an equal number of $CD4^+$ T cells and $CD8^+$ T cells expressing $\alpha E\beta 7$. Collazo et al. [19] reported that expression of SH2 domain–containing inositol 5-phosphatase (SHIP) is required for robust expansion of donor $\alpha E\beta 7^+$ $CD4^+$ T cells during graft-versus-host and host-versus-graft responses by $CD4^+$ T cell and limits their immunoregulatory capacity. These observations on the role of $\alpha E\beta 7$ in allograft rejection and GVHD identify a potential opportunity for therapeutic intervention using inhibitors specific for this integrin.

Separation of deleterious GVHD pathology from beneficial graft-versus-leukemia (GVL) responses following bone marrow transplantation (BMT) remains a major challenge in the treatment of hematologic malignancies by allogeneic hematopoietic cell transplantation (HCT). Liu et al. [62] used αE null/null mice to show that $\alpha E\beta 7$ expression by $CD8^+$ T cells is required for the former but not the latter process, identifying $\alpha E\beta 7$ blockade as an improved strategy for GVHD prophylaxis. Li et al. [61] showed that preconditioning of host mice with anti-CD3 mAb also separates GVHD and GVL effects, and does so by reducing the number of $\alpha E\beta 7$ expressing dendritic cells in the mesenteric lymph nodes.

7.6.3 Tumor Immunology

Le Floc'h et al. [29, 57, 58] reported that αEβ7 expression by $CD8^+$ CTL clones in tumors can be induced by TGF-β expression within the tumor. These studies also showed that αEβ7 can participate in formation of the immunological synapse between the CTL and the tumor target, and that interaction with E-cadherin expressed by the tumor target is required for polarization and subsequent release of cytotoxic granules. Subsequent studies showed that interaction of αEβ7 with E-cadherin, but not αLβ2 with ICAM-1, acts at the level of the immunologic synapse formed between tumor-infiltrating lymphocytes and tumor cells to promote CCR5-dependent retention of CTL [30], that interaction of αEβ7 with E-cadherin promotes the phosphorylation of the ERK1/2 kinases and Phospholipase C-γ1 (PLC-γ1), which is sufficient to induce the polarization of cytolytic granules [57], and that interaction between CTL and epithelial tumor cells is regulated by αE expression at the immune synapse which can profoundly influence effector functions of CD8 T cells [29]. Thus, αEβ7 potentially plays a role in tumor elimination through interaction with E-cadherin. These findings raise the exciting possibility that the characteristic loss of E-cadherin expression and gain in invasiveness by metastatic epithelial tumors exhibited by many neoplastic epithelial cells [97] might, in part, reflect CTL selection. That said, the frequency of tumor-reactive CTL clones that express αEβ7 remains a matter of speculation. Nonetheless, together, these studies provide novel insight onto the role of αEβ7 in CTL function. Also, of relevance to the field of tumor immunotherapy, Trinite et al. [101] reported that immature ($CD4^-$ $\alpha E\beta 7^+$) rat dendritic cells can induce rapid caspase-independent apoptosis-like cell death and subsequent phagocytosis of tumor targets. Both of these sets of findings have spurred interest in the development of novel immunotherapeutic strategies to combat cancer.

7.6.4 T Regulatory Cells

The role of αEβ7 in Treg function is controversial and highly dependent on the model employed. However, there is evidence that αEβ7 plays an important role in promoting both the function and localization of Treg cells, and even that αEβ7 marks Tregs with the most potent immunosuppressive properties. McHugh et al. [68] and Lehmann et al. [60] both reported that that the $\alpha E\beta 7^+$ population was more efficient at suppressing anti-CD3 stimulated proliferation of $CD4^+$ CD25-cells than the αEβ7-subset. TGF-β plays a role in the development and function of Treg cells, and the presence of αEβ7 on the surface of a major subpopulation of Treg cells argues, at least, that these cells have recently been exposed to TGF-β. However, such expression may be misleading and it remains to be determined if a direct role of αEβ7 on Treg is always relevant. As described below, αEβ7 expressing dendritic cell subsets can also control the suppressive function of Tregs.

More recently, Suffia et al. [94] have shown that αEβ7 plays an essential role in retention of Treg and control of Leishmania major infection, and that targeted disruption of the αE gene renders mice susceptible to Leishmania infection, a result that could be reversed by transfer of αEβ7 expressing Tregs from wild type mice. In contrast, the Powrie group has reported that targeted disruption of αE has no effect on the suppressive capacity of Tregs in a mouse model of colitis [3]. Rather, expression of αEβ7 by dendritic cells was found to be necessary for Treg function (see below). Van et al. [106] showed that CD47 controls the in vivo proliferation and homeostasis of the αEβ7 expressing subset of peripheral Tregs. There is also evidence that αEβ7 expressing CD8 T cells can be suppressive. For example, Uss et al. provided evidence that $\alpha E\beta 7^+$ CD8 T cells can be potently immunosuppressive in vitro [103], effectively functioning as T regs.

7.6.5 Dendritic Cells

While the precise function of each dendritic cell (DC) subset remains to be clearly defined, it is clear that expression of $\alpha E\beta 7$ allows DCs to control the balance of effector responses to foreign antigens. Annaker et al. reported that $\alpha E\beta 7$ expression by DCs is required for the induction of Tregs to suppress intestinal bowel disease [3]. In this model, $\alpha E\beta 7^-$ DCs promoted mainly effector cytokine IFN-γ production by the responding T cells whereas $\alpha E\beta 7^+$ DCs enhanced immune protection by inducing the gut homing receptor CCR9 on responding T cells. These data indicated that $\alpha E\beta 7$ can control the balance of effector vs regulatory T cell activity in the intestine. Indeed, Coombes et al. have shown that mucosal $\alpha E\beta 7^+$ DC induce $Foxp3^+$ Treg by a TGF-β and retinoic acid-dependent pathway [20]. Subsequent studies confirmed that retinoic acid is centrally involved in regulating this pathway [53], and that human $\alpha E\beta 7^+$ DC share this ability to induce T reg [108]. Choi et al. [18] reported that DC are dominant in normal aortic intima and, in contrast to macrophages which promote atherosclerosis, the $\alpha E\beta 7^+$ DC subset was associated with protection from atherosclerosis. Weiner et al. reviewed the existing literature on oral tolerance and also concluded that $\alpha E\beta 7^+$ DCs induce T regs [109].

There is also evidence that $\alpha E\beta 7^+$ DC subsets can indirectly *promote* immune responses. For example, $\alpha E\beta 7^+$ DC appear adept at generating gut-tropic effector CD8 T cells [43]. Recent studies provide further insight into the antigen-presenting qualities of $\alpha E\beta 7^+$ dendritic cells. Bedoui et al. [6] reported that $\alpha E\beta 7^+$ DCs in non-lymphoid tissues are specialized in the cross-presentation of cell-associated antigens and are essential for inducing proliferation of CD8 T cells, a finding that appears consistent with recent work in human DC [108].

7.6.6 Innate Immune Responses

McCarty et al. [66] reported that circulating Vδ2 T cells display enhanced gut-homing potential upon microbial activation and populate the human intestinal mucosa, generating functionally distinct $\alpha E\beta 7^+$ and $\alpha E\beta 7^-$ subsets that promoted inflammation by colonic $\alpha\beta$ T cells. Further evidence that $\alpha E\beta 7$ functions in innate immune responses is provided by the findings of Kinnebrew et al. [52] who reported that $\alpha E\beta 7^+$ CD11b $(\alpha M\beta 2)^+$ DCs in the lamina propria, in addition to promoting long-term tolerance to ingested antigens, also rapidly produce IL-23 in response to detection of flagellin in the lamina propria. Flores-Langarica et al. [28] showed that systemic flagellin immunization can induce mucosal immune responses.

7.7 Conclusions

Integrin $\alpha E\beta 7$ has proved to be enigmatic and tantalising. Considerable efforts to define its true significance in vivo have met with mixed fortunes. Whilst this integrin undoubtedly contributes to mucosal specific retention of diverse leukocyte subpopulations there are valid grounds in the future to seek deeper significance in its signaling capacity, especially in relation to cross-talk with the epithelium. Studies of αE knockout mice have clearly identified an important role for this integrin in allograft rejection and also have provided a glimpse of its possible significance in immunoregulation. Resonance with the finding that $\alpha E\beta 7$ is expressed by major leukocyte subsets is striking and the functional relationships between these observations provide fertile ground for further investigation. It is evident, for example, that while $\alpha E\beta 7$ expressing mouse dendritic cells are important, the molecular function of $\alpha E\beta 7$ in this context, and on T_{rm} in the brain, is less clear. The role of $\alpha E\beta 7$ on similar cells in humans also invites further study. TGF-β signaling to both the $\alpha E\beta 7$ expressing leukocyte and its target (if any), and the significance of cell surface-bound TGF-β, merit further attention. In mice, the αE integrin gene locus is sandwiched between the Th2 cytokine gene cluster (IL-4, IL-5 and IL-13) and a cluster of chemokine genes (eotaxin, TCA-3, MCP-1, 3, 5, MIP-1α and 1β, RANTES). In future studies to address the role of $\alpha E\beta 7$ in immunoregulation it will be essential to utilize αE null/null and control mice that are congenic at the Th2 and chemokine loci, and to use conditional knockout mice with disruption of the gene targeted to specific leukocyte subsets.

References

1. Allez M, Brimnes J, Dotan I, Mayer L (2002) Expansion of CD8+ T cells with regulatory function after interaction with intestinal epithelial cells. Gastroenterology 123:1516–1526
2. Andrew DP, Rott LS, Kilshaw PJ, Butcher EC (1996) Distribution of alpha 4 beta 7 and alpha E beta 7 integrins on thymocytes, intestinal epithelial lymphocytes and peripheral lymphocytes. Eur J Immunol 26:897–905
3. Annacker O, Coombes JL, Malmstrom V, Uhlig HH, Bourne T, Johansson-Lindbom B et al (2005) Essential role for CD103 in the T cell-mediated regulation of experimental colitis. J Exp Med 202:1051–1061
4. Annunziato F, Cosmi L, Liotta F, Lazzeri E, Romagnani P, Angeli R et al (2006) CXCR3 and alpha E beta 7 integrin identify a subset of CD8+ mature thymocytes that share phenotypic and functional properties with CD8+ gut intraepithelial lymphocytes. Gut 55:961–968
5. Baker E, Sutherland GR, Jiang WM, Yuan Q, Leung E, Watson JD et al (1992) Mapping of the human integrin beta 7 gene (ITG beta 7) to 12q13.13 by non-isotopic in situ hybridization. Mamm Genome 2:272–273
6. Bedoui S, Whitney PG, Waithman J, Eidsmo L, Wakim L, Caminschi I et al (2009) Cross-presentation of viral and self antigens by skin-derived CD103+ dendritic cells. Nat Immunol 10:488–495
7. Brenan M, Puklavec M (1992) The MRC OX-62 antigen: a useful marker in the purification of rat veiled cells with the biochemical properties of an integrin. J Exp Med 175:1457–1465
8. Brenan M, Rees DJ (1997) Sequence analysis of rat integrin alpha E1 and alpha E2 subunits: tissue expression reveals phenotypic similarities between intraepithelial lymphocytes and dendritic cells in lymph. Eur J Immunol 27:3070–3079
9. Brew R, West DC, Burthem J, Christmas SE (1995) Expression of the human mucosal lymphocyte antigen, HML-1, by T cells activated with mitogen or specific antigen in vitro. Scand J Immunol 41:553–562
10. Brown DW, Furness J, Speight PM, Thomas GJ, Li J, Thornhill MH et al (1999) Mechanisms of binding of cutaneous lymphocyte-associated antigen-positive and alphaebeta7-positive lymphocytes to oral and skin keratinocytes. Immunology 98:9–15
11. Butcher EC, Williams M, Youngman K, Rott L, Briskin M (1999) Lymphocyte trafficking and regional immunity. Adv Immunol 72:209–253
12. Cepek KL, Parker CM, Madara JL, Brenner MB (1993) Integrin alpha E beta 7 mediates adhesion of T lymphocytes to epithelial cells. J Immunol 150:3459–3470
13. Cepek KL, Shaw SK, Parker CM, Russell GJ, Morrow JS, Rimm DL et al (1994) Adhesion between epithelial cells and T lymphocytes mediated by E-cadherin and the alpha E beta 7 integrin. Nature 372:190–193
14. Cerf-Bensussan N, Begue B, Gagnon J, Meo T (1992) The human intraepithelial lymphocyte marker HML-1 is an integrin consisting of a beta 7 subunit associated with a distinctive alpha chain. Eur J Immunol 22:885
15. Cerf-Bensussan N, Guy-Grand D, Lisowska-Grospierre B, Griscelli C, Bhan AK (1986) A monoclonal antibody specific for rat intestinal lymphocytes. J Immunol 136:76–82
16. Cerf-Bensussan N, Jarry A, Brousse N, Lisowska-Grospierre B, Guy-Grand D, Griscelli C (1987) A monoclonal antibody (HML-1) defining a novel membrane molecule present on human intestinal lymphocytes. Eur J Immunol 17:1279–1285
17. Chatenoud L, Salomon B, Bluestone JA (2001) Suppressor T cells–they're back and critical for regulation of autoimmunity! Immunol Rev 182:149–163
18. Choi JH, Cheong C, Dandamudi DB, Park CG, Rodriguez A, Mehandru S et al (2011) Flt3 signaling-dependent dendritic cells protect against atherosclerosis. Immunity 35:819–831
19. Collazo MM, Wood D, Paraiso KH, Lund E, Engelman RW, Le CT et al (2009) SHIP limits immunoregulatory capacity in the T-cell compartment. Blood 113:2934–2944
20. Coombes JL, Siddiqui KR, Arancibia-Carcamo CV, Hall J, Sun CM, Belkaid Y et al (2007) A functionally specialized population of mucosal CD103+ DCs induces Foxp3+ regulatory T cells via a TGF-beta and retinoic acid-dependent mechanism. J Exp Med 204:1757–1764
21. Corbi AL, Garcia-Aguilar J, Springer TA (1990) Genomic structure of an integrin alpha subunit, the leukocyte p150,95 molecule. J Biol Chem 265:2782–2788
22. Cresswell J, Robertson H, Neal DE, Griffiths TR, Kirby JA (2001) Distribution of lymphocytes of the alpha(E)beta(7) phenotype and E-cadherin in normal human urothelium and bladder carcinomas. Clin Exp Immunol 126:397–402
23. Csencsits KL, Walters N, Pascual DW (2001) Cutting edge: dichotomy of homing receptor dependence by mucosal effector B cells: alpha(E) versus L-selectin. J Immunol 167:2441–2445
24. El-Asady R, Yuan R, Liu K, Wang D, Gress RE, Lucas PJ et al (2005) TGF-{beta}-dependent CD103 expression by CD8(+) T cells promotes

selective destruction of the host intestinal epithelium during graft-versus-host disease. J Exp Med 201:1647–1657
25. Farstad IN, Halstensen TS, Lien B, Kilshaw PJ, Lazarovits AI, Brandtzaeg P (1996) Distribution of beta 7 integrins in human intestinal mucosa and organized gut-associated lymphoid tissue. Immunology 89:227–237
26. Feng Y, Wang D, Yuan R, Parker CM, Farber DL, Hadley GA (2002) CD103 expression is required for destruction of pancreatic islet allografts by CD8(+) T cells. J Exp Med 196:877–886
27. Fleming JC, Pahl HL, Gonzalez DA, Smith TF, Tenen DG (1993) Structural analysis of the CD11b gene and phylogenetic analysis of the alpha-integrin gene family demonstrate remarkable conservation of genomic organization and suggest early diversification during evolution. J Immunol 150:480–490
28. Flores-Langarica A, Marshall JL, Hitchcock J, Cook C, Jobanputra J, Bobat S et al (2012) Systemic flagellin immunization stimulates mucosal CD103+ dendritic cells and drives Foxp3+ regulatory T cell and IgA responses in the mesenteric lymph node. J Immunol 189:5745–5754
29. Franciszkiewicz K, Le Floc'h A, Boutet M, Vergnon I, Schmitt A, Mami-Chouaib F (2013) CD103 or LFA-1 engagement at the immune synapse between cytotoxic T cells and tumor cells promotes maturation and regulates T-cell effector functions. Cancer Res 73:617–628
30. Franciszkiewicz K, Le Floc'h A, Jalil A, Vigant F, Robert T, Vergnon I et al (2009) Intratumoral induction of CD103 triggers tumor-specific CTL function and CCR5-dependent T-cell retention. Cancer Res 69:6249–6255
31. Gabor MJ, Sedgwick JD, Lemckert FA, Godfrey DI, Korner H (2001) Lymphotoxin controls alphaEbeta7-integrin expression by peripheral CD8+ T cells. Immunol Cell Biol 79:323–331
32. Gavin MA, Clarke SR, Negrou E, Gallegos A, Rudensky A (2002) Homeostasis and anergy of CD4(+)CD25(+) suppressor T cells in vivo. Nat Immunol 3:33–41
33. Grueter B, Petter M, Egawa T, Laule-Kilian K, Aldrian CJ, Wuerch A et al (2005) Runx3 regulates integrin alpha E/CD103 and CD4 expression during development of CD4-/CD8+ T cells. J Immunol 175:1694–1705
34. Hadley GA, Bartlett ST, Via CS, Rostapshova EA, Moainie S (1997) The epithelial cell-specific integrin, CD103 (alpha E integrin), defines a novel subset of alloreactive CD8 + CTL. J Immunol 159:3748–3756
35. Hadley GA, Charandee C, Weir MR, Wang D, Bartlett ST, Drachenberg CB (2001) CD103+ CTL accumulate within the graft epithelium during clinical renal allograft rejection. Transplantation 72:1548–1555
36. Hadley GA, Rostapshova EA, Gomolka DM, Taylor BM, Bartlett ST, Drachenberg CI et al (1999) Regulation of the epithelial cell-specific integrin, CD103, by human CD8+ cytolytic T lymphocytes. Transplantation 67:1418–1425
37. Higgins JM (2001) The Haspin gene: location in an intron of the integrin alphaE gene, associated transcription of an integrin alphaE-derived RNA and expression in diploid as well as haploid cells. Gene 267:55–69
38. Higgins JM, Cernadas M, Tan K, Irie A, Wang J, Takada Y et al (2000) The role of alpha and beta chains in ligand recognition by beta 7 integrins. J Biol Chem 275:25652–25664
39. Higgins JM, Mandlebrot DA, Shaw SK, Russell GJ, Murphy EA, Chen YT et al (1998) Direct and regulated interaction of integrin alphaEbeta7 with E-cadherin. J Cell Biol 140:197–210
40. Hoffkes HG, Schmidtke G, Uppenkamp M, Schmucker U (1996) Multiparametric immunophenotyping of B cells in peripheral blood of healthy adults by flow cytometry. Clin Diagn Lab Immunol 3:30–36
41. Jenkinson SE, Whawell SA, Swales BM, Corps EM, Kilshaw PJ, Farthing PM (2011) The alphaE(CD103)beta7 integrin interacts with oral and skin keratinocytes in an E-cadherin-independent manner*. Immunology 132:188–196
42. Jiang WM, Jenkins D, Yuan Q, Leung E, Choo KH, Watson JD et al (1992) The gene organization of the human beta 7 subunit, the common beta subunit of the leukocyte integrins HML-1 and LPAM-1. Int Immunol 4:1031–1040
43. Johansson-Lindbom B, Svensson M, Pabst O, Palmqvist C, Marquez G, Forster R et al (2005) Functional specialization of gut CD103+ dendritic cells in the regulation of tissue-selective T cell homing. J Exp Med 202:1063–1073
44. Karecla PI, Bowden SJ, Green SJ, Kilshaw PJ (1995) Recognition of E-cadherin on epithelial cells by the mucosal T cell integrin alpha M290 beta 7 (alpha E beta 7). Eur J Immunol 25:852–856
45. Karecla PI, Green SJ, Bowden SJ, Coadwell J, Kilshaw PJ (1996) Identification of a binding site for integrin alphaEbeta7 in the N-terminal domain of E-cadherin. J Biol Chem 271:30909–30915
46. Kilshaw PJ (1993) Expression of the mucosal T cell integrin alpha M290 beta 7 by a major subpopulation of dendritic cells in mice. Eur J Immunol 23:3365–3368
47. Kilshaw PJ, Baker KC (1988) A unique surface antigen on intraepithelial lymphocytes in the mouse. Immunol Lett 18:149–154
48. Kilshaw PJ, Higgins JM (2002) Alpha E: no more rejection? J Exp Med 196:873–875
49. Kilshaw PJ, Murant SJ (1991) Expression and regulation of beta 7(beta p) integrins on mouse lymphocytes: relevance to the mucosal immune system. Eur J Immunol 21:2591–2597

50. Kilshaw PJ, Murant SJ (1990) A new surface antigen on intraepithelial lymphocytes in the intestine. Eur J Immunol 20:2201–2207
51. Kim SK, Schluns KS, Lefrancois L (1999) Induction and visualization of mucosal memory CD8 T cells following systemic virus infection. J Immunol 163:4125–4132
52. Kinnebrew MA, Buffie CG, Diehl GE, Zenewicz LA, Leiner I, Hohl TM et al (2012) Interleukin 23 production by intestinal CD103(+)CD11b(+) dendritic cells in response to bacterial flagellin enhances mucosal innate immune defense. Immunity 36:276–287
53. Klebanoff CA, Spencer SP, Torabi-Parizi P, Grainger JR, Roychoudhuri R, Ji Y et al (2013) Retinoic acid controls the homeostasis of pre-cDC-derived splenic and intestinal dendritic cells. J Exp Med 210:1961–1976
54. Krissansen GW, Print CG, Prestidge RL, Hollander D, Yuan Q, Jiang WM et al (1992) Immunologic and structural relatedness of the integrin beta 7 complex and the human intraepithelial lymphocyte antigen HML-1. FEBS Lett 296:25–28
55. Kunisawa J, Kurashima Y, Higuchi M, Gohda M, Ishikawa I, Ogahara I et al (2007) Sphingosine 1-phosphate dependence in the regulation of lymphocyte trafficking to the gut epithelium. J Exp Med 204:2335–2348
56. Kutlesa S, Wessels JT, Speiser A, Steiert I, Muller CA, Klein G (2002) E-cadherin-mediated interactions of thymic epithelial cells with CD103 + thymocytes lead to enhanced thymocyte cell proliferation. J Cell Sci 115:4505–4515
57. Le Floc'h A, Jalil A, Franciszkiewicz K, Validire P, Vergnon I, Mami-Chouaib F (2011) Minimal engagement of CD103 on cytotoxic T lymphocytes with an E-cadherin-Fc molecule triggers lytic granule polarization via a phospholipase Cgamma-dependent pathway. Cancer Res 71:328–338
58. Le Floc'h A, Jalil A, Vergnon I, Le Maux Chansac B, Lazar V, Bismuth G et al (2007) Alpha E beta 7 integrin interaction with E-cadherin promotes antitumor CTL activity by triggering lytic granule polarization and exocytosis. J Exp Med 204:559–570
59. Lefrancois L, Barrett TA, Havran WL, Puddington L (1994) Developmental expression of the alpha IEL beta 7 integrin on T cell receptor gamma delta and T cell receptor alpha beta T cells. Eur J Immunol 24:635–640
60. Lehmann J, Huehn J, de la Rosa M, Maszyna F, Kretschmer U, Krenn V et al (2002) Expression of the integrin alpha Ebeta 7 identifies unique subsets of CD25+ as well as CD25-regulatory T cells. Proc Natl Acad Sci USA 99:13031–13036
61. Li N, Chen Y, He W, Yi T, Zhao D, Zhang C et al (2009) Anti-CD3 preconditioning separates GVL from GVHD via modulating host dendritic cell and donor T-cell migration in recipients conditioned with TBI. Blood 113:953–962
62. Liu K, Anthony BA, Yearsly MM, Hamadani M, Gaughan A, Wang JJ et al (2011) CD103 deficiency prevents graft-versus-host disease but spares graft-versus-tumor effects mediated by alloreactive CD8 T cells. PLoS ONE 6:e21968
63. Mackay LK, Stock AT, Ma JZ, Jones CM, Kent SJ, Mueller SN et al (2012) Long-lived epithelial immunity by tissue-resident memory T (TRM) cells in the absence of persisting local antigen presentation. Proc Natl Acad Sci USA 109:7037–7042
64. Maloy KJ, Powrie F (2001) Regulatory T cells in the control of immune pathology. Nat Immunol 2:816–822
65. Maric I, Holt PG, Perdue MH, Bienenstock J (1996) Class II MHC antigen (Ia)-bearing dendritic cells in the epithelium of the rat intestine. J Immunol 156:1408–1414
66. McCarthy NE, Bashir Z, Vossenkamper A, Hedin CR, Giles EM, Bhattacharjee S et al (2013) Proinflammatory Vdelta2+ T cells populate the human intestinal mucosa and enhance IFN-gamma production by colonic alphabeta T cells. J Immunol 191:2752–2763
67. McFarland RD, Douek DC, Koup RA, Picker LJ (2000) Identification of a human recent thymic emigrant phenotype. Proc Natl Acad Sci USA 97:4215–4220
68. McHugh RS, Whitters MJ, Piccirillo CA, Young DA, Shevach EM, Collins M et al (2002) CD4(+)CD25(+) immunoregulatory T cells: gene expression analysis reveals a functional role for the glucocorticoid-induced TNF receptor. Immunity 16:311–323
69. McLellan AD, Kapp M, Eggert A, Linden C, Bommhardt U, Brocker EB et al (2002) Anatomic location and T-cell stimulatory functions of mouse dendritic cell subsets defined by CD4 and CD8 expression. Blood 99:2084–2093
70. Micklem KJ, Dong Y, Willis A, Pulford KA, Visser L, Durkop H et al (1991) HML-1 antigen on mucosa-associated T cells, activated cells, and hairy leukemic cells is a new integrin containing the beta 7 subunit. Am J Pathol 139:1297–1301
71. Moldenhauer G, Mielke B, Dorken B, Schwartz-Albiez R, Moller P (1990) Identity of HML-1 antigen on intestinal intraepithelial T cells and of B-ly7 antigen on hairy cell leukaemia. Scand J Immunol 32:77–82
72. Moller P, Mielke B, Moldenhauer G (1990) Monoclonal antibody HML-1, a marker for intraepithelial T cells and lymphomas derived thereof, also recognizes hairy cell leukemia and some B-cell lymphomas. Am J Pathol 136:509–512
73. Nelson DJ, McMenamin C, McWilliam AS, Brenan M, Holt PG (1994) Development of the airway intraepithelial dendritic cell network in the rat from

class II major histocompatibility (Ia)-negative precursors: differential regulation of Ia expression at different levels of the respiratory tract. J Exp Med 179:203–212
74. Noti JD, Gordon M, Hall RE (1992) Human p150,95 alpha-subunit: genomic organization and analysis of the 5'-flanking region. DNA Cell Biol 11:123–138
75. Nunes-Cabaco H, Caramalho I, Sepulveda N, Sousa AE (2011) Differentiation of human thymic regulatory T cells at the double positive stage. Eur J Immunol 41:3604–3614
76. Parker CM, Cepek KL, Russell GJ, Shaw SK, Posnett DN, Schwarting R et al (1992) A family of beta 7 integrins on human mucosal lymphocytes. Proc Natl Acad Sci USA 89:1924–1928
77. Quayle AJ, Pudney J, Munoz DE, Anderson DJ (1994) Characterization of T lymphocytes and antigen-presenting cells in the murine male urethra. Biol Reprod 51:809–820
78. Ribi E, Granger DL, Milner KC, Yamamoto K, Strain SM, Parker R et al (1982) Induction of resistance to tuberculosis in mice with defined components of mycobacteria and with some unrelated materials. Immunology 46:297–305
79. Riedl E, Stockl J, Majdic O, Scheinecker C, Rappersberger K, Knapp W et al (2000) Functional involvement of E-cadherin in TGF-beta 1-induced cell cluster formation of in vitro developing human Langerhans-type dendritic cells. J Immunol 165:1381–1386
80. Rihs S, Walker C, Virchow JC Jr, Boer C, Kroegel C, Giri SN et al (1996) Differential expression of alpha E beta 7 integrins on bronchoalveolar lavage T lymphocyte subsets: regulation by alpha 4 beta 1-integrin crosslinking and TGF-beta. Am J Respir Cell Mol Biol 15:600–610
81. Roberts AI, O'Connell SM, Ebert EC (1993) Intestinal intraepithelial lymphocytes bind to colon cancer cells by HML-1 and CD11a. Cancer Res 53:1608–1611
82. Roberts K, Kilshaw PJ (1993) The mucosal T cell integrin alpha M290 beta 7 recognizes a ligand on mucosal epithelial cell lines. Eur J Immunol 23:1630–1635
83. Robertson H, Wong WK, Burt AD, Mohamed MA, Talbot D, Kirby JA (2001) Relationship between TGFbeta(1), intratubular CD103 positive T cells and acute renal allograft rejection. Transplant Proc 33:1159
84. Robertson H, Wong WK, Talbot D, Burt AD, Kirby JA (2001) Tubulitis after renal transplantation: demonstration of an association between CD103+ T cells, transforming growth factor beta1 expression and rejection grade. Transplantation 71:306–313
85. Robinson PW, Green SJ, Carter C, Coadwell J, Kilshaw PJ (2001) Studies on transcriptional regulation of the mucosal T-cell integrin alphaEbeta7 (CD103). Immunology 103:146–154
86. Rostapshova EA, Burns JM, Bartlett ST, Hadley GA (1998) Integrin-mediated interactions influence the tissue specificity of CD8+ cytolytic T lymphocytes. Eur J Immunol 28:3031–3039
87. Schieferdecker HL, Ullrich R, Weiss-Breckwoldt AN, Schwarting R, Stein H, Riecken EO et al (1990) The HML-1 antigen of intestinal lymphocytes is an activation antigen. J Immunol 144:2541–2549
88. Schon MP, Arya A, Murphy EA, Adams CM, Strauch UG, Agace WW et al (1999) Mucosal T lymphocyte numbers are selectively reduced in integrin alpha E (CD103)-deficient mice. J Immunol 162:6641–6649
89. Shaw SK, Cepek KL, Murphy EA, Russell GJ, Brenner MB, Parker CM (1994) Molecular cloning of the human mucosal lymphocyte integrin alpha E subunit. Unusual structure and restricted RNA distribution. J Biol Chem 269:6016–6025
90. Shibahara T, Si-Tahar M, Shaw SK, Madara JL (2000) Adhesion molecules expressed on homing lymphocytes in model intestinal epithelia. Gastroenterology 118:289–298
91. Siddiqui KR, Laffont S, Powrie F (2010) E-cadherin marks a subset of inflammatory dendritic cells that promote T cell-mediated colitis. Immunity 32:557–567
92. Smith TJ, Ducharme LA, Shaw SK, Parker CM, Brenner MB, Kilshaw PJ et al (1994) Murine M290 integrin expression modulated by mast cell activation. Immunity 1:393–403
93. Strauch UG, Mueller RC, Li XY, Cernadas M, Higgins JM, Binion DG et al (2001) Integrin alpha E(CD103)beta 7 mediates adhesion to intestinal microvascular endothelial cell lines via an E-cadherin-independent interaction. J Immunol 166:3506–3514
94. Suffia I, Reckling SK, Salay G, Belkaid Y (2005) A role for CD103 in the retention of CD4+ CD25+ Treg and control of Leishmania major infection. J Immunol 174:5444–5455
95. Suvas PK, Dech HM, Sambira F, Zeng J, Onami TM (2007) Systemic and mucosal infection program protective memory CD8 T cells in the vaginal mucosa. J Immunol 179:8122–8127
96. Suzuki R, Nakao A, Kanamaru Y, Okumura K, Ogawa H, Ra C (2002) Localization of intestinal intraepithelial T lymphocytes involves regulation of alphaEbeta7 expression by transforming growth factor-beta. Int Immunol 14:339–345
97. Takeichi M (1993) Cadherins in cancer: implications for invasion and metastasis. Curr Opin Cell Biol 5:806–811
98. Taraszka KS, Higgins JM, Tan K, Mandelbrot DA, Wang JH, Brenner MB (2000) Molecular basis for leukocyte integrin alpha(E)beta(7) adhesion to epithelial (E)-cadherin. J Exp Med 191:1555–1567
99. Teraki Y, Shiohara T (2002) Preferential expression of alphaEbeta7 integrin (CD103) on CD8+ T cells in the psoriatic epidermis: regulation by

interleukins 4 and 12 and transforming growth factor-beta. Br J Dermatol 147:1118–1126
100. Tiisala S, Paavonen T, Renkonen R (1995) Alpha E beta 7 and alpha 4 beta 7 integrins associated with intraepithelial and mucosal homing, are expressed on macrophages. Eur J Immunol 25:411–417
101. Trinite B, Chauvin C, Peche H, Voisine C, Heslan M, Josien R (2005) Immature CD4-CD103+ rat dendritic cells induce rapid caspase-independent apoptosis-like cell death in various tumor and nontumor cells and phagocytose their victims. J Immunol 175:2408–2417
102. Uchida Y, Kawai K, Ibusuki A, Kanekura T (2011) Role for E-cadherin as an inhibitory receptor on epidermal gammadelta T cells. J Immunol 186:6945–6954
103. Uss E, Rowshani AT, Hooibrink B, Lardy NM, van Lier RA, ten Berge IJ (2006) CD103 is a marker for alloantigen-induced regulatory CD8+ T cells. J Immunol 177:2775–2783
104. Van den Bossche J, Bogaert P, van Hengel J, Guerin CJ, Berx G, Movahedi K et al (2009) Alternatively activated macrophages engage in homotypic and heterotypic interactions through IL-4 and polyamine-induced E-cadherin/catenin complexes. Blood 114:4664–4674
105. Van den Bossche J, Malissen B, Mantovani A, De Baetselier P, Van Ginderachter JA (2012) Regulation and function of the E-cadherin/catenin complex in cells of the monocyte-macrophage lineage and DCs. Blood 119:1623–1633
106. Van VQ, Darwiche J, Raymond M, Lesage S, Bouguermouh S, Rubio M et al (2008) Cutting edge: CD47 controls the in vivo proliferation and homeostasis of peripheral CD4 + CD25 + Foxp3 + regulatory T cells that express CD103. J Immunol 181:5204–5208
107. Wakim LM, Woodward-Davis A, Bevan MJ (2010) Memory T cells persisting within the brain after local infection show functional adaptations to their tissue of residence. Proc Natl Acad Sci USA 107:17872–17879
108. Watchmaker PB, Lahl K, Lee M, Baumjohann D, Morton J, Kim SJ et al (2014) Comparative transcriptional and functional profiling defines conserved programs of intestinal DC differentiation in humans and mice. Nat Immunol
109. Weiner HL, da Cunha AP, Quintana F, Wu H (2011) Oral tolerance. Immunol Rev 241:241–259
110. Wong DA, Davis EM, LeBeau M, Springer TA (1996) Cloning and chromosomal localization of a novel gene-encoding a human beta 2-integrin alpha subunit. Gene 171:291–294
111. Wright SH, Brown J, Knight PA, Thornton EM, Kilshaw PJ, Miller HR (2002) Transforming growth factor-beta1 mediates coexpression of the integrin subunit alphaE and the chymase mouse mast cell protease-1 during the early differentiation of bone marrow-derived mucosal mast cell homologues. Clin Exp Allergy 32:315–324
112. Yarmus M, Woolf E, Bernstein Y, Fainaru O, Negreanu V, Levanon D et al (2006) Groucho/transducin-like Enhancer-of-split (TLE)-dependent and -independent transcriptional regulation by Runx3. Proc Natl Acad Sci USA 103:7384–7389
113. Yu CI, Becker C, Wang Y, Marches F, Helft J, Leboeuf M et al (2013) Human CD1c+ dendritic cells drive the differentiation of CD103 + CD8 + mucosal effector T cells via the cytokine TGF-beta. Immunity 38:818–830
114. Yuan J, Zhang G, Yang X, Liu K, Wang F (2013) Transplantation of allograft transforming growth factor-beta1 transfected CD103(+) lamina propria dendritic cells could effectively induce antigen-specific regulatory T cells in vivo. Transplant Proc 45:3408–3413
115. Yuan Q, Jiang WM, Hollander D, Leung E, Watson JD, Krissansen GW (1991) Identity between the novel integrin beta 7 subunit and an antigen found highly expressed on intraepithelial lymphocytes in the small intestine. Biochem Biophys Res Commun 176:1443–1449
116. Yuan Q, Kozak CA, Jiang WM, Hollander D, Watson JD, Krissansen GW (1992) Genetic mapping of the gene coding for the integrin beta 7 subunit to the distal part of mouse chromosome 15. Immunogenetics 35:403–407
117. Zelenika D, Adams E, Humm S, Graca L, Thompson S, Cobbold SP et al (2002) Regulatory T cells overexpress a subset of Th2 gene transcripts. J Immunol 168:1069–1079
118. Zhou S, Ueta H, Xu XD, Shi C, Matsuno K (2008) Predominant donor CD103 + CD8 + T cell infiltration into the gut epithelium during acute GvHD: a role of gut lymph nodes. Int Immunol 20:385–394

Structural Aspects of Integrins

8

Robert C. Liddington

Abstract

Structural studies on integrins have recently made great strides in recent years. Crystal structures of the complete extracellular fragments of three integrins in open and closed conformations, 6 α-I domains in complex with ligands, and at least 20 intracellular proteins in complex with cytosolic tails have been obtained; and several transmembrane and cytosolic complexes have been determined by NMR. High resolution EM studies complement these atomic resolution techniques by studying the integrin in different activation states. Although we still have only a few experimental examples among integrin family members, the high level of sequence homology between integrins means that reliable models can be built for the other members of the integrin family. These structures make sense of a lot of preceding biochemical, biophysical and mutagenesis studies, and generate many new testable hypotheses of integrin function. This chapter emphasizes new structural insights applicable to all integrins, with an emphasis on those integrins that contain an α-I domain. The structural data reinforce the notion of the integrin as a molecule in dynamic equilibrium at the cell surface, regulated by binding both to extracellular and intracellular ligands.

Keywords

Integrins · Structure · Mechanism · Allostery

8.1 Overall Structure

Integrins are $\alpha\beta$ heterodimers, consisting of a head domain from which emerge two legs, one from each subunit, ending in a pair of single-pass transmembrane helices and short cytoplasmic tails,

R. C. Liddington (✉)
Sanford-Burnham Medical Research Institute, 10901 North Torrey Pines Road, La Jolla, CA 92037, USA
e-mail: rlidding@sanfordburnham.org

D. Gullberg (ed.), *I Domain Integrins*, Advances in Experimental Medicine and Biology 819,
DOI: 10.1007/978-94-017-9153-3_8,

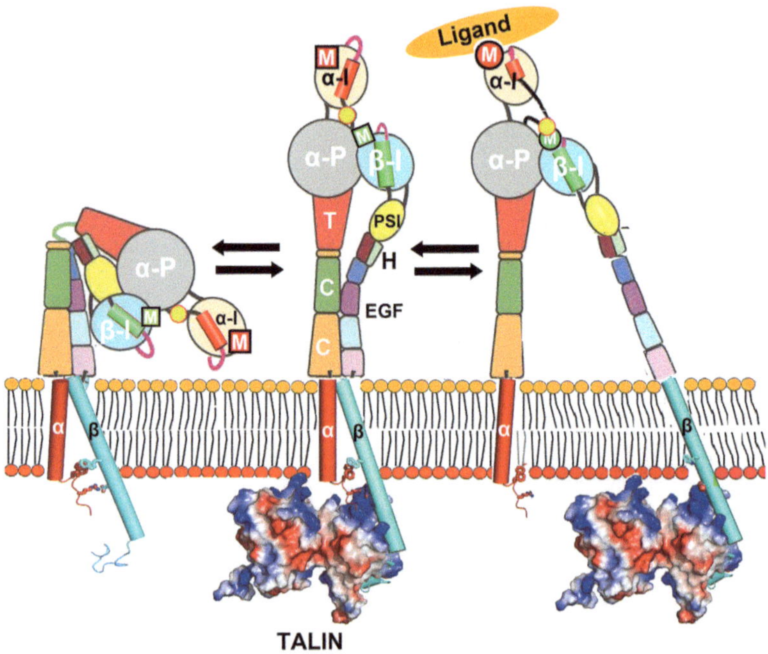

Fig. 8.1 Cartoon of the $\alpha X\beta 2$ structure derived by crystallography and EM studies. At *left*, the bent, low affinity integrin stabilized by bonds between the head, legs and cytoplasmic tails. At *center*, an unknown trigger causes the integrin to "stand up", while maintaining most of its low affinity bonds. At *right*, binding of activated talin and/or binding of an extracellular ligand, trigger an open, high affinity form of the integrin, with TM helices separated

except for $\alpha 6\beta 4$ (Fig. 8.1). The integrin "head" comprises a seven-bladed propeller from the α-subunit that makes an intimate contact with the β-I domain. Nine α-subunits (α1, α2, α10, α11, αD, αE, αL, αM and αX) contain an additional domain, the α-I domain, that is inserted between two loops on the upper surface of the propeller, where it plays a central role in ligand binding [27, 41, 58]. The α-I domain contains an invariant ligand binding site called MIDAS, for Metal Ion-Dependent Adhesion Site [34], in which a metal ion is coordinated by three loops from the I domain, and a glutamic or aspartic acid from the ligand completes an octahedral coordination sphere around the metal. In those integrins that lack an α-I domain, the β-I domain and propeller form the major ligand recognition sites; in the α-I domain integrins, the β-I domain plays a regulatory role.

In the absence of ligand, bonds between the legs, tails and head are believed to hold the head in an "inactive" or "resting" conformation that has low affinity for ligand [20, 56, 61]. Recent structural data suggest that integrins possess three global conformations (see Fig. 8.1): a bent conformation in which the head adopts a "closed", low affinity conformation and the cytoplasmic tails form an inhibitory complex; an extended conformation of the head that retains its low ligand affinity; and a high affinity form in which the legs and tails separate, and the "hybrid" domain, which is part of the head, swings away from the β-I domain, propeller and α-I domain, promoting conformational changes that create high affinity binding sites on both the head and tail [54]. During "outside-in" signaling, the head binds to ECM proteins or counter-receptors on other cells, triggering conformational changes that propagate down the "legs" and through the plasma membrane, leading to a reorganization of the C-terminal tails that allows them to bind intracellular proteins [46]. During "inside-out" signaling, cytosolic proteins bind and sequester one or both of the cytoplasmic tails, triggering conformational changes in the head that promote a high affinity "active" integrin [19, 60], in which the integrin "stands up".

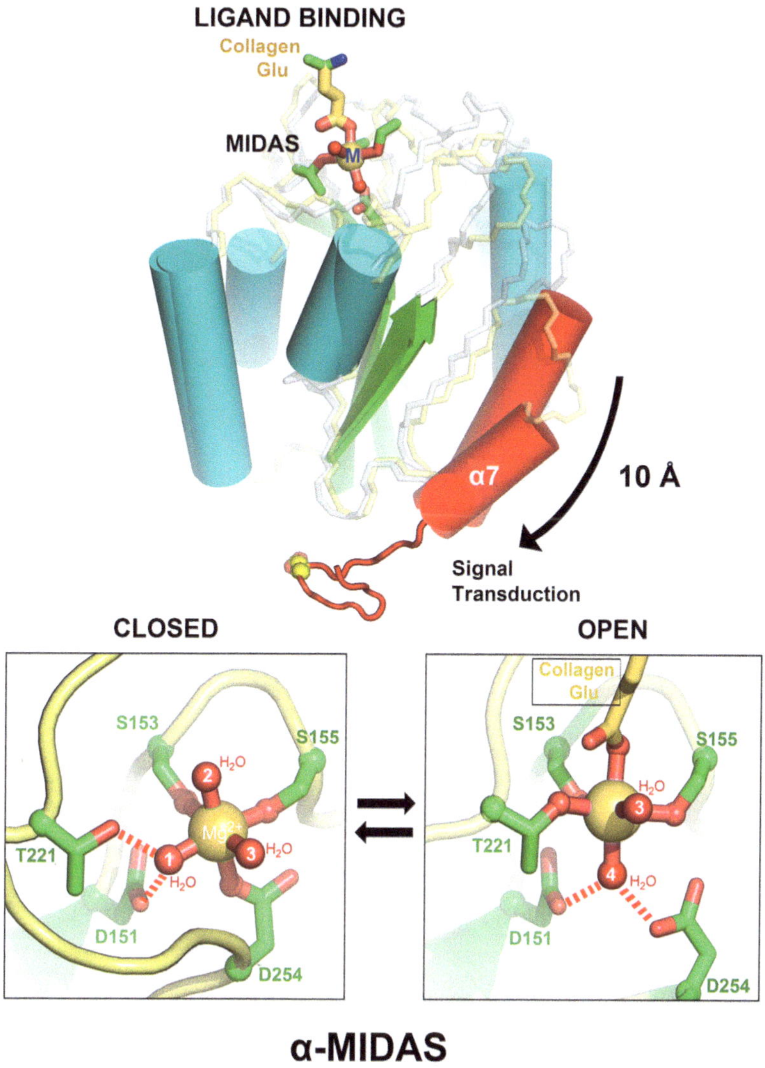

Fig. 8.2 Two conformations of the α-I domain. *Upper panel* conformational changes in the α2-I domain on binding collagen. *Lower panel* Conformational changes in the α-MIDAS motif upon binding ligand. At *left*, the "closed" conformation observed in the absence of ligand; at *right*, the "open" conformation seen when ligand is bound. These changes are mechanically linked to the tertiary changes in the domain. It is very likely that all α-I domains undergo the same switch

8.2 The α-I Domain

The first crystal structure of an α-I domain revealed a compact domain comprising a central mostly parallel β-sheet surrounded on both sides by amphipathic α-helices [34] (Fig. 8.2). Subsequent crystal structures of recombinant αL, α1 and α2 I domains display the same three-dimensional fold, as expected given their reasonable sequence similarity [15, 43, 45]. The MIDAS motif lies at the C-terminal end of the central β-sheet, with three loops contributing sidechains that coordinate the metal ion (Fig. 8.2 Lower panel). The metal-coordinating MIDAS residues are invariant among α-I domains, and mutagenesis of any of these residues abrogates ligand binding. Surface-exposed sidechains surrounding the MIDAS motif are more variable; they provide additional ligand contact points and hence ligand specificity [26, 41, 52].

The structures of 6 ligand-bound α-I domains have now been determined. The first was the α2-I domain bound to a collagen-like triple helix [17]. More recently, the structures of the αL-I domain in complex with homologous fragments

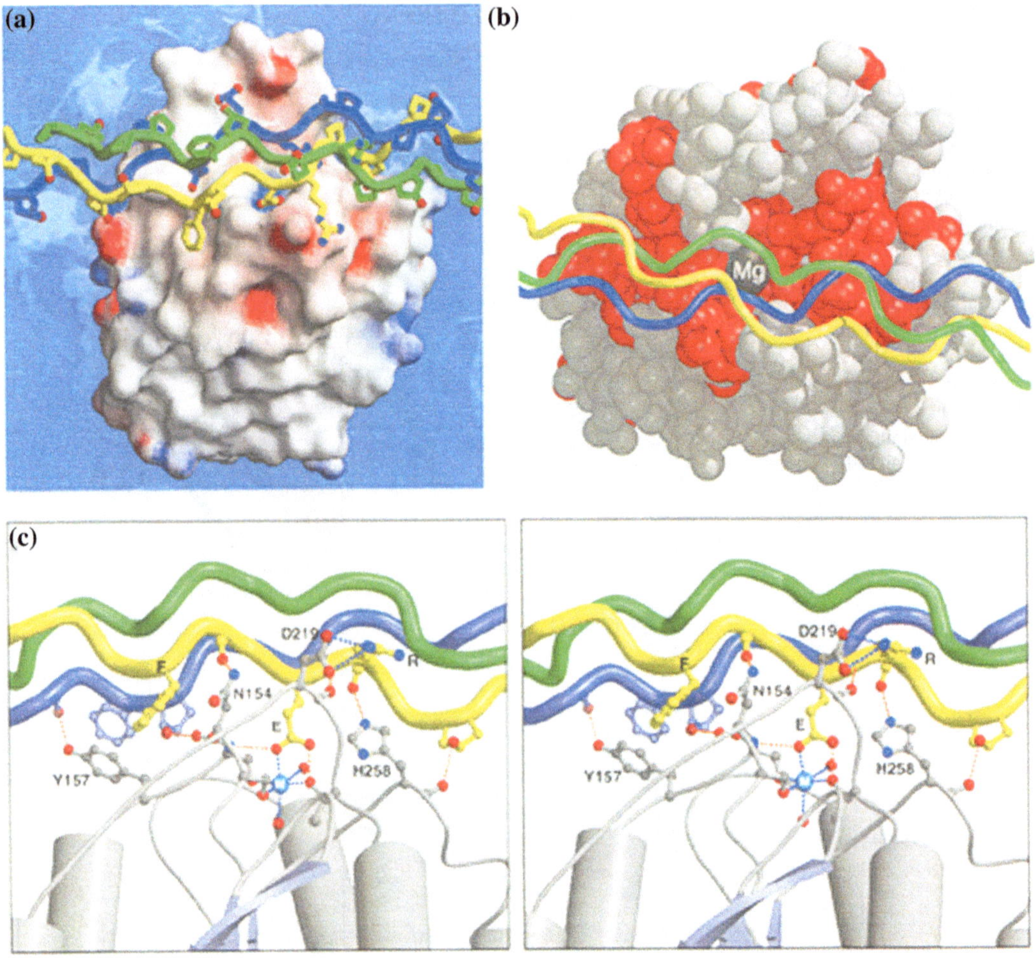

Fig. 8.3 Collagen binding to the α2-I domain. **a** Surface model of the α2-I domain colored by surface charge (*red* = negative, *blue* = positive) with a triple helical fragment of collagen bound. **b** Space-filling model of the complex (rotated about a horizontal axis compared with **a**), showing residues (in *red*) that are invariant in the collagen-binding integrins, α1β1, α2β1 and α10β1. The strong conservation of the binding surface suggests that these integrins will engage collagen in the same fashion. **c** Stereo close-up image of the α2-I:collagen complex

of ICAM-1 [51], ICAM-3 [53] and ICAM-5 [67] have been determined. Recently, the first authentic complex of an αM-I domain bound to ligand (the C3d domain of complement C3) [2] validates the earlier structure of the αM I domain bound to a "ligand-mimetic" crystal contact [33, 34]. They all demonstrate that ligand binding triggers a profound conformational switch in the α-I domain that underlies affinity regulation and signal transduction. The conformational switch is essentially identical in all these examples, strongly suggesting that all α-I domains will undergo the same switch.

8.3 Structural Determinants of Collagen Binding

Recombinant α2-I domain was crystallized as a complex with a homotrimer of a 21-mer peptide containing a critical GFOGER (where O is hydroxyproline) motif [17, 30]. The peptide closely resembles the structure of uncomplexed collagen-like peptides [16], and has the properties of a folded protein domain (i.e., stable secondary and tertiary structure). Three loops on the upper surface of the α2-I domain that comprise the MIDAS motif also engage the collagen, with a collagen glutamate completing the coordination sphere of the metal (Fig. 8.3). The critical roles of both the MIDAS and surrounding residues have been confirmed by mutagenesis [52].

The buried surface area on complex formation ($\sim$1,200 A^2) is at the lower limit of known protein-protein interfaces (the value is almost identical for the αL-I:ICAM complex), especially given the fact that some of the binding energy must be expended in switching the conformation of the I domain from closed to open. The quite reasonable affinity of the interaction (Kd = 35–90 nM) [24] reflects the unusually strong bonds formed by the glutamate-metal-I domain bridge, which has been estimated to contribute $\sim$5 kcal/mol. This bridge is indeed critical, since the conservative substitution of collagen Glu to Asp in the GFOGER motif eliminates binding [31], presumably because the aspartic acid is too short to reach down from the rigid collagen triple helix to bind to the partly buried metal ion. However, in the case of the αM-C3d interaction, Asp is the preferred residue, perhaps because it lies on a flexible loop at the end of a helical segment [2].

The MIDAS motif and much of the collagen-binding surface are strictly invariant among the collagen-binding I domain integrins (αl, α2, α10 and α11), suggesting that these integrins will all engage collagen in a similar fashion, with a strict requirement for glutamate in the collagen motif. The periphery of the binding surface is more variable, however (Fig. 8.3b), which would explain their collagen type preferences. The recent structure of the α1-I domain bound to collagen containing the closely related motif, GLOGEN, confirms this [11]. In addition, a gain-of-function point mutation in the α2-I domain (i.e. one that favors the open conformation) [10] displays relaxed specificity and alternate binding modes to the GFOGER motif. Given the special nature of collagen (see Chap. 3 by Zutter and Santoro), this observation may point to profound consequences for collagen recognition by activated cells.

8.4 The Integrin α-I Domain and the von Willebrand Factor (vWF) A Domain: A Caveat

The integrin α-I domain is generally categorized as a member of the vWF-A domain superfamily, based on sequence similarity and a highly conserved overall 3-dimensional structure. However, since MIDAS- and non-MIDAS containing vWFA-domains have distinct ligand-binding and allosteric properties, this author believes that much confusion could be avoided if the family was reclassified into two sub-families: Two examples illustrate my point. First, the eponymous vWF-A1 and vWF-A3 domains lack at least one of the acidic residues of authentic MIDAS motifs, and so do not bind metal; moreover, they bind ligands via distinct surfaces, and conformational changes are not induced [21]. In fact, vWF-A3, like integrin α2β1, binds triple-helical collagen, but it utilizes a different surface (one side of its β-sheet) [7]. Second, the "vWF-A domain" of Factor B *does* contain a functional MIDAS motif, and binds an acidic moiety in its ligand, complement iC3b, in the canonical integrin fashion; in this case, the metal ion engages the C-terminal carboxylate iC3b, triggering integrin-like conformational changes [18], suggesting that it should be classed as an I domain. Indeed, a genome-wide collection of vWF-A domains has been compiled [62] and have been these subdivided into I-like and A-like domains based on the conservation of key MIDAS residues.

8.5 Conformational Changes in the α-I Domain

Ligand binding alters the conformation of α-I domains in the same way in the three cases studied thus far (α2, αM and αL), as well as in a subset of "vWF-A domain" (as noted above) and the matrix receptor, TEM8 (in complex with pathogen; see below). Binding of an acidic residue to the α-MIDAS causes a switch in in Mg^{2+}

coordination in which a direct bond to a MIDAS threonine is gained while a direct bond to an aspartic acid is lost (Fig. 8.2b, c). These subtle changes in metal coordination are linked to extensive secondary and tertiary changes that create a complementary surface for binding ligand and generate a 10 Å downward movement of the C-terminal helix, $\alpha 7$. The helix movement links the change in the upper ligand-binding surface to the lower surface of the domain. The shift of the helix $\alpha 7$ is highly significant in the context of the whole integrin, since the helix is packed against the propeller and β-I domain (see Sect. 8.8).

The close similarity between the structural changes seen in the three α-I domains and a subset of A domains suggests that there are just two principal conformations for I domains, "open" and "closed" (Fig. 8.2). The "open" conformation is seen in the presence of ligand or ligand mimetic, while the "closed" conformation is seen in the absence of ligand. It therefore appears to be the formation of a strong ligand-metal bond, requiring a change in metal coordination, that triggers the conformational switch. Springer's group has engineered disulfide-linked αL-I domains with intermediate affinity and packing of the C-terminal helix, and suggested the existence of an intermediate state [51]. However, in these structures the MIDAS motif exists in only two conformations, and it remains to be seen whether the intermediate conformation has biological relevance or is an artifact of the engineered disulfide. It should be noted that it is not necessary to invoke an intermediate tertiary conformation in order to explain an intermediate affinity. In principle, a shift in the position of the equilibrium between two states is sufficient [38].

Various studies have now been published in support of the hypothesis that the open and closed conformations of the α-I domain equate with high and low affinity states. Thus, mutants of the αM-I domain that are predicted to destabilize the closed conformation and favor the open conformation increase the affinity for the ligand iC3b [41]. The epitope for an antibody that binds only to the high affinity form of the $\alpha M\beta 2$ integrin maps to a region that undergoes extensive conformational changes between the closed and open forms [35].

Disulfide engineering studies on recombinant I domains and full-length integrins, which lock the domain either into the open or closed state, also support the hypothesis [39, 49, 50]. So does the structure of the αL-I domain in complex with the inhibitor lovastatin [25], which reveals allosteric inhibition by binding between the β-sheet and the C-terminal helix, preventing the helical shift.

It should also be appreciated that pathogens often utilize integrin α-I domains for cell entry, and there is evidence that many bind across the MIDAS motif. However, in general they bind preferentially to the (default) closed conformation, sometimes involving direct bonds to the MIDAS, but they do not induce conformational changes [4]. One counter-example is anthrax toxin, which utilizes a glutamate to engage the *bona fide* MIDAS motif of the "vWF A domain" of the collagen receptor, TEM8, in its open conformation [6]. There is also one clear example of gene transfer, in which the Gram-positive pathogen, *Streptococcus pneumoniae*, has an α-I domain inserted into the tip of its pilus, perhaps to act as a shear stress-activated adhesin for attachment to host cells [23, 36].

8.6 The β-I Domain and the Integrin Headpiece

The existence of a β-I domain was initially predicted based on the conserved and critical MIDAS-like sequence, DxSxS, and hydropathy plot comparisons with the α-I domain [34]; and later from more sophisticated sequence analysis [59]. The structure of the β3-I domain, contained within the $\alpha V\beta 3$ crystal structure [64], confirmed that the basic fold and topology are very similar to the α-I domain, albeit with many large insertion/loops between β-strands, which had confounded conventional sequence alignment algorithms.

In contrast to α-I, the β-I domain is not folded independently, but packs rigidly against the α-subunit propeller, with the major ligand

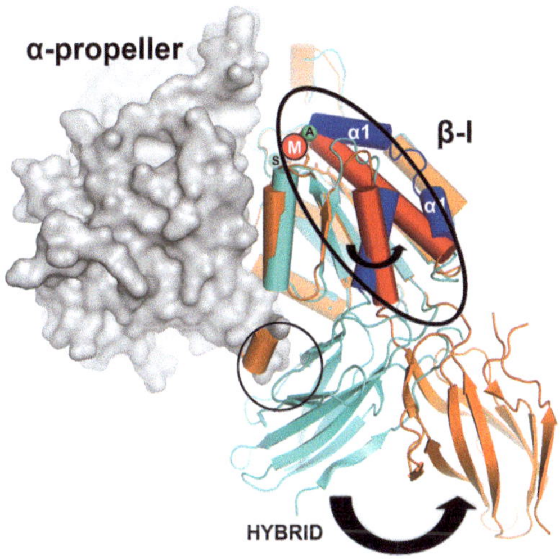

Fig. 8.4 Tertiary and quaternary changes triggered by ligand binding in integrins that lack an α-I domain. Ligand binding to the β-MIDAS motif (M) causes a shift of helix α1, which generates a rotation of α7 helix (*black arrow* within region *circled* in *black*) and a loosening of the contacts between the β-hybrid domain and the propeller. The β-hybrid domain is then free to swing by as much as 60° away from the α-propeller

recognition elements lying at the interface (see Fig. 8.4) [42, 64]. The β-MIDAS is similar to the α-MIDAS, except that the α-MIDAS threonine is replaced by glutamate. This difference likely explains the different cation specificities—in α-MIDAS, the smaller Mg^{2+} ion favors ligands lacking a formal charge; while in β-MIDAS, the larger Ca^{2+} ion favors multiple acidic ligands. The structure of αVβ3 in complex with an RGD-style peptide shows that the Asp sidechain completes the coordination sphere of the MIDAS metal ion [65], as predicted. There are also further metal-binding sites adjacent to the β-MIDAS (the "ADMIDAS" and "SyMBS") that play important structural and possibly regulatory roles in ligand-binding and regulation [68].

Although Xiong et al. initially proposed the opposite, the conformation of the β-I domain in their unliganded crystal structure [64] corresponds to the closed conformation of the α-I domain. Soaking of RGD ligand into preformed crystals induced small changes within the β-I domain, but these were not propagated to the rest of the headpiece; i.e., they were frustrated by the constraints of the closed quaternary structure [37]. This situation is typical in crystallography: either the ligand binds and induces small changes constrained by the lattice, or it induces large changes that destroy the lattice.

However, Springer's group has recently discovered a rare exception to this rule, and report a crystal form of the αIIbβ3 headpiece with large solvent channels in which the lattice tolerates (and/or adjusts to) a switch from the closed to the open conformation, involving an outward swing of the hybrid domain by ~40°. Preformed crystals were simply soaked with different concentrations/durations of an RGD ligand mimetic and different Ca^{2+}/Mg^{2+} ratios [69] (Fig. 8.4). This remarkable observations settles many questions with regard to head-opening, although the crystal structure of the headpiece in complex with a non-peptidic ligand is still lacking.

8.7 Quaternary Regulation in Integrins Lacking an α-I Domain

Takagi et al. [57] showed that the inactive (resting) form of the integrin αVβ3, observed in physiological concentrations of Ca^{2+} and Mg^{2+}, is largely bent, and closely resembles the crystal structure, in which the C-termini of both chains are closely apposed. Based on the one case studied of an α-I domain integrin, the αXβ2 ectodomain, it also adopts a similar (although distinct) bent default conformation [63]. Other integrins tested had a lower propensity to adopt the bent conformation; however, the experiments were performed with extracellular heterodimers truncated near the plasma membrane, so that they lacked the transmembrane helices and cytoplasmic tails that are known to contribute critically to the stability of the inactive conformation. By engineering a disulfide link between the α-subunit propeller and the EGF4 domain of the β-subunit (which are 4 Å apart in the bent (crystal) structure, but would be very far apart in the "standing-up" conformation), Takagi et al. further showed that integrin expressed on the cell surface was in a low affinity state and could only be activated under reducing conditions.

The current model for integrins invokes a minimum of three distinct states: (i) bent, low affinity; (ii) standing-up, legs together, low affinity; and (iii) standing-up, legs apart, high affinity (see Fig. 8.1). The position of equilibrium depends on the concentrations and activation status of extracellular and intracellular ligands, as well as divalent cations. At the heart of the switch is an outward swing of the β-hybrid domain with respect to the β-I domain, by as much as 60° (Fig. 8.4). In αIIbβ3, the primary response to extracellular ligand binding is a concerted reorganization of the N-terminal helix (attached directly to the β-MIDAS) and the adjacent C-terminal helix. Rather than translate downward (as in the case of the α-I domain), the principle motion of α7 is a rotation about an axis close to the β-MIDAS, which is linked to the rotation of the β-hybrid domain. Thus, although some details may differ, the data support the prediction that the trigger for the integrin switch is similar in integrins that contain or lack and α-I domain: i.e., a subtle change in metal coordination at the MIDAS motif is linked to a reorganization of the I domain architecture that leads to quaternary changes toward an open, high-affinity state [33].

As noted above, these experiments were performed with truncated integrins and small peptide ligands. The nature of the trigger in the integrin head seems secure, but it remains to be seen how the quaternary changes are promulgated across the plasma membrane. Recent studies have shown that full-length integrin can be reconstituted into lipid nanodiscs and visualized by high resolution Electron Microscopy [12], so we should soon have an answer.

8.8 Quaternary Changes in Integrins Containing an α-I Domain

As noted above, in integrins that lack an α-I domain, the β-I domain and α-subunit propeller are the major recognition elements [42]. However, in integrins that contain an α-I domain, the β-I and α-propeller do not play direct roles in ligand recognition; instead they play important regulatory roles. This concept initially caused some confusion: thus, mutation of the β-MIDAS motif led to loss of iC3b binding to αMβ2 [3] which was initially interpreted as evidence for a direct role for the β-I domain in ligand; it now seems clear, however, that the mutation works allosterically, by preventing conformational changes in the α-I domain.

How does the quaternary organization of the integrin regulate the affinity of the α-I domain? We know that regulation occurs allosterically (rather than by steric masking of the binding site), since the α-I domain is a major antibody epitope. Hypotheses focused on the loss-of-function effect of mutating a Glu residue within a conserved ΦEGT motif (where Φ is any hydrophobic residue) at the end of the α-I domain C-terminal (α7) helix [1, 22, 66]; and it was suggested that the Glu could act as an intradimer ligand by completing the coordination sphere of the β-MIDAS motif. The first crystal structures of the αXβ2 headpiece (from Xie et al. [63]) were inconclusive: they showed the α-I domain in the closed conformation, but rather loosely attached to the rest of the headpiece. However, a recent structure of the αXβ2 ectodomain displays an activated α-I domain by virtue of a fortuitous crystal contact [47]. Although the rest of the $\alpha\beta$ headpiece is in the closed conformation, the predicted "internal ligand", Glu318, is observed coordinating the β-MIDAS motif with only minor compensatory movements in the β-I domain (Fig. 8.5). By contrast, the α-I domain adopts a fully "open" conformation, with the MIDAS threonine directly coordinating the metal and (what appears to be) a chloride ion completing the coordination sphere. The first half of the α-I domain α7 helix has shifted by ~10 Å, as expected, but the remainder is unwound, thereby switching the orientation with respect to the headpiece. Thus, the crystalline environment seems to have created a hybrid molecule with a fully active α-I domain in the context of an inactive headpiece. It is possible that such a

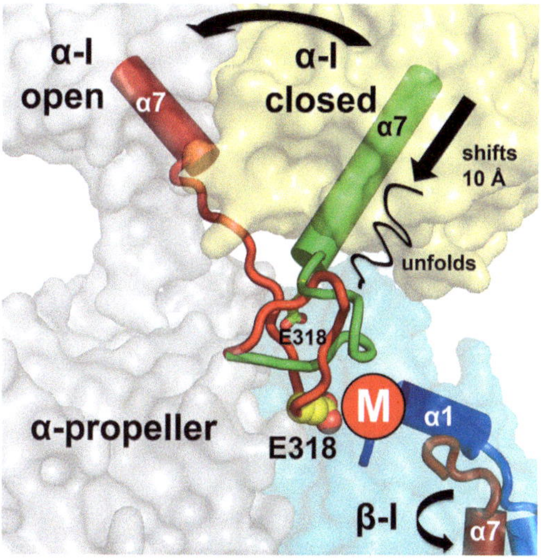

Fig. 8.5 Close-up comparison of the α-I domain-containing integrin head of αXβ2. In the open conformation, Glu318 acts as an internal ligand to the β-MIDAS that generates a 10 Å shift in the first half of helix α7 of the αI domain, while the second half of the helix unwinds, leading to a 30–40° rotation of the α-I domain about the propeller and β-I domain. The open α-I domain is stabilized by a crystal contact, and the the β-I domain remains principally in the closed conformation

hybrid state could exist in vivo, providing long-range, flexible and rapid (non-equilibrium) responses to the presence of ligand and/or mechanical stress; with a slow switch to the overall open (equilibrium) conformation occurring if the signal persisted.

8.9 Transmembrane (TM) and Tail Interactions

There are abundant biochemical and genetic data supporting the notion that interactions between integrin α- and β-TM helices and cytosolic tails help to hold the resting integrin in a low affinity conformation. In a classic study by Ginsberg and colleagues, a salt bridge between αIIb Arg995 and β3 Asp723 was shown to be necessary and sufficient to hold the integrin in its resting state [20]. A definitive structure of the αIIbβ3 tails in bicelles [32] reveals a remarkably stable conformation, in which the two helices pack closely together at the extracellular end; at the intracellular end they diverge, but the void is effectively filled by a highly conserved aromatic triplet that breaks the α-subunit helix and turns inward (Fig. 8.6). Isolated α and β subunit helices studies in bicelles show a remarkably well-conserved structure: the α-subunit is always orthogonal to the membrane while the β-subunit helix is always tilted. Recently, Ginsberg has shown that Lys 716β, whose Cα is buried in the membrane, can "snorkel" to the hydrophilic headgroups by extension of the Lys sidechain, and moreover that this residue is essential for maintaining the tilted helix and TM signaling [28, 55]. Recent studies on α-I domain integrins have yielded consistent results for isolated TM regions, and structures of αβ complexes are in progress [13]. The switch to the "open" conformation may entail a simple separation of the tails, which maintain their structural integrity and reassemble rapidly when the integrin returns to the low affinity state.

The penultimate question is how cytosolic proteins interact with the cytoplasmic tails of integrins, and the number of structural examples of protein domains bound to tail peptides (mostly β, but some α) has grown rapidly in recent years (see Table 8.1). It is clear that some proteins bind strongly enough to the β-tail to promote integrin activation. Talin was the first such molecule to be thus characterized, and remains the central player [8], although the number of additional contributors, such as kindlin and filamin [9], is growing fast. A model of talin activation is presented in Fig. 8.6. Talin sequesters the β-tail, breaking the critical R995α-D723β bond. It is also clear that phosphorylation of the β-tails provides a rapid means of switching between binding partners, and thus between cell migration and adhesion [14, 40, 44].

The final question is how the cytosolic activators generate force across the membrane. In the case of talin, recent work by Ginsberg's group suggests that dissociation of the tails, which have flexible linkages to the extracellular domains, is sufficient [29]. Key to this process is talin's ability to bind simultaneously to the integrin β-tail and membrane (Fig. 8.6), the latter providing a pivot point to force the two helices apart.

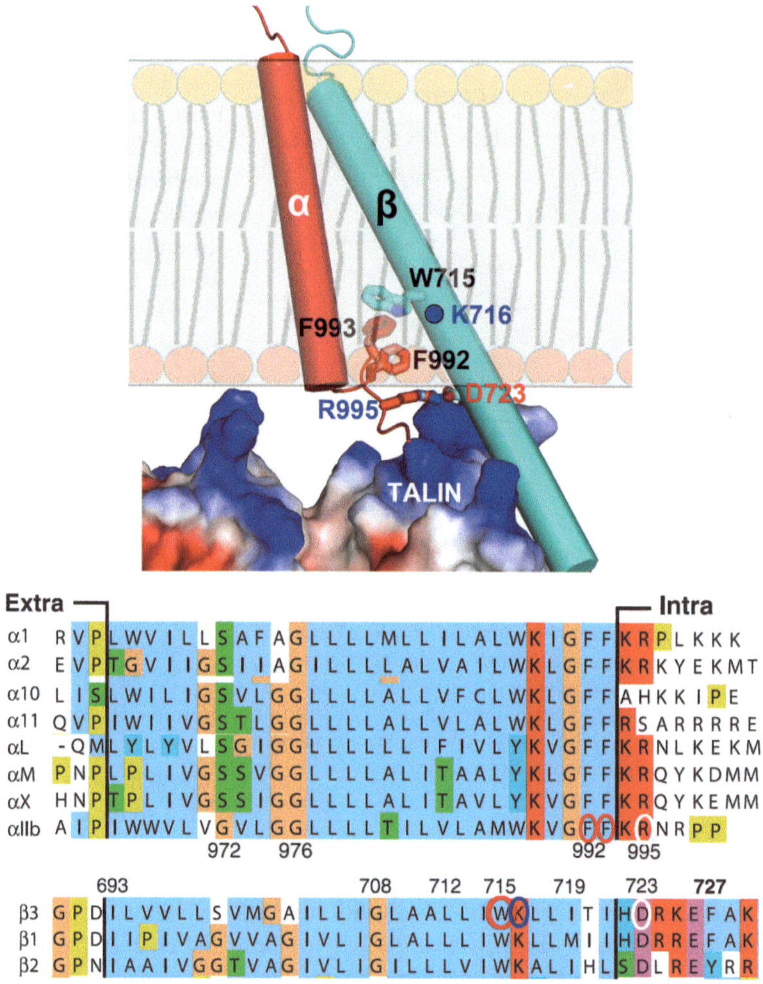

Fig. 8.6 Integrin-tail interactions. *Upper panel* Atomic interactions between the αIIb and β3 tails as revealed by NMR, melded to the complex of Talin2 and β1D tail. *Lower panel* Aligned sequences of α and β TM segments. Important residues in αIIbβ3 (and conserved in α-I domain integrins) are circled. See text for details

8.10 Perspectives

Structural and structure-function studies have revealed many of the major paradigms of integrin allostery underlying affinity regulation and bi-directional signal transduction. Notably lacking is the structure of an intact integrin bound to a physiological ligand in a membrane environment that would reveal the true "active" conformation of the molecule. EM studies show the greatest promise here, most likely using nanodisks. We are beginning to understand the structures of the TM helices and their cytosolic extensions, but the biophysics of inside-out signaling in particular requires further study. The role of mechanical force, whether of intracellular (actomyosin motors) or extracellular (shear flow in the vasculature) origin has not been discussed here, but its interplay with the chemical forces that attract cognate molecules is a fascinating field for current and future study [48]. Finally, this chapter has addressed the structural basis of affinity changes within

Table 8.1 Reported structures of integrins, 1996–present

Year	Protein	PDB code	Notes
Wild-type α-I domains			
1996	αM-I, Mg^{2+}	1IDO	
1996	αL-I, Mn^{2+}	1LFA	
1996	αL-I, Mg^{2+}	1ZOO, 1ZOP	
1996	αL-I metal-free	1ZON	
1997	αM-I, Mn^{2+}	1JLM	
1998	αM-I, soaks	1BHO, 1BHQ, 1IDN	Mg^{2+}, Mn^{2+}, free
1998	α2-I	1AOX	
2000	α1-I	1QC5	
2000	α1-I	1CK4	Rat
2000	αL-I	1DGQ	NMR structure
2003	αX-I	1N3Y	
2003	αL-I	1MQ9	High affinity form
I-like domains			
2004	A domain, Factor B	1Q0P	
2010	*Haemophilus* pilus	2WW8	
I domain-ligand complexes			
2000	α2-Collagen	1DZI	
2003	αL-ICAM1	1MQ8	
2005	αL-ICAM3	1T0P	
2008	αL-ICAM5	3BN3	
2013	αM-C3d	4M76	
2013	α1-Collagen	2M32	NMR/HADDOCK model
Engineered I domains/complexes			
2002	αM-I	1MIU, 1MQA	Ile switch
2003	αL-I	1MJN	Intermediate affinity
2003	αM-I	1MF7,1N9Z,1NA5	Modulatory mutants
2009	αL-I	3HI6	Disulfide-bonded intermediate
2011	α1-I	4A0Q	Activating mutation
2011	αL-ICAM-1	3TCX	Mutant high affinity I domain
2013	α2-Collagen	4BJ3	Mutant high affinity I domain
I domain-small molecule/FAB complexes			
2001	αL-LOVASTATIN	1CQP	
2004-2014	αL modulators	1RD4,1XDD,1XDG,1XUO,2ICA, 2O7N 3BQM, 3BQN,3E2M,4IXD,3F74,3F78	
2009/2010	αL-EFALIZUMAB	3EOA,3EOB,3M6F	

(continued)

Table 8.1 (continued)

Year	Protein	PDB code	Notes
2011	αM-FAB	3Q3G,3QA3	
Cytoplasmic Tail-protein complexes			
2003	β3-Talin (chimera)	1MIZ,1MK7,1MK9	
2005	β3-PIP-kinase	1Y19	
2005	αIIb-Filamin	2BP3	
2006	β7-Filamin	2BRQ	
2007	β3-TalinF3	2H7E	NMR
2007	β3-PIP-kinase	2H7D	NMR
2008	β2-Filamin	2JF1	
2008	β3-DOK1	2V76	
2008	β2-P-14-3-3	2V7D	P = Phosphorylation
2009	β1D-Talin2	3V9W	
2010	β3-shc-P	2L1C	
2012	αIIb-CIB1	2LM5	
2012	β1-Acap1	3T9K	
2012	β3-Src	4HXJ	
2013	β4-14-3-3	4HKC	
2013	β1-Acap1	4DX9	
Transmembrane Domains			
2008	β3	2RMZ,2RN0	
2008	αIIb	2K1A	
2009	αIIbβ3	2K9J	
2009	αIIbβ3	2KNC	
2011	β3	2KV9	S–S linked
2012	α2	2L8S	Detergent micelles
2014	αLβ2	2M3E	
2014	β3	2L91	
Cytoplasmic Domains			
2000	αIIb mutant	1DPQ	
2000	αIIb	1DPK	
2002	αIIbβ3	1M8O	
2002	αIIbβ3	1KUP,1KUZ	
2004	αIIbβ3	1S4W,1S4X	Micelles
2008	αL	2K8O	NMR
2011	P-β3	2LJF	Aqueous
2011	αMβ2	2LKE,2LKJ	
2011	P-β3	2LJD,LJE	
2012	αXβ2	2LUV	
α6β4 Intracellular domains/complexes			
1999	β4-FibIII pair	1QG3	
2008	β4-FibIII	2YRZ	NMR

(continued)

Table 8.1 (continued)

Year	Protein	PDB code	Notes
2009	β4-FibIII	3F7Q,3F7R	
2009	β4 Calx	3FQ4,3FSO,3H6A	
2009	β4-Plectin complex	3F7P	
Full-length integrins			
2014	αIIbβ3	4CAK	EM
2009	αVβ3	3IJE	Chimera?
Headpiece/ectodomain			
2001	αVβ3	1JV2	
2002	αVβ3-RGD	1L5G	
2002	αVβ3-Mn^{2+}	1M1X	
2004	αIIbβ3-inhibitor	1TYE	
2008	αIIbβ3 rerefined	2VC2,2VDK,2VDL,2VDM,2VDN,2VDO (supercede 1TY7,1TY3,1TXV,1TY5,1TY6)	
2008	αIIbβ3-Fibrinogen	2VDO,2VDP,2VDQ,2VDR	
2009	αIIbβ3	3FCS	"Complete ectodomain"
2009	αIIbβ3	3FCU	Open conformation
2010	αXβ2	3K6S,3K7L,3K72	
2011	αIIbβ3	3NID,3NIF,3NIG	Antagonist
2012	αVβ3	4DX9	Coiled-coil
2012	αVβ3	4G1M	Rerefined
2012	α5β1-RGD	3VI3,3VI4	
2012	α4β7	3V4P,3V4V	$\alpha\beta$ complex
2013	αIIbβ3-RGD	3ZDX,3ZDY,3ZDZ,3ZE0,3ZE1,3ZE2	
2013	α4β7	4IRZ	
2014	αXβ2	4NEH, 4NEN	

individual integrins. Lateral association (clustering) of integrins in the plasma membrane at sites of ECM contact also plays a major role in integrin signaling, and we still know little about its structural basis and regulation [5].

References

1. Alonso JL, Essafi M, Xiong JP, Stehle T, Arnaout MA (2002) Does the integrin αA domain act as a ligand for its βA domain? Curr Biol 12:R340–R342
2. Bajic G, Yatime L, Sim RB, Vorup-Jensen T, Andersen GR (2013) Structural insight on the recognition of surface-bound opsonins by the integrin I domain of complement receptor 3. Proc Natl Acad Sci USA 110:16426–16431
3. Bajt ML, Loftus JC, Gawaz MP, Ginsberg MH (1992) Characterization of a gain of function mutation of integrin αIIb β3 (platelet glycoprotein IIb-IIIa). J Biol Chem 267:22211–22216
4. Bergelson JM, Chan BM, Finberg RW, Hemler ME (1993) The integrin VLA-2 binds echovirus 1 and extracellular matrix ligands by different mechanisms. J Clin Invest 92:232–239
5. Boettiger D (2012) Mechanical control of integrin-mediated adhesion and signaling. Curr Opin Cell Biol 24:592–599
6. Bradley KA, Mogridge J, Jonah G, Rainey A, Batty S, Young JA (2003) Binding of anthrax toxin to its receptor is similar to α integrin-ligand interactions. J Biol Chem 278:49342–49347
7. Brondijk TH, Bihan D, Farndale RW, Huizinga EG (2012) Implications for collagen I chain registry from the structure of the collagen von Willebrand factor A3 domain complex. Proc Natl Acad Sci USA 109:5253–5258

8. Calderwood DA (2004) Talin controls integrin activation. Biochem Soc Trans 32:434–437
9. Calderwood DA, Campbell ID, Critchley DR (2013) Talins and kindlins: partners in integrin-mediated adhesion. Nat Rev Mol Cell Biol 14:503–517
10. Carafoli F, Hamaia SW, Bihan D, Hohenester E, Farndale RW (2013) An activating mutation reveals a second binding mode of the integrin α2 I domain to the GFOGER motif in collagens. PLoS ONE 8:e69833
11. Chin YK, Headey SJ, Mohanty B, Patil R, McEwan PA, Swarbrick JD et al (2013) The structure of integrin α1I domain in complex with a collagen-mimetic peptide. J Biol Chem 288:36796–36809
12. Choi WS, Rice WJ, Stokes DL, Coller BS (2013) Three-dimensional reconstruction of intact human integrin αIIbβ3: new implications for activation-dependent ligand binding. Blood 122:4165–4171
13. Chua GL, Tang XY, Patra AT, Tan SM, Bhattacharjya S (2012) Structure and binding interface of the cytosolic tails of αXβ2 integrin. PLoS ONE 7:e41924
14. Deshmukh L, Meller N, Alder N, Byzova T, Vinogradova O (2011) Tyrosine phosphorylation as a conformational switch: a case study of integrin β3 cytoplasmic tail. J Biol Chem 286:40943–40953
15. Emsley J, King SL, Bergelson JM, Liddington RC (1997) Crystal structure of the I domain from integrin α2β1. J Biol Chem 272:28512–28517
16. Emsley J, Knight CG, Farndale RW, Barnes MJ (2004) Structure of the integrin α2β1-binding collagen peptide. J Mol Biol 335:1019–1028
17. Emsley J, Knight CG, Farndale RW, Barnes MJ, Liddington RC (2000) Structural basis of collagen recognition by integrin α2β1. Cell 101:47–56
18. Forneris F, Ricklin D, Wu J, Tzekou A, Wallace RS, Lambris JD et al (2010) Structures of C3b in complex with factors B and D give insight into complement convertase formation. Science 330:1816–1820
19. Garcia-Alvarez B, de Pereda JM, Calderwood DA, Ulmer TS, Critchley D, Campbell ID et al (2003) Structural determinants of integrin recognition by talin. Mol Cell 11:49–58
20. Hughes PE, Diaz-Gonzalez F, Leong L, Wu C, McDonald JA, Shattil SJ et al (1996) Breaking the integrin hinge. A defined structural constraint regulates integrin signaling. J Biol Chem 271:6571–6574
21. Huizinga EG, Tsuji S, Romijn RA, Schiphorst ME, de Groot PG, Sixma JJ et al (2002) Structures of glycoprotein Ibα and its complex with von Willebrand factor A1 domain. Science 297:1176–1179
22. Huth JR, Olejniczak ET, Mendoza R, Liang H, Harris EA, Lupher ML Jr et al (2000) NMR and mutagenesis evidence for an I domain allosteric site that regulates lymphocyte function-associated antigen 1 ligand binding. Proc Natl Acad Sci USA 97:5231–5236
23. Izore T, Contreras-Martel C, El Mortaji L, Manzano C, Terrasse R, Vernet T et al (2010) Structural basis of host cell recognition by the pilus adhesin from *Streptococcus pneumoniae*. Structure 18:106–115
24. Jung SM, Moroi M (1998) Platelets interact with soluble and insoluble collagens through characteristically different reactions. J Biol Chem 273:14827–14837
25. Kallen J, Welzenbach K, Ramage P, Geyl D, Kriwacki R, Legge G et al (1999) Structural basis for LFA-1 inhibition upon lovastatin binding to the CD11a I-domain. J Mol Biol 292:1–9
26. Kamata T, Puzon W, Takada Y (1994) Identification of putative ligand binding sites within I domain of integrin α2β1 (VLA-2, CD49b/CD29). J Biol Chem 269:9659–9663
27. Kamata T, Takada Y (1994) Direct binding of collagen to the I domain of integrin α2β1 (VLA-2, CD49b/CD29) in a divalent cation-independent manner. J Biol Chem 269:26006–26010
28. Kim C, Schmidt T, Cho EG, Ye F, Ulmer TS, Ginsberg MH (2012) Basic amino-acid side chains regulate transmembrane integrin signalling. Nature 481:209–213
29. Kim C, Ye F, Hu X, Ginsberg MH (2012) Talin activates integrins by altering the topology of the β transmembrane domain. J Cell Biol 197:605–611
30. Knight CG, Morton LF, Onley DJ, Peachey AR, Messent AJ, Smethurst PA et al (1998) Identification in collagen type I of an integrin α2β1-binding site containing an essential GER sequence. J Biol Chem 273:33287–33294
31. Knight CG, Morton LF, Peachey AR, Tuckwell DS, Farndale RW, Barnes MJ (2000) The collagen-binding A-domains of integrins α1β1 and α2β1 recognize the same specific amino acid sequence, GFOGER, in native (triple-helical) collagens. J Biol Chem 275:35–40
32. Lau TL, Kim C, Ginsberg MH, Ulmer TS (2009) The structure of the integrin αIIbβ3 transmembrane complex explains integrin transmembrane signalling. EMBO J 28:1351–1361
33. Lee JO, Bankston LA, Arnaout MA, Liddington RC (1995) Two conformations of the integrin A-domain (I-domain): a pathway for activation? Structure 3:1333–1340
34. Lee JO, Rieu P, Arnaout MA, Liddington R (1995) Crystal structure of the A domain from the α subunit of integrin CR3 (CD11b/CD18). Cell 80:631–638
35. Li R, Rieu P, Griffith DL, Scott D, Arnaout MA (1998) Two functional states of the CD11b A-domain: correlations with key features of two Mn^{2+}-complexed crystal structures. J Cell Biol 143:1523–1534
36. Liddington R (2010) Catching pneumonia. Structure 18:6–8

37. Liddington RC (2002) Will the real integrin please stand up? Structure 10:605–607
38. Loftus JC, Liddington RC (1997) Cell adhesion in vascular biology. New insights into integrin-ligand interaction. J Clin Invest 99:2302–2306
39. McCleverty CJ, Liddington RC (2003) Engineered allosteric mutants of the integrin αMβ2 I domain: structural and functional studies. Biochem J 372:121–127
40. McCleverty CJ, Lin DC, Liddington RC (2007) Structure of the PTB domain of tensin1 and a model for its recruitment to fibrillar adhesions. Protein Sci 16:1223–1229
41. Michishita M, Videm V, Arnaout MA (1993) A novel divalent cation-binding site in the A domain of the β2 integrin CR3 (CD11b/CD18) is essential for ligand binding. Cell 72:857–867
42. Mould AP, Askari JA, Aota S, Yamada KM, Irie A, Takada Y et al (1997) Defining the topology of integrin α5β1-fibronectin interactions using inhibitory anti-α5 and anti-β1 monoclonal antibodies. Evidence that the synergy sequence of fibronectin is recognized by the amino-terminal repeats of the α5 subunit. J Biol Chem 272:17283–17292
43. Nolte M, Pepinsky RB, Venyaminov S, Koteliansky V, Gotwals PJ, Karpusas M (1999) Crystal structure of the α1β1 integrin I-domain: insights into integrin I-domain function. FEBS Lett 452:379–385
44. Oxley CL, Anthis NJ, Lowe ED, Vakonakis I, Campbell ID, Wegener KL (2008) An integrin phosphorylation switch: the effect of β3 integrin tail phosphorylation on Dok1 and talin binding. J Biol Chem 283:5420–5426
45. Qu A, Leahy DJ (1995) Crystal structure of the I-domain from the CD11a/CD18 (LFA-1, αLβ2) integrin. Proc Natl Acad Sci USA 92:10277–10281
46. Schwartz MA, Ginsberg MH (2002) Networks and crosstalk: integrin signalling spreads. Nat Cell Biol 4:E65–E68
47. Sen M, Yuki K, Springer TA (2013) An internal ligand-bound, metastable state of a leukocyte integrin, αXβ2. J Cell Biol 203:629–642
48. Seong J, Tajik A, Sun J, Guan JL, Humphries MJ, Craig SE et al (2013) Distinct biophysical mechanisms of focal adhesion kinase mechanoactivation by different extracellular matrix proteins. Proc Natl Acad Sci USA 110:19372–19377
49. Shimaoka M, Shifman JM, Jing H, Takagi J, Mayo SL, Springer TA (2000) Computational design of an integrin I domain stabilized in the open high affinity conformation. Nat Struct Biol 7:674–678
50. Shimaoka M, Takagi J, Springer TA (2002) Conformational regulation of integrin structure and function. Annu Rev Biophys Biomol Struct 31:485–516
51. Shimaoka M, Xiao T, Liu JH, Yang Y, Dong Y, Jun CD et al (2003) Structures of the αL I domain and its complex with ICAM-1 reveal a shape-shifting pathway for integrin regulation. Cell 112:99–111
52. Smith C, Estavillo D, Emsley J, Bankston LA, Liddington RC, Cruz MA (2000) Mapping the collagen-binding site in the I domain of the glycoprotein Ia/IIa (integrin α2β1). J Biol Chem 275:4205–4209
53. Song G, Yang Y, Liu JH, Casasnovas JM, Shimaoka M, Springer TA et al (2005) An atomic resolution view of ICAM recognition in a complex between the binding domains of ICAM-3 and integrin αLβ2. Proc Natl Acad Sci USA 102:3366–3371
54. Springer TA, Dustin ML (2012) Integrin inside-out signaling and the immunological synapse. Curr Opin Cell Biol 24:107–115
55. Surya W, Li Y, Millet O, Diercks T, Torres J (2013) Transmembrane and juxtamembrane structure of αL integrin in bicelles. PLoS ONE 8:e74281
56. Takagi J, Erickson HP, Springer TA (2001) C-terminal opening mimics 'inside-out' activation of integrin α5β1. Nat Struct Biol 8:412–416
57. Takagi J, Petre BM, Walz T, Springer TA (2002) Global conformational rearrangements in integrin extracellular domains in outside-in and inside-out signaling. Cell 110:599–611
58. Tuckwell D, Calderwood DA, Green LJ, Humphries MJ (1995) Integrin α2 I-domain is a binding site for collagens. J Cell Sci 108:1629–1637
59. Tuckwell DS, Humphries MJ (1997) A structure prediction for the ligand-binding region of the integrin β subunit: evidence for the presence of a von Willebrand factor A domain. FEBS Lett 400:297–303
60. Vinogradova O, Haas T, Plow EF, Qin J (2000) A structural basis for integrin activation by the cytoplasmic tail of the αIIb-subunit. Proc Natl Acad Sci USA 97:1450–1455
61. Vinogradova O, Velyvis A, Velyviene A, Hu B, Haas T, Plow E et al (2002) A structural mechanism of integrin αIIbβ3 "inside-out" activation as regulated by its cytoplasmic face. Cell 110:587–597
62. Whittaker CA, Hynes RO (2002) Distribution and evolution of von Willebrand/integrin A domains: widely dispersed domains with roles in cell adhesion and elsewhere. Mol Biol Cell 13:3369–3387
63. Xie C, Zhu J, Chen X, Mi L, Nishida N, Springer TA (2010) Structure of an integrin with an αI domain, complement receptor type 4. EMBO J 29:666–679
64. Xiong JP, Stehle T, Diefenbach B, Zhang R, Dunker R, Scott DL et al (2001) Crystal structure of the extracellular segment of integrin αVβ3. Science 294:339–345
65. Xiong JP, Stehle T, Zhang R, Joachimiak A, Frech M, Goodman SL et al (2002) Crystal structure of the extracellular segment of integrin αVβ3 in complex with an Arg-Gly-Asp ligand. Science 296:151–155

66. Yang W, Shimaoka M, Salas A, Takagi J, Springer TA (2004) Intersubunit signal transmission in integrins by a receptor-like interaction with a pull spring. Proc Natl Acad Sci USA 101:2906–2911
67. Zhang H, Casasnovas JM, Jin M, Liu JH, Gahmberg CG, Springer TA et al (2008) An unusual allosteric mobility of the C-terminal helix of a high-affinity αL integrin I domain variant bound to ICAM-5. Mol Cell 31:432–437
68. Zhang K, Chen J (2012) The regulation of integrin function by divalent cations. Cell Adh Migr 6:20–29
69. Zhu J, Zhu J, Springer TA (2013) Complete integrin headpiece opening in eight steps. J Cell Biol 201:1053–1068

Integrin Recognition Motifs in the Human Collagens

9

Samir Hamaia and Richard W. Farndale

Abstract

The best-known (fibrillar) collagens support cellular adhesion primarily through a subset of collagen-binding integrins, $\alpha1\beta1$, $\alpha2\beta1$, $\alpha10\beta1$ and $\alpha11\beta1$, which have been shown to recognise a series of similar sequences. These contain Gxx′GEx″motifs (where x is a hydrophobic residue, x′ is usually O (hydroxyproline) and x″ is often R). Here, we review the variations within such sequences that support integrin reactivity, and their distribution across the 28 human collagens. The main basis for our understanding is the use of triple-helical, homotrimeric collagen peptides, but this work is far from exhaustive, and there is good evidence that heterotrimeric collagens where the sequence of interest occurs in two or even just a single chain may still support integrin binding. The fibrillar collagens I, II and III are rich in GxOGER motifs, whereas GxOGEK is more widely distributed, and less frequent in these three archetypal fibrillar collagens.

Keywords

Integrin • Collagen • Recognition motif • Peptide

9.1 Introduction

In this chapter, we review the interactions of the collagen-binding integrins, $\alpha1\beta1$, $\alpha2\beta1$, $\alpha10\beta1$ and $\alpha11\beta1$, with the triple-helical domains of the collagen. Much detailed data has been accumulated from the use of the most abundant fibrillar collagens, types I, II and III, and from work in this and other laboratories where triple-helical synthetic peptides have been used to investigate the structure of complexes between collagens and integrins. The use of such (generally) homotrimeric peptides has established the rules that govern these interactions. An underlying assumption is that the integrin-binding motifs discovered in this way may be relevant to all homotrimeric, and some hetero-trimeric triple-helical collagen molecules where the same sequence occurs. The general structure

S. Hamaia · R. W. Farndale (✉)
Department of Biochemistry, Cambridge University, Downing Site, Cambridge CB2 1QW, UK
e-mail: rwf10@cam.ac.uk

D. Gullberg (ed.), *I Domain Integrins*, Advances in Experimental Medicine and Biology 819,
DOI: 10.1007/978-94-017-9153-3_9, © Springer Science+Business Media Dordrecht 2014

of the collagen family has been described clearly by Ricard-Blum [52], who also tabulates collagen-like proteins that contain triple-helical domains, but go by other names.

9.2 Collagen Structure

The collagens are defined by their repeated Gxx′ sequences, which can support assembly as a right-handed superhelix, the COL domain. Each of the three constituent polypeptide chains adopts a left-handed helical conformation, the polyproline II helix, with three residues per turn. Viewed axially, there is little space in the centre of the superhelix. This constraint dictates that glycine, lacking a bulky sidechain, should occupy every third position in the primary sequence, reviewed in [45]. Hence, both x and x′ sidechains are exposed on the superhelix surface where they may be free to interact with other molecules. The capacity to adopt the polyproline helical conformation arises from the distortion of the α chain backbone introduced by the cyclic iminoacids proline (P) and hydroxyproline (O) in the x and x′ positions respectively. Twenty eight human collagens exist, assembled from about 45 gene products, the family expressing surprising structural diversity [27]. The presence of a triple helix, as either a continuous COL domain or with non-helical interruptions, is their common defining feature. The collagens fall into several groups, with the fibrillar collagens, comprising types I, II, III, V, XI, XXIV and XXVII, being the prime example and the most abundant proteins within the vertebrate organism. Their fibrillar structure is achieved by the packing of the cord-like trimeric tropocollagen monomers in the typical quarter-staggered array, stabilised by electrostatic interactions between the sidechains of adjacent triple helices. The structure of the different collagen types, both as monomers and supramolecular assemblies, is nicely portrayed and reviewed by Ricard-Blum [52].

The fibrillar collagen genes encode precursors, procollagens, from which N- and C-terminal propeptides are trimmed by specific enzymes as the translated triple helix is secreted from the cell and assembles into the fibril. Processing leaves short unstructured telopeptide extensions at both the N- and C-terminal ends of the (generally single) COL domain. In contrast, the non-fibrillar collagens often contain several different classes of non-collagenous domains, notably VWF A, fibronectin III and thrombospondin domains, each with the capacity to support complex two- or three-dimensional network assembly. We have not attempted a detailed review of integrin binding to the non-helical domains of the non-fibrillar collagens, but there are many reports of such activity, such as within the C-terminal propeptide of collagen I which binds both $\alpha 1\beta 1$ and $\alpha 2\beta 1$ [14, 72], which have yet to be elucidated. Similarly endostatin, the C-terminal domain of collagen XVIII, interacts with $\alpha 5\beta 1$ by an unknown mechanism [51, 73].

Collagens as a group necessarily undergo post-translational modification, especially the hydroxylation of x′ proline and a proportion of x′ lysine residues. Subsequent crosslinking between triple helices occurs by condensation of hydroxylysine and lysine sidechains, adding to the stability of the fibril, whilst hydroxyproline supports hydrogen bonding through water molecules within the hydration shell surrounding the collagen monomer, sufficient to stabilise the initial assembly of both the helix itself and the fibril [49].

The collagen triple helices may be interrupted by stretches of non-helical sequence, separating the triple helix into multiple serial COL domains, conventionally numbered from the C-terminus. In the pro-form of the fibrillar collagens, such COL2 domains are found in collagens I, III and $\alpha 2$(V), whilst the remaining members of the sub-group also contain a smaller COL3 domain. In general, these are lost during procollagen cleavage, leaving just the COL1 domain and its telopeptide extensions, but this process is incomplete in collagen III, so that some unprocessed molecules occur on the surface of fibres where the COL2 domain may impede further fibre growth. Collagen IV and the fibril-associated collagens with interrupted triple-helices (FACIT) are more extreme forms of such structures. Interruptions may take the form of extended insertions, which might be expected to contain secondary structure, or of small non-canonical insertions (interruptions or imperfections) into the COL domain of maybe three or four residues lacking glycine, or a deletion with just two intervening residues between conventional Gxx′ triplets.

The triple helix assembles with a one-residue offset between strands. As a consequence, three possible isoforms occur in heterotrimeric collagens, with, in principle, the single α2(I) chain of collagen I able to occupy the leading, middle or trailing position, and so that the trimer can present three different surfaces to protein binding partners such as integrins, as will be discussed further below. Firm knowledge of the order of strands in the heterotrimeric collagens is scarce. Whether the order is maintained through longer helical interruptions is not known, and in principle, needs to be defined case-by-case. From NMR experiments on a Gly/Ser substituted peptide [35], it seems likely that the order persists through short insertions or deletions with only slight perturbation resulting from the interruption. In the case of α1(VIII), for example, the COL1 domain contains eight occurrences of a single-residue deletion (resulting in a "two-residue triplet"), and the corresponding α2(VIII) contains eight equivalent compensating deletions usually four residues towards the C-terminus, that will allow the maintenance of an essentially linear COL domain with only minor deviation. In contrast, the FACIT collagen IX has unequal longer insertions in its α-chains that will introduce a flexible kink between the COL2 and COL3 domains.

9.3 Integrin I-Domain Structure

The vertebrate integrins are a family of 24 heterodimeric trans-membrane cellular adhesion molecules, with carefully regulated expression and affinity, reviewed in [4]. Integrins mediate either cell–cell interactions, binding counter-receptors such as ICAM and VCAM, or cell–matrix interactions [24]. The two matrix molecules best-known in the latter context are fibronectin and collagen. Fibronectin is a ligand for both β1 and β3 integrins, with α5β1 and αvβ3 as its most widespread receptors, whereas collagen is considered to interact directly only with that subset of four β1 integrins, α1β1, α2β1, α10β1 and α11β1, which is distinguished by the presence within the α-subunit of an inserted, or I, domain [4]. These integrins are discussed in detail in Chaps. 2, 3, 4 and 5 of this volume. The α-I domain adopts the dinucleotide-binding, or Rossman, fold [17], and its evolution is discussed in Chap. 1. This structure, the prototype for which is the von Willebrand factor A domain, is found in intracellular signalling species such as G protein α-subunits and in extracellular adhesive proteins including the terminal extensions of the non-fibrillar collagens [13]. Under physiological conditions, the integrin α-I domains, though not necessarily all other A-domains [6], constitutively co-ordinate a divalent cation, Mg^{2+}, in their metal ion dependent adhesion site (MIDAS) which is the focus of their interaction with collagens [18]. Of the four collagen-binding integrins, α1β1 and α2β1 have been studied for almost three decades whilst the properties of both α10β1, which was purified more recently using its capacity to bind collagen II [5, 7, 8] and α11β1 (reviewed in Chap. 5) are still not fully investigated. Although α1β1 and α2β1 share most sequence identity, the binding properties of α1β1 and α10β1 appear most similar, but distinct from those of α2β1 and α11β1, which also appear to share some functional redundancy [4]. Integrin α1β1 is considered a selective ligand for collagen IV, and α2β1 for collagen I [67, 68]. This indicates differential ability to interact with specific collagen types, and by inference therefore, with different motifs within the collagens. Crucially, the tissue distribution and temporal expression of the integrins may differ significantly, with α10β1 being most abundant in cartilage, for example. Transcripts of all four are increasingly being detected in diverse settings, where functional significance is as yet unknown. In this volume, Chap. 10, Heino summarises the integrin-reactivity of the different collagen types.

The possibility that leukocyte integrins (also with I domain-containing α subunits) can bind collagen arises from time to time. Collagen is featured in "Integrins at a glance" as a ligand for αxβ2 [24], and recently Lahti et al. showed binding of leukocytes and recombinant I domains to collagens and to a GFOGER-containing collagenous peptide [33]. In our hands, these I domains do indeed bind to such peptides, but weakly, and non-specifically: where cation dependence occurs in our data, similar binding occurs to the control peptide, GPP-10, which lacks any integrin motif (Fig. 9.1). To our minds, therefore, this may represent a different mode of binding than that regarded as the canonical mechanism [18].

9.4 Regions of Collagen That Bind Integrins

In this chapter, we will focus upon the binding of motifs within the COL domains to integrins, where triple-helical structure is a pre-requisite for binding activity.

Michael Barnes [20, 21, 30, 37, 39, 66] and Sam Santoro [15, 55–57, 62] were the most prominent workers in the field during the last 15 years of the twentieth century. Both laid the foundations for the present understanding, and used the human platelet collagen receptors as an established and accessible cellular model for collagen receptors in general, together with a human fibrosarcoma cell line, HT1080, that is still considered to express integrin α2β1 as its major functional collagen-binding integrin. The three other family members may occur at low copy number in both cell types, but the cellular adhesion to collagen is virtually abolished by α2 subunit-specific blocking antibodies.

Separately, Barnes [38] and Santoro [63] each mapped α2β1 onto the cyanogen bromide peptide, α1(I) CB3, identifying α2β1 as a key receptor that supported strong adhesion of platelets to collagen, but which was not sufficient to activate the platelet. Santoro found integrin-binding activity to be restricted to CB3 [63], whilst Barnes was able to show sites within several other CB peptides, although CB3 contained the highest affinity ligand for α2β1. Santoro used chemical derivatisation to show that specific reactivities within the intact collagens were responsible for binding to α2β1 and to an activatory platelet receptor, and subsequently to propose a linear tetrapeptide, DGEA, as an α2β1 recognition motif [61]. In turn, Barnes extended his reductive use of CB peptides, which yielded low-resolution maps of integrin binding sites in collagens I and III, to the development of a small library of overlapping, triple-helical synthetic peptides which encompassed the whole of α1(I)CB3 [30, 31]. By this method Barnes identified the sequence GFOGER as what remains the highest-affinity triple-helical ligand for α2β1 discovered to date.

These two contrasting reactivities remain in the literature: DGEA, reported to be active in linear form [61], and GFOGER, active as a triple-helix but not as a shorter, linear peptide [31]. In our hands, neither linear nor triple-helical DGEA bound the resting platelet, a point to be discussed further below. Nonetheless, using either DGEA or GFOGER as a title keyword to search PubMed returns ~50 papers for each attesting to their integrin-reactivity, but to date no complex of an integrin I domain with DGEA has been deposited in the Protein Data Bank, although three complexes between integrin I domains and GFOGER or a similar ligand are present (1DZI, 4BJ3 and 2M32).

9.5 Collagen Sequences That Bind Integrins

9.5.1 GxOGER and Related Ligands in the Fibrillar Collagens

Both Barnes and Emsley promptly recognised GFOGER as just one of a series of integrin-binding motifs occurring in the fibrillar collagens [18, 31], as did Hook's group, who reported

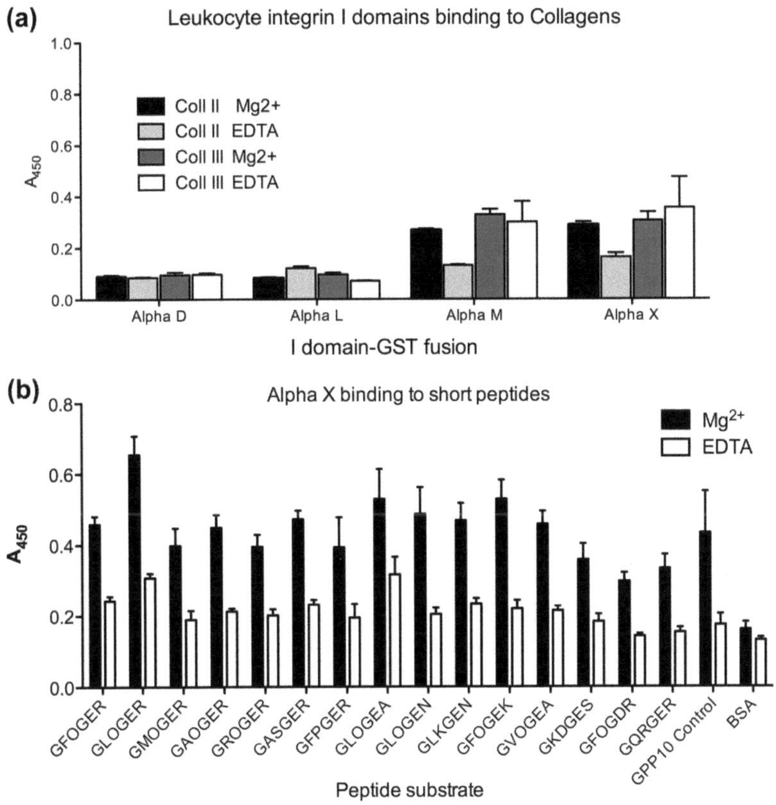

Fig. 9.1 **a** The ability of the recombinant I domains of leukocyte integrins to bind to collagen II and III monomers in ELISA-like assays is shown. With the exception of αM and αX applied to collagen II, there was no diminution of binding in the presence of EDTA. **b** Explores the capacity of triple-helical peptide motifs, in GPC–[GPP]$_5$–Gxx′GEx″–[GPP]$_5$–GPC format, to support leukocyte integrin I domain binding. Note that these values are low, and the Control peptide, GPP10, displays almost as much activity as the authentic motifs, and is also EDTA-sensitive. Compare amplitudes with the data in Fig. 9.2, where α2 I domain is used under the same conditions

a weakly-binding GASGER in collagen I and subsequently GROGER in collagen III [29, 74]. These motifs generally have a hydrophobic residue at position x in the GxOGER generic sequence. Siljander summarised their occurrence at specific loci within the D-periods of a wide range of fibrillar collagens, (D1: GLOGER; D2: GAOGER; D3: GFOGER in collagens I and II, GAOGER in III; GMOGER in collagens I, II and III) [59]. Beyond D3, conservation is less good, with weakly-binding motifs such as GAOGER present in some collagens and GQRGER or GLSGER in others. The activity of several of these sequences has been tested using synthetic peptides, and subsequently, Raynal extended this approach by synthesising the 27-residue peptide library, Collagen Toolkit III, and was able to confirm the activity of GROGER and to discover GLOGEN as an integrin-binding motif, later shown to be a preferred ligand for $\alpha1\beta1$ [50].

GROGER, quite a good ligand for $\alpha2\beta1$, unexpectedly contains the positively-charged arginine in place of the bulky hydrophobic residue that was thought to be essential. It seems quite likely that the three-carbon stem of its sidechain might fulfil the same function, supporting hydrophobic contact with the I domain surface, but this requires confirmation through structural study.

GLOGER was identified as a good ligand for $\alpha1\beta1$, as good as GFOGER, whilst GLOGEA seems similarly effective. It is therefore not surprising that GLOGEN proved to be a higher-affinity selective ligand for $\alpha1\beta1$, recently identified in this context, along with GVOGEA, a weaker but specific $\alpha1\beta1$ ligand [23]. Subsequently, a low-resolution SAXS structure of a complex between $\alpha1$ I domain and a GLOGEN-containing peptide has been reported, with the interesting demonstration of a 2-to-1 I domain-to-peptide complex [12]. To achieve this outcome, an activated form of $\alpha1$ I, E317A, was used, similar to the equivalent active form of $\alpha2$ I domain, E318W [2], also shown previously to form a 2-to-1 $\alpha2$ I-to-GFOGER crystal complex [9]. Both I domains supported 2-to-1 complexes in solution, suggesting that this is not merely a crystallisation artefact. Whether such complexes can occur in nature is debatable, but might represent a means of cell–cell adhesion, using a single strand of collagen as an intermediate bridging ligand. Such single triple helices might be found in non-fibrillar collagens such as collagen IV, or after dispersion of collagen fibres during tissue resorption or remodeling during wound repair. Whatever the significance of these 2-to-1 structures, they reveal a lower affinity mode of binding that may become operative when an integrin is activated, discussed further below.

Inspection of sequence, based on knowledge accrued since the first reports in 2000, has proved valuable in identifying potential integrin-binding sites in the collagen family at large, sometimes supported by synthesis of corresponding triple-helical peptides. The fibrillar collagens, I, II and III have attracted most attention, not least because the development of the peptide Toolkits renders their study straightforward. No such reagents exist at present for other collagens. The sequences of the fibrillar collagens, V, XI, XXIV and XXVII, are much less rich in GXOGER motifs than the more familiar I, II and III (see Table 9.1), reflecting their different evolutionary paths after the emergence of the ancestral collagen I, and the restricted expression of collagen XXVII to embryogenesis [47].

9.5.2 GxOGEK

Conservative substitution of K for R within triple-helical integrin ligand peptides revealed a similar motif, GFOGEK, as a useful ligand for $\alpha1\beta1$, $\alpha2\beta1$ and $\alpha11\beta1$ [23, 75], suggesting that it might prove to be representative of a series of GxOGEK homologues. Such sequences, like GxOGER, are widespread within the collagens as a whole.

GxOGEK motifs are rare in collagens I, II and III, with single occurrences in $\alpha2$(I) and at a discrete conserved locus in $\alpha1$(II) and $\alpha1$(III), seven triplets before the collagenase cleavage site. Collagen V lacks established GxOGER motifs, except for GMOGER in $\alpha2$(V), at loci conserved with integrin sites in D1 and D3 of collagens I, II and III. At the nearby locus in D3 where GFOGER occurs in collagens I and II, collagen V contains GNOGER, that deserves investigation. Collagen V also contains several GLOGEx″ motifs that might offer a site to $\alpha1\beta1$, including GLOGEK at a locus conserved in its $\alpha1$ and $\alpha3$ chains. Few other of the OGEK motifs that are relatively abundant appear to be promising integrin sites, lacking hydrophobic side-chains at the x position. GLOGEK in $\alpha2$(V) and GIOGEK in $\alpha3$(V) are the exceptions. Collagen V has a role in pericellular collagen fibrillogenesis in various tissues [60, 64, 71], and can support $\alpha11\beta1$-mediated cell migration [48].

Collagen XI, a co-constituent of cartilage fibres along with collagen II, is more promising, with a GFOGER site aligned in its $\alpha1$ and $\alpha2$ chains at a central D3 locus conserved with that in collagens I and II. GLOGEA in $\alpha1$(XI), that may bind $\alpha1\beta1$, aligns with GLOGES in $\alpha2$(XI), located in D1 near the GLOGEN and GLOGER sites in collagens I, II and III. This will most likely reconstitute an integrin site similar to the GFOGER/GPOGES site in $\alpha1$(I) and $\alpha2$(I). In Chap. 4, Lundgren–Åkerland describes binding of $\alpha10\beta1$-expressing cells to collagens II and XI, as well as to GFOGER and GLOGER-containing peptides.

The remaining fibrillar collagens, XXIV and XXVII, lack established integrin-binding motifs

Table 9.1 Integrin recognition motifs in all human collagens

	Chain	GER	GEK	GLOGE	RGD	Known and prospective integrin motifs (listed from N-terminus)
Fibrillar	Iα1	11	1	1	2	GROGER, GLOGER, [a]GFOGER, GMOGER, GQRGER, GASGER
	Iα2	8	3	2	2	GROGER, GLOGER, GLOGER
	II	11	4	1	3	GLOGER, GVOGEA, [a]GFOGER, GMOGER, GQRGER
	III	14	4	1	1	GROGER, GLOGEN, [a]GAOGER, GMOGER, GLSGER
	Vα1	7	14	3	2	*GLOGEK*, *GLOGEO*, *GLOGEG*
	Vα2	10	3	1	7	GMOGER, [a]*GNOGER*, GMOGER, GQRGER
	Vα3	7	14	0	3	
	XIα1	6	14	1	0	GLOGEA, [a]GFOGER
	XIα2	8	14	2	3	[a]GFOGER
	XXIV	5	7	4	0	*GNOGER*, [a]*GLOGEO*
	XXVII	3	4	2	3	[a]*GLOGEO*, GLOGEA
Network	IVα1	6	21	3	3	GFOGER, **GDQ**, *GLOGEK*, *GLOGEK*, *GLOGEK*
	IVα2	6	6	3	9	*GLOGEM*, **GRA**, *GLOGEV*, GAOGER, *GLOGEK*
	IVα3	4	9	1	6	*GLOGES*, GFOGER
	IVα4	5	11	1	8	GLOGEA, GFOGER, GFOGER
	IVα5	6	14	2	0	*GLOGEK*, GFOGER, *GLOGEO*
	IVα6	3	10	1	3	*GLOGEK*, *GLOGEL*
	VIIIα1	0	1	0	0	
	VIIIα2	1	0	0	2	*GVOGER*
	X	5	1	0	0	*GKOGER*, GFOGEK, GROGER
FACIT	IXα1	3	3	1	0	*GLOGEL*
	IXα2	2	4	2	0	*GLOGEI*, *GLOGEK*
	IXα3	4	3	0	2	GMOGER
	XII	1	1	0	2	*GLOGEK*
	XIV	3	2	0	1	
	XXVI	7	17	0	1	
	XIX	6	7	1	1	*GLOGEH*, *GIOGEK*
	XX	2	2	0	0	
	XXI	1	4	0	0	
	XXII	8	14	4	2	*GLOGEV*, *GLOGEI*, *GNOGER*, GLOGEN
TM	XIII	4	10	0	0	
	XVII	1	3	0	0	
	XXIII	0	1	0	2	
	XXV	3	10	0	0	
Multiplexins	XV	1	9	0	0	
	XVIII	3	4	1	1	*GLOGEO*

(continued)

Table 9.1 (continued)

	Chain	GER	GEK	GLOGE	RGD	Known and prospective integrin motifs (listed from N-terminus)
Other	VIα1	4	6	1	3	*GLOGEK*, GAOGER
	VIα2	1	4	0	5	
	VIα3	4	3	0	5	GFOGEK, GAOGER
	VIα5	0	3	0	1	
	VIα6	2	3	1	1	*GLOGEM*
	VII	22	17	2	3	GAOGER, GLOGER, GFOGER, GROGER, GLOGER, GAOGER
	XXVI	1	2	1	0	*GLOGEM*
	XXVIII	6	2	1	1	

[a] Indicates conserved high-affinity locus in D-period 3 of the fibrillar collagens. Alignment of other sites is approximate. **Bold** indicates the composite site in α1(IV) and α2(IV). *Italics* indicates prospective sites, not tested to date. In the fibrillar collagens, some known low-affinity sites (GAOGER) are omitted to conserve space

such as GxOGER, except the untested GNOGER in α1(XXIV). Plausible GLOGEx″ α1β1-selective motifs are also found in α1(XXIV), with Q, V, O and D as x″. Similarly, collagen XXVII contains single occurrences of GLOGEO and GLOGEA, prospective α1β1 ligands.

9.6 Sites in the Non-fibrillar Collagens

9.6.1 Network-Forming Collagens, IV, VIII and X

The non-fibrillar collagens are much less rich in defined GxOGER integrin-binding sites, and where such sites have been proposed, their exact location within the triple-helical domains is more often by inference than experiment. Parkin et al. have prepared a collagen IV interactome [44], and mapped putative integrin sites within the three different heterotrimers of collagen IV (α1α1α2, α3α4α5 and α5α5α6), using the location of GxOGEx″ motifs as a guide. A point of interest is that few motifs are aligned in all three constituent chains. One exception, in a region described as an endothelial cell binding domain in the most abundant form of basal lamina collagen IV, α1α1α2, contains a GFOGER/GFOGER/GLOGEM locus near to the disperse α1β1 site identified by Kuhn's group, discussed further below. Other potential sites, GLOGEx″ being a prime example that is represented in all six α chains, often occur unsupported in the other chains of the heterotrimer, but nonetheless form credible sites for both α1β1 and α2β1; critical interactions are thought to involve just one GxOGEx″ glutamate within a triple helix, with crucial ancillary stabilisation from hydrophobic x or positively-charged x″ residues. In α3α4α5, both GLOGEx″ and GFOGER occur unsupported in different loci in each α chain, but the adjacent chains contain x hydrophobic residues that may contribute to binding. OGEK triplets are sparse, with GLOGEK and GIOGEK in α5(IV) the most promising, although GEK occurs many times. These, along with GFOGER, form promising loci in the α5α5α6 form of collagen IV, but the α6 chain contributes just one each of GLOGEK, GFOGEK and GLOGEx″ as putative integrin sites.

Kuhn's group reported and have researched extensively a disperse site for α1β1, comprising GPOGDQ triplets in α1(IV) and the aligned GAKGRA triplet in α2(IV). The nature of this site remains enigmatic: one of the α1 chain aspartate residues co-ordinates the metal ion in the α1 I domain MIDAS, yet a competent recombinant α1 I domain is unable to bind a model homotrimeric peptide containing the sequence GFOGDR [23]. The critical but ancillary role of the hydrophobic F residue, important in defining the affinity of peptides for

$\alpha 2\beta 1$, seems insufficient to support the GDR triplet in binding $\alpha 1\beta 1$, despite GFOGER being a moderately good ligand for $\alpha 1\beta 1$. In the native $\alpha 2$(IV) collagen chain, the lysine of the GAK triplet preceding GRA has been proposed to have a critical role, by forming a salt bridge with the I domain surface [53].

Chin models a slightly different peptide-I domain relationship [12], using low resolution SAXS- and NMR-derived structures to direct the docking programme HADDOCK, and suggest that the peptide, triple-helical GLOGEN, sits more centrally within the binding trench in $\alpha 1$ than in $\alpha 2$, where the corresponding GFOGER is described as binding to the edge of the equivalent trench in the $\alpha 2$ I domain. Part of the rationale for this may be a steric clash of the bulkier sidechain of the x″ arginine in GER-containing motifs, explaining why GEN may be a preferred ligand for $\alpha 1\beta 1$. In line with these ideas, Seo modelled and expressed an unnatural sequence, GFPGEN, that they found to be selective for $\alpha 1\beta 1$ [58]. Emsley described the $\alpha 1$ I domain as having a flatter, more open MIDAS, more readily able to receive a short co-ordinating aspartate than $\alpha 2$ I domain [18]. Further work to extend these concepts to include $\alpha 10\beta 1$ is called for.

Both the $\alpha 1$ and $\alpha 2$ chains of collagen VIII are devoid of defined integrin motifs, although GVOGER in its $\alpha 2$ chain, by analogy with GVOGEA in collagen II, presents a possible binding motif. Turner showed that endothelial cell attachment to collagen $\alpha 2$(VIII) homotrimers was partially mediated by $\alpha 2\beta 1$, using anti-$\alpha 2$ blocking antibodies [69]. Anti-$\beta 1$ completely inhibited adhesion, implying the presence of other collagen-binding integrins. Of general note, a cyclic RGD peptide had no effect on cell adhesion to this collagen preparation despite the presence of two RGD motifs within the COL domain. Here, as in other collagens, triple-helical RGD was not recognised by the relevant integrins although the latter were present and competent to bind fibronectin. Adiguzel note the upregulation of collagen VIII in atherosclerotic tissue [1], where it may support vascular smooth muscle cell migration, and identify $\beta 1$ integrin-mediated signalling to the small GTPase, RhoA.

Collagen X contains one known and one possible GxOGER motif, GRO and GKOGER, along with the proven GFOGEK.

9.6.2 FACIT Collagens, IX, XII, XIV, XVI, XIX, XX, XXI and XXII

The heterotrimeric collagen IX, a FACIT collagen associated with collagen II fibres, was reported to express strong integrin-binding activity, tested on all four collagen binding I domains, compared to other collagens [28] and showing similar high affinity for $\alpha 1$, $\alpha 2$ and $\alpha 11$, but rather weaker for $\alpha 10$. (See also Chap. 4 in this volume.) This might be attributed to its aligned GLOGEL and GLOGEI motifs which occur in the (N-terminal) COL3 domain of the $\alpha 1$(IX) and $\alpha 2$(IX) chains. Collagen IX is not rich in established motifs; the COL2 domain contains GLOGEK in $\alpha 2$(IX) and GMOGER in $\alpha 3$(IX). It may be that these GLOGEx″ motifs are selective for $\alpha 1\beta 1$. Whilst the patency of GFOGER in collagen I (and by inference, in collagen II) has been questioned by Orgel's group, GFOGER being proposed to be systematically buried within the complex twisted structure of the fibre [46], the presence of strongly integrin-binding FACIT collagens may compensate for this effect, if it does indeed occur.

Collagen XII presents only the untested GLOGEK. Although it is proposed that integrin-mediated tension may stimulate collagen XII expression, its capacity to bind integrins directly is unknown [65]. It is worth noting that the non-collagenous domain 3 (NC3) of collagen XII contains several VWF A domain structures with intact DxSxS MIDAS motifs. This raises the interesting possibility that the affinity of the FACIT collagen XII for the surface of a collagen fibre may include a cation-dependent interaction between the its VWF A domains and integrin-binding motifs in the fibrillar collagen. A similar VWF A domain-mediated mechanism of collagen–collagen interaction (dimerisation) has been described for collagen VI [3] in this case utilising the COL domain sequence GSOGER as a counter-ligand. The ability of collagen XII to regulate the biomechanical properties of associated fibres has been attributed to its NC3 domain [40].

Collagen XIV contains no defined integrin motif, although GMOGEK and GTOGER represent untested possibilities. Collagen XVI, although rich in GEK and GER triplets, lacks any OGEK or OGER motifs, but Eble et al. [16], propose GLQGER and GIKGER, but not GGKGER and GKAGER, as integrin-binding in collagen XVI. Two of its interrupting sequences contain RGD motifs, raising the possibility of $\alpha5\beta1$- or $\alpha v\beta3$-mediated adhesion to these less-structured NC domains. However, Eble concluded that these motifs, along with one RGD triplet that lies firmly within a COL domain, do not contribute to cell-binding activity, reinforcing the conclusion that RGD is cryptic in collagens when located in their COL domains.

Collagen XIX presents two possible motifs, GLOGEH and GIOGEK, again untested.

Collagens XX and XXI, both designated FACITs, contain four or five of either GER or GEK motifs, none preceded by O, and therefore lack obvious integrin-reactivity. In contrast, collagen XXII, with a much longer interrupted triple-helix, contain 14 GEK motifs, possible integrin-binders, including GROGEK, and eight GERs, including GEOGER and GNOGER.. Most telling, four GLOGEx″ motifs occur, including GLOGEN, now shown to be an $\alpha1\beta1$-selective motif [23]. Koch applied HACAT keratinocytes ($\alpha2\beta1$-expressing) and WI-26 fibroblasts (which express both $\alpha1\beta1$ and $\alpha2\beta1$) to collagen XXII, the latter binding more efficiently [32]. It is tempting to conclude that the sequence, GLOGEN was central to this outcome.

9.6.3 "Other" Collagens

Amongst the collagens VI, VII, XXVI and XXVIII (undesignated by Ricard-Blum), only collagen VII is as rich as the archetypal fibrillar collagens in GxOGER motifs, where x is A and L (twice each), F or R. This suggests a marked propensity to bind integrins. Saelman et al. report $\alpha2\beta1$-dependent adhesion of platelets to collagens including VI and VII [54]. However, Chen et al. reported $\alpha2\beta1$-mediated attachment to the NC1 domain of VII that was RGD-independent and survived denaturation [11], and Liebert et al. [36] observe co-localisation of collagen VII with $\alpha6\beta4$, not known as a collagen receptor. This interaction may regulate laminin-332 organisation during wound healing [43]. Whether the collagen VII COL domain, although an attractive candidate, expresses the anticipated strong integrin reactivity remains to be established.

Of the others, the short ($\sim$100 triplet) $\alpha1$(VI) and $\alpha3$(VI) COL domains each contain only GAOGER and either GLOGEK or GFOGEK, whilst XXVI contains a single GLOGEM motif and XXVIII contains a single GVOGER motif, mentioned elsewhere above.

9.6.4 Multiplexins, Collagens XV and XVIII

Collagen XV lacks obvious motifs, although contains several GEK triplets, whilst collagen XVIII contains a single GLOGEO motif. Consistent with this, collagen XV has been found not to support cell adhesion [26], but instead to bind other matrix proteins (FN, LN and VN) that will offer RGD-dependent cell-binding activity. Halfter et al. report little cell binding activity of collagen XVIII [22].

9.6.5 Transmembrane Collagens, XIII, XVII, XXIII and XXV

The transmembrane collagens are intriguing, since they offer novel and barely-explored opportunities for cell–cell interaction. Collagen XIII lacks obvious integrin sites, with four GER and several GEK triplets, but none preceded by

O. Nonetheless, $\alpha 1\beta 1$, though not $\alpha 2\beta 1$, binding is reported [42]. Integrin reactivity has also been identified for collagen XVII [41], which when transfected into suitable cell lines, also binds the non-integrin immune collagen receptor, LAIR1 [34, 42]. Again, obvious integrin-binding motifs are absent.

Collagen XXIII contains the low-affinity motif, GASGER, but no other obvious integrin-binding activity amongst the sundry GEK triplets, suggesting weak or absent integrin reactivity. Veit et al., using bacterially expressed foldon-peptide collagenous materials, suggest that the integrin reactivity they observed in collagen XXIII does not reside in GTSGER or GEKGER [70].

Collagen XXV, similar in most respects, lacks even low affinity GER motifs.

9.7 Recognition of Integrins by Multiple Collagen α-Chains

The first collagen peptide-I domain complex [18] showed that two strands of a homotrimeric GFOGER peptide, the leading and the middle strand, were involved in integrin binding. The middle strand was crucial, containing the glutamate responsible for much of the binding energy. As described, however, the contribution of the hydrophobic phenylalanine was also critical, and the identity of x in GxOGER-containing peptides defines an affinity series at the level of binding assays [23, 50, 59], and in determining the ability of a cell to migrate across a peptide coated surface [19]. The positively charged arginine residue also forms an important salt bridge with corresponding negative charges on the I domain surface, in $\alpha 2$ I domain at least. The leading strand, however, makes several significant contributions, including further hydrophobic bonding and a less intimate charge–charge interaction with its F and R residues respectively. These interactions have been detailed by Emsley and others and need little reiteration here. From this body of work it seems unlikely that a heterotrimer containing a GxOGER motif in just a single strand would support high-affinity interaction. This remains to be explored experimentally.

9.8 Effect of Cellular Activation and Integrin Binding

The recent publication of a new 2-to-1 I domain-to-peptide structure [9, 12] shows that a second binding mode is possible. In this receptor–ligand complex of $\alpha 2$ I with GFOGER, one copy of an activated form of $\alpha 2$ I (E318W) adopts a relationship to two strands of the peptide that is essentially identical to that described by Emsley, although in this iteration, the trailing strand takes the primary role by housing the crucial glutamate. Other equivalent supporting interactions are provided by the middle strand. The second copy of the I domain, however, binds mainly to just one strand, the leading strand, with its glutamate conventionally co-ordinating the MIDAS. There is no arginine available in an adjacent strand to meet the requirement for salt-bridging, and, by the same token, no phenylalanine to offer additional hydrophobic bond support. Proline in the peptide flanking sequence fulfils this function. The 2-to-1 complex was sufficiently robust to survive gel filtration, but not when a lower-affinity peptide, GMOGER, was used. Crucially, an active mutant of $\alpha 2$ I, E318W, was used to prepare the complex; wild type would not support stable complex formation in solution with either peptide. A similar complex was published subsequently between the preferred ligand for $\alpha 1\beta 1$, GLOGEN, and the corresponding E317A form of $\alpha 1$ domain, where, presumably, similar considerations apply [12].

This work suggests that, upon activation, interactions of integrins with lower affinity sequences can become useful. For $\alpha 2\beta 1$, the affinity series, $x = F > L \geq R > M > A$ in GxOGER-containing motifs appears to hold well in different cell adhesion studies, e.g. using platelets and HT1080 cells [19, 59]. With the latter, although resting cell adhesion was virtually negligible in relatively stringent static

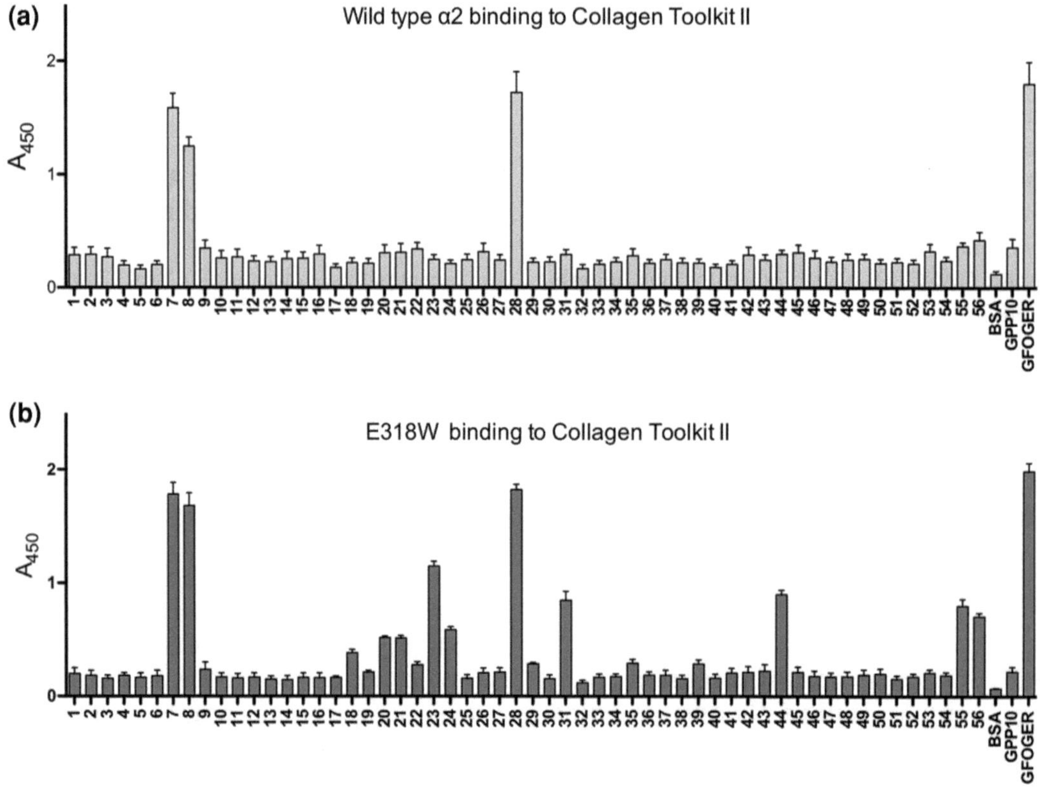

Fig. 9.2 **a** The ability of wild type α2 I domains to bind to Toolkit II is shown. Wild-type binds well to just three peptides, II-7, II-8 and II-28. GLOGER occurs in the overlap between II-7 and II-8; GFOGER in II-28. **b** Shows the equivalent experiment using the more active I domain, E318W, which, in addition, binds well to established motifs, GMOGER in II-31, GQRGER in II-44, and less well to GAOGER in II-18. Other binding activity remains undefined, but it may be significant that II-23 and II-24 share GKAGEK in their overlapping sequence. All other weakly-binding peptides include GEx″ motifs. (Reproduced from Carafoli et al. [9] under the creative commons license. See http://creativecommons.org/licenses/by/3.0/)

adhesion assays (with multiple wash steps) the capacity of cells to migrate upon GAOGER surfaces was much greater than on GFOGER. Thus, a tight-binding surface may not be what is needed if cells are to invade, repopulate and repair a wound quickly and successfully.

The use of activated I domains provides other useful insights; lower-affinity peptides, with corresponding lower-affinity motifs, appear positive on Toolkit maps, (see Fig. 9.2) but binding curves (Fig. 9.3) indicate that the affinity of such motifs may be one or two orders of magnitude lower than the best motif, GFOGER [9]. Nonetheless, these weak interactions may become more important in the context of an activated cell. It is important to note that whilst the selectivity of the activated form of the I domain is decreased (more Toolkit peptides become positive), binding affinity for the established peptides increases. This indicates that the activated I domain surface is able to make a greater number of contacts with the peptides than the resting integrin, suggesting greater plasticity of its binding surface. (As an aside, DGEA, which occurs in Toolkit peptide II-03, does not bind either wild type or E318W I domain.)

Taken together, these data support the idea that weaker motifs, or motifs present in only a single strand of a heterotrimeric collagen, may be valuable in the context of cellular behaviour that requires weaker interactions, and where

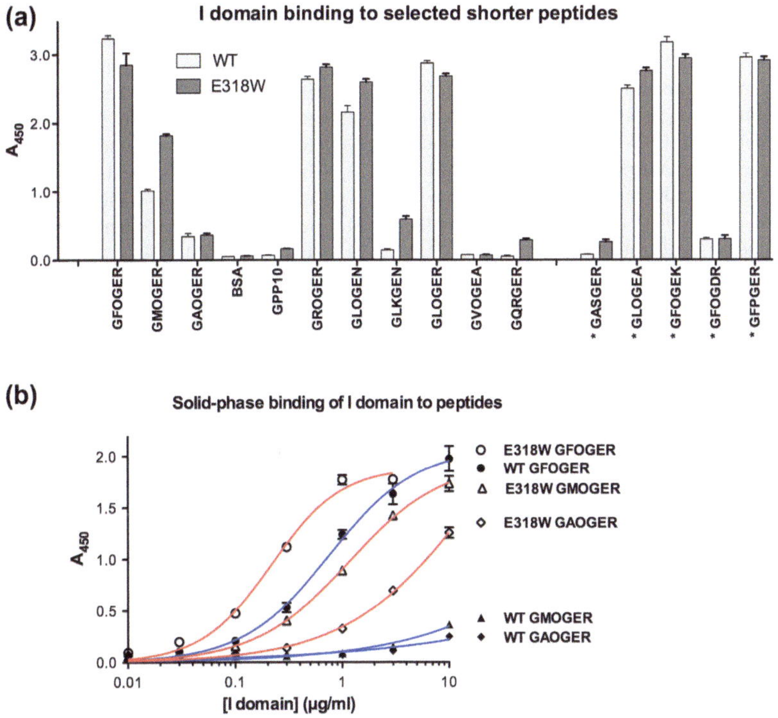

Fig. 9.3 **a** The ability of wild type and E318W $\alpha 2$ I domain to bind to shorter triple-helical peptides. Peptides were presented as in Fig. 9.1, and show that the activating mutation leads to improved binding to the moderate affinity motif, GMOGER, and to some low affinity peptides. **b** Quantitates this effect, and shows that the activating mutation shifts the binding curve threefold or so to the left for GFOGER and GMOGER. For the low affinity GAOGER, binding becomes measurable for E318W compared with wild type $\alpha 2$ I domain. (Reproduced from Carafoli et al. [9] under the creative commons license. See http://creativecommons.org/licenses/by/3.0/)

cells may be activated, perhaps by local cytokine levels in an inflammatory lesion.

9.9 Engineered Collagens as Integrin-Binding Proteins

Finally, we mention the potential for the use of integrin-binding motifs in tissue engineering applications. This concept arose in part from the location of an integrin-binding site in the *Streptococcus pyogenes* adhesin, SCL1, within its 213 residue COL domain [25], subsequently identified as the non-hydroxylated analogue of mammalian binding motifs, GLPGER [10]. Such motifs are rare in bacterial collagens, perhaps unique, although a number of bacteria express collagen-like proteins. Höök's group subsequently designed a sequence, GFPGEN, that was suitable for bacterial expression and that displayed $\alpha 1\beta 1$-selectivity [58]. Bacterial expression of accurately folded collagens becomes a real possibility. The combination of exploration of integrin selectivity with re-engineering into bacterial expression systems makes the production of collagen-like biomaterials by fermentation an exciting prospect.

Acknowledgments Work described here from the authors' laboratory was supported by grants from the Medical Research Council, British Heart Foundation and the Wellcome Trust.

References

1. Adiguzel E, Hou G, Sabatini PJ, Bendeck MP (2013) Type VIII collagen signals via beta1 integrin and RhoA to regulate MMP-2 expression and smooth muscle cell migration. Matrix Biol 32(6):332–341
2. Aquilina A, Korda M, Bergelson JM, Humphries MJ, Farndale RW, Tuckwell D (2002) A novel gain-of-function mutation of the integrin [alpha]2 VWFA domain. Eur J Biochem 269(4):1136–1144
3. Ball S, Bella J, Kielty C, Shuttleworth A (2003) Structural basis of type VI collagen dimer formation. J Biol Chem 278(17):15326–15332
4. Barczyk M, Carracedo S, Gullberg D (2010) Integrins. Cell Tissue Res 339(1):269–280
5. Bengtsson T, Camper L, Schneller M, Lundgren-Akerlund E (2001) Characterization of the mouse integrin subunit alpha10 gene and comparison with its human homologue. Genomic structure, chromosomal localization and identification of splice variants. Matrix Biol 20(8):565–576
6. Bienkowska J, Cruz M, Atiemo A, Handin R, Liddington R (1997) The von Willebrand factor A3 domain does not contain a metal ion-dependent adhesion site motif. J Biol Chem 272:25162–25167
7. Camper L, Hellman U, Lundgren-Akerlund E (1998) Isolation, cloning, and sequence analysis of the integrin subunit alpha10, a beta1-associated collagen binding integrin expressed on chondrocytes. J Biol Chem 273(32):20383–20389
8. Camper L, Holmvall K, Wangnerud C, Aszodi A, Lundgren-Akerlund E (2001) Distribution of the collagen-binding integrin alpha10beta1 during mouse development. Cell Tissue Res 306(1):107–116
9. Carafoli F, Hamaia SW, Bihan D, Hohenester E, Farndale RW (2013) An activating mutation reveals a second binding mode of the integrin alpha2 I domain to the GFOGER motif in collagens. PLoS One 8(7):e69833
10. Caswell CC, Barczyk M, Keene DR, Lukomska E, Gullberg DE, Lukomski S (2008) Identification of the first prokaryotic collagen sequence motif that mediates binding to human collagen receptors, integrins alpha2beta1 and alpha11beta1. J Biol Chem 283(52):36168–36175
11. Chen M, O'Toole EA, Li YY, Woodley DT (1999) Alpha 2 beta 1 integrin mediates dermal fibroblast attachment to type VII collagen via a 158-amino-acid segment of the NC1 domain. Exp Cell Res 249(2):231–239
12. Chin YK, Headey SJ, Mohanty B, Patil R, McEwan PA, Swarbrick JD, Mulhern TD, Emsley J, Simpson JS, Scanlon MJ (2013) The structure of integrin alpha1I domain in complex with a collagen-mimetic peptide. J Biol Chem 288(52):36796–36809
13. Colombatti A, Bonaldo P, Doliana R (1993) Type A modules: interacting domains found in several non-fibrillar collagens and in other extracellular matrix proteins. Matrix 13:297–306
14. Davies D, Tuckwell DS, Calderwood DA, Weston SA, Takigawa M, Humphries MJ (1997) Molecular characterisation of integrin-procollagen C-propeptide interactions. Eur J Biochem 246(2):274–282
15. Dickeson SK, Walsh JJ, Santoro SA (1997) Contributions of the I and EF hand domains to the divalent cation-dependent collagen binding activity of the $\alpha2\beta1$ integrin. J Biol Chem 272:7661–7668
16. Eble JA, Kassner A, Niland S, Morgelin M, Grifka J, Grassel S (2006) Collagen XVI harbors an integrin alpha1 beta1 recognition site in its C-terminal domains. J Biol Chem 281(35):25745–25756
17. Emsley J, King SL, Bergelson JM, Liddington RC (1997) Crystal structure of the I-domain from integrin $\alpha2\beta1$. J Biol Chem 272:28512–28517
18. Emsley J, Knight CG, Farndale RW, Barnes MJ, Liddington RC (2000) Structural basis of collagen recognition by integrin alpha2beta1. Cell 101(1):47–56
19. Farndale RW, Lisman T, Bihan D, Hamaia S, Smerling CS, Pugh N, Konitsiotis A, Leitinger B, de Groot PG, Jarvis GE, Raynal N (2008) Cell–collagen interactions: the use of peptide Toolkits to investigate collagen–receptor interactions. Biochem Soc Trans 36(Pt 2):241–250
20. Fitzsimmons CM, Cawston TE, Barnes MJ (1986) The platelet reactivity of collagen type I: evidence for multiple platelet-reactive sites in the type I collagen molecule. Thromb Haemost 56:95–99
21. Gullberg D, Gehlsen KR, Turner DC, Ahlen K, Zijenah LS, Barnes MJ, Rubin K (1992) Analysis of $\alpha1\beta1$, $\alpha2\beta1$ and $\alpha3\beta1$ integrins in cell–collagen interactions: identification of conformation dependent $\alpha1\beta1$ binding sites in collagen type 1. EMBO J 11:3865–3873
22. Halfter W, Dong S, Schurer B, Cole GJ (1998) Collagen XVIII is a basement membrane heparan sulfate proteoglycan. J Biol Chem 273(39):25404–25412
23. Hamaia SW, Pugh N, Raynal N, Nemoz B, Stone R, Gullberg D, Bihan D, Farndale RW (2012) Mapping of potent and specific binding motifs, GLOGEN and GVOGEA, for integrin alpha1beta1 using Collagen Toolkits II and III. J Biol Chem 287(31):26019–26028
24. Humphries JD, Byron A, Humphries MJ (2006) Integrin ligands at a glance. J Cell Sci 119(Pt 19):3901–3903
25. Humtsoe JO, Kim JK, Xu Y, Keene DR, Hook M, Lukomski S, Wary KK (2005) A streptococcal collagen-like protein interacts with the alpha2beta1 integrin and induces intracellular signaling. J Biol Chem 280(14):13848–13857
26. Hurskainen M, Ruggiero F, Hagg P, Pihlajaniemi T, Huhtala P (2010) Recombinant human collagen XV regulates cell adhesion and migration. J Biol Chem 285(8):5258–5265
27. Kadler KE, Baldock C, Bella J, Boot-Handford RP (2007) Collagens at a glance. J Cell Sci 120(Pt 12):1955–1958

28. Kapyla J, Jaalinoja J, Tulla M, Ylostalo J, Nissinen L, Viitasalo T, Vehvilainen P, Marjomaki V, Nykvist P, Saamanen AM, Farndale RW, Birk DE, Ala-Kokko L, Heino J (2004) The fibril-associated collagen IX provides a novel mechanism for cell adhesion to cartilaginous matrix. J Biol Chem 279(49):51677–51687
29. Kim JK, Xu Y, Xu X, Keene DR, Gurusiddappa S, Liang X, Wary KK, Hook M (2005) A novel binding site in collagen type III for the integrins, alpha 1beta 1 and alpha 2beta 1. J Biol Chem 280:32512–32520
30. Knight CG, Morton LF, Onley DJ, Peachey AR, Messent AJ, Smethurst PA, Tuckwell DS, Farndale RW, Barnes MJ (1998) Identification in collagen type I of an integrin alpha2 beta1-binding site containing an essential GER sequence. J Biol Chem 273(50):33287–33294
31. Knight CG, Morton LF, Peachey AR, Tuckwell DS, Farndale RW, Barnes MJ (2000) The collagen-binding A-domains of integrins alpha(1)beta(1) and alpha(2)beta(1) recognize the same specific amino acid sequence, GFOGER, in native (triple-helical) collagens. J Biol Chem 275(1):35–40
32. Koch M, Schulze J, Hansen U, Ashwodt T, Keene DR, Brunken WJ, Burgeson RE, Bruckner P, Bruckner-Tuderman L (2004) A novel marker of tissue junctions, collagen XXII. J Biol Chem 279(21):22514–22521
33. Lahti M, Heino J, Kapyla J (2013) Leukocyte integrins alphaLbeta2, alphaMbeta2 and alphaXbeta2 as collagen receptors–receptor activation and recognition of GFOGER motif. Int J Biochem Cell Biol 45(7):1204–1211
34. Lebbink RJ, de Ruiter T, Adelmeijer J, Brenkman AB, van Helvoort JM, Koch M, Farndale RW, Lisman T, Sonnenberg A, Lenting PJ, Meyaard L (2006) Collagens are functional, high affinity ligands for the inhibitory immune receptor LAIR-1. J Exp Med 203:1419–1425
35. Li Y, Brodsky B, Baum J (2009) NMR conformational and dynamic consequences of a Gly to Ser substitution in an osteogenesis imperfecta collagen model peptide. J Biol Chem 284(31):20660–20667
36. Liebert M, Washington R, Wedemeyer G, Carey TE, Grossman HB (1994) Loss of co-localization of alpha 6 beta 4 integrin and collagen VII in bladder cancer. Am J Pathol 144(4):787–795
37. Morton LF, Peachey AR, Barnes MJ (1989) Platelet-reactive sites in collagens type I and type III. Evidence for separate adhesion and aggregatory sites. BJ 258:157–163
38. Morton LF, Peachey AR, Zijenah LS, Goodall AH, Humphries MJ, Barnes MJ (1994) Conformation-dependent platelet adhesion to collagen involving integrin alpha 2 beta 1-mediated and other mechanisms: multiple alpha 2 beta 1-recognition sites in collagen type I. Biochem J 299(Pt 3):791–797
39. Morton LF, Zijenah LS, McCulloch IY, Knight CG, Humphries MJ, Barnes MJ (1991) Integrin-dependent platelet recognition sites in collagen: identification of a short platelet-reactive sequence in the type III-derived fragment alpha 1(III) CB3. Biochem Soc Trans 19(4):439S
40. Nishiyama T, McDonough AM, Bruns RR, Burgeson RE (1994) Type XII and XIV collagens mediate interactions between banded collagen fibers in vitro and may modulate extracellular matrix deformability. J Biol Chem 269(45):28193–28199
41. Nykvist P, Tasanen K, Viitasalo T, Kapyla J, Jokinen J, Bruckner-Tuderman L, Heino J (2001) The cell adhesion domain of type XVII collagen promotes integrin-mediated cell spreading by a novel mechanism. J Biol Chem 276(42):38673–38679
42. Nykvist P, Tu H, Ivaska J, Kapyla J, Pihlajaniemi T, Heino J (2000) Distinct recognition of collagen subtypes by alpha(1)beta(1) and alpha(2)beta(1) integrins. Alpha(1)beta(1) mediates cell adhesion to type XIII collagen. J Biol Chem 275(11):8255–8261
43. Nystrom A, Velati D, Mittapalli VR, Fritsch A, Kern JS, Bruckner-Tuderman L (2013) Collagen VII plays a dual role in wound healing. J Clin Invest 123(8):3498–3509
44. Parkin JD, San Antonio JD, Pedchenko V, Hudson B, Jensen ST, Savige J (2011) Mapping structural landmarks, ligand binding sites, and missense mutations to the collagen IV heterotrimers predicts major functional domains, novel interactions, and variation in phenotypes in inherited diseases affecting basement membranes. Hum Mutat 32(2):127–143
45. Persikov AV, Ramshaw JAM, Kirkpatrick A, Brodsky B (2000) Amino acid propensities for the collagen triple-helix. Biochemistry 39: 14960–14967
46. Perumal S, Antipova O, Orgel JP (2008) Collagen fibril architecture, domain organization, and triple-helical conformation govern its proteolysis. Proc Natl Acad Sci USA 105(8):2824–2829
47. Plumb DA, Dhir V, Mironov A, Ferrara L, Poulsom R, Kadler KE, Thornton DJ, Briggs MD, Boot-Handford RP (2007) Collagen XXVII is developmentally regulated and forms thin fibrillar structures distinct from those of classical vertebrate fibrillar collagens. J Biol Chem 282(17):12791–12795
48. Popova SN, Rodriguez-Sanchez B, Liden A, Betsholtz C, Van Den Bos T, Gullberg D (2004) The mesenchymal alpha11beta1 integrin attenuates PDGF-BB-stimulated chemotaxis of embryonic fibroblasts on collagens. Dev Biol 270(2):427–442
49. Prockop DJ, Kivirikko KI, Tuderman L, Guzman NA (1979) The biosynthesis of collagen and its disorders (Pt 2). NEJM 301:77–85
50. Raynal N, Hamaia SW, Siljander PR, Maddox B, Peachey AR, Fernandez R, Foley LJ, Slatter DA, Jarvis GE, Farndale RW (2006) Use of synthetic peptides to locate novel integrin alpha2beta1-binding motifs in human collagen III. J Biol Chem 281(7):3821–3831

51. Rehn M, Veikkola T, Kukk-Valdre E, Nakamura H, Ilmonen M, Lombardo C, Pihlajaniemi T, Alitalo K, Vuori K (2001) Interaction of endostatin with integrins implicated in angiogenesis. Proc Natl Acad Sci USA 98(3):1024–1029
52. Ricard-Blum S (2011) The Collagen Family. Cold Spring Harb Perspect Biol 3(1):a004978
53. Sacca B, Sinner EK, Kaiser J, Lubken C, Eble JA, Moroder L (2002) Binding and docking of synthetic heterotrimeric collagen type IV peptides with alpha1beta1 integrin. ChemBioChem 3(9): 904–907
54. Saelman EU, Nieuwenhuis HK, Hese KM, de Groot PG, Heijnen HF, Sage EH, Williams S, McKeown L, Gralnick HR, Sixma JJ (1994) Platelet adhesion to collagen types I through VIII under conditions of stasis and flow is mediated by GPIa/IIa (alpha 2 beta 1-integrin). Blood 83(5):1244–1250
55. Santoro SA (1986) Identification of a 160000 dalton platelet membrane protein that mediates the initial divalent cation-dependent adhesion of platelets to collagen. Cell 46:913–920
56. Santoro SA (1988) Molecular basis of platelet adhesion to collagen. In: Jamieson GA (ed.) Platelet membrane receptors: molecular biology, immunology, biochemistry, and pathology. Alan R. Liss, New York
57. Santoro SA, Cunningham L (1980) Collagen-mediated platelet aggregation: the role of multiple interactions between the platelet surface and collagen. Thromb Haemost 43:158–162
58. Seo N, Russell BH, Rivera JJ, Liang X, Xu X, Afshar-Kharghan V, Hook M (2010) An engineered alpha1 integrin-binding collagenous sequence. J Biol Chem 285(40):31046–31054
59. Siljander PR, Hamaia S, Peachey AR, Slatter DA, Smethurst PA, Ouwehand WH, Knight CG, Farndale RW (2004) Integrin activation state determines selectivity for novel recognition sites in fibrillar collagens. J Biol Chem 279(46):47763–47772
60. Smith SM, Zhang G, Birk DE (2014) Collagen V localizes to pericellular sites during tendon collagen fibrillogenesis. Matrix Biol 33:47–53
61. Staatz WD, Fok KF, Zutter MM, Adams SP, Rodriguez BA, Santoro SA (1991) Identification of a tetrapeptide recognition sequence for the a2b1-binding sites in collagen I. JBC 266:7363–7367
62. Staatz WD, Rajpara SM, Wayner EA, Carter WG, Santoro SA (1989) The membrane glycoprotein Ia-IIa (VLA_2) complex mediates the Mg^{2+}-dependent adhesion of platelets to collagen. JCB 108:1917–1924
63. Staatz WD, Walsh JJ, Pexton T, Santoro SA (1990) The alpha 2 beta 1 integrin cell surface collagen receptor binds to the alpha 1 (I)-CB3 peptide of collagen. J Biol Chem 265(9):4778–4781
64. Sun M, Chen S, Adams SM, Florer JB, Liu H, Kao WW, Wenstrup RJ, Birk DE (2011) Collagen V is a dominant regulator of collagen fibrillogenesis: dysfunctional regulation of structure and function in a corneal-stroma-specific Col5a1-null mouse model. J Cell Sci 124(Pt 23):4096–4105
65. Trachslin J, Koch M, Chiquet M (1999) Rapid and reversible regulation of collagen XII expression by changes in tensile stress. Exp Cell Res 247(2):320–328
66. Tuckwell DS, Reid KBM, Barnes MJ, Humphries MJ (1996) The A-domain of integrin alpha 2 binds specifically to a range of collagens but is not a general receptor for the collagenous motif. Eur J Biochem 241(3):732–739
67. Tulla M, Lahti M, Puranen JS, Brandt AM, Kapyla J, Domogatskaya A, Salminen TA, Tryggvason K, Johnson MS, Heino J (2008) Effects of conformational activation of integrin alpha 1I and alpha 2I domains on selective recognition of laminin and collagen subtypes. Exp Cell Res 314(8):1734–1743
68. Tulla M, Pentikainen OT, Viitasalo T, Kapyla J, Impola U, Nykvist P, Nissinen L, Johnson MS, Heino J (2001) Selective binding of collagen subtypes by integrin alpha 1I, alpha 2I, and alpha 10I domains. J Biol Chem 276(51):48206–48212
69. Turner NJ, Murphy MO, Kielty CM, Shuttleworth CA, Black RA, Humphries MJ, Walker MG, Canfield AE (2006) Alpha2(VIII) collagen substrata enhance endothelial cell retention under acute shear stress flow via an alpha2beta1 integrin-dependent mechanism: an in vitro and in vivo study. Circulation 114(8):820–829
70. Veit G, Zwolanek D, Eckes B, Niland S, Kapyla J, Zweers MC, Ishada-Yamamoto A, Krieg T, Heino J, Eble JA, Koch M (2011) Collagen XXIII, novel ligand for integrin alpha2beta1 in the epidermis. J Biol Chem 286(31):27804–27813
71. Wenstrup RJ, Florer JB, Davidson JM, Phillips CL, Pfeiffer BJ, Menezes DW, Chervoneva I, Birk DE (2006) Murine model of the Ehlers-Danlos syndrome. Col5a1 haploinsufficiency disrupts collagen fibril assembly at multiple stages. J Biol Chem 281(18):12888–12895
72. Weston SA, Hulmes DJS, Mould AP, Watson RB, Humphries MJ (1994) Identification of integrin $\alpha_2\beta_1$ as cell surface receptor for the carboxy-terminal propeptide of type I procollagen. J Biol Chem 269:20982–20986
73. Wickstrom SA, Alitalo K, Keski-Oja J (2002) Endostatin associates with integrin alpha5beta1 and caveolin-1, and activates Src via a tyrosyl phosphatase-dependent pathway in human endothelial cells. Cancer Res 62(19):5580–5589
74. Xu Y, Gurusiddappa S, Rich RL, Owens RT, Keene DR, Mayne R, Hook A, Hook M (2000) Multiple binding sites in collagen type I for the integrins $\alpha1\beta1$ and $\alpha2\beta1$. JBC 275:38981–38989
75. Zhang WM, Kapyla J, Puranen JS, Knight CG, Tiger CF, Pentikainen OT, Johnson MS, Farndale RW, Heino J, Gullberg D (2003) Alpha 11beta 1 integrin recognizes the GFOGER sequence in interstitial collagens. J Biol Chem 278(9):7270–7277

Cellular Signaling by Collagen-Binding Integrins

10

Jyrki Heino

Abstract

The four collagen-binding αI domain integrins form their own subgroup among cell adhesion receptors. The signaling functions of α1β1 and α2β1 integrins have been analyzed in many experimental models, whereas less studies are available about the more recently found α10β1 and α11β1 heterodimers. Interestingly, collagen binding by α1β1 and α2β1 often generates opposite cellular responses. For example α1β1 has often been reported to promote cell proliferation and to suppress collagen synthesis, whereas α2β1 can in many model systems inhibit growth and promote collagen synthesis. There are obviously cell type dependent factors modifying the signaling. Additionally the structure and the organization of collagenous matrix play a critic role. Many recent studies have also stressed the importance of the crosstalk between the integrins and other cell surface receptors.

Keywords

Integrins · Collagen · Signaling

10.1 Introduction

The members of the collagen receptor subgroup of the integrins recognize their ligands using an inserted domain in their α subunit (αI domain, often called as αA domain). Four heterodimers belong to this category, namely α1β1, α2β1, α10β1 and α11β1 [8, 11, 77, 102]. While α1β1 and α2β1 are very abundantly expressed on distinct cell types, the tissue distribution of α10β1 is mainly limited to cartilage. Integrin α11β1 is found on mesenchymal cells, e.g. fibroblasts. The four receptors have differences in their ability to recognize extracellular matrix (ECM) and other ligands. Table 10.1 collects the published information about recognition of distinct collagen subtypes by integrins. Furthermore, the fact that the cytoplasmic domains of the α subunits are different [8, 11, 77, 102] suggests that they also generate unique intracellular signals. Numerous cell type and tissue

J. Heino (✉)
Department of Biochemistry, University of Turku, 20014, Turku, Finland
e-mail: jyrki.heino@utu.fi

D. Gullberg (ed.), *I Domain Integrins*, Advances in Experimental Medicine and Biology 819,
DOI: 10.1007/978-94-017-9153-3_10,

Table 10.1 The recognition of distinct collagen subtypes

Fibril-forming collagens (I, II, III, V, XI, XXIV, XXVIII)
Collagens I, II: α1β1/α1I domain, α2β1/α2I domain, α10β1/α10I domain, α11β1/α11I domain
Collagens III, V: α1β1/α1I domain, α2β1/α2I domain, α10β1/α10I domain
Collagen XI: α2I domain [77]
The approximated avidity of α1β1 binding to fibril-forming collagens is lower than that of α2β1. References for α10β1/α10I and α11β1/α11I [110, 76]
Network forming collagens (IV, VIII, X)
Collagen IV: Integrin α1β1/α1I domain, α2β1/α2I domain, α10I domain. Integrin α1β1 seems to prefer type IV over fibril-forming collagens, opposite to α2β1 integrin [47, 76]
Collagen VIII: Platelet binding is mediated by α2β1 [77]
Collagen X: α2β1 [59]
Beaded-filaments forming collagen (VI)
Collagen VI: α1β1/α1I, α10I domain. Binding by α2β1/α2I domain is much weaker [96]
Anchoring fibrils forming collagen (VII)
Collagen VII: NC1 domain in is recognized by α2β1 on human fibroblasts [15]. Platelet binding is mediated by α2β1 [77]
FACIT collagens (IX, XII, XIV, XVI, XIX, XX, XXI, XXII)
Collagen IX: α1β1/α1I domain, α2β1/α2I domain, α10β1/α10 I domain, α11I domain [46]
Collagen XIV: CD44, unlike α1β1 or α2β1 [26, 48]
Collagen XVI: α1β1/α1I domain, α2β1/α2I domain. Binging by α1β1 is stronger [25]
Transmembrane collagens (XIII, XVII, XXIII, XXV)
Collagen XIII: α1β1 integrin/α1I domain. Binding by α2β1/α2I domain much weaker [69]
Collagen XVII: The largest collagenous domain (COL 15) cannot be recognized by the collagen receptors. However, when denatured the multiple KGD motifs can be used by α5β1 and αV-integrins [70]
Collagen XXIII: α2β1 [76]
Multiplexins (XV, XVIII)
Collagen XVIII: α1β1 [24]. Endostatin, the C-terminal cleavage product is recognized by α5β1 and αV-integrins [76]

(Note that here collagens XXVI and XXVIII have not been listed to any of the subgroups)

specific factors, e.g. interplay with other cellular receptors, also modify the signaling by the collagen receptors [42].

This chapter is focused on the four members of the collagen receptor subgroup of the integrins. However, also other integrins have been reported to function as collagen receptors. Integrin α3β1, generally known as a laminin receptor, may also act as an assisting collagen IV receptor [23]. Similarly, the leukocyte αI domain integrins can bind to various collagen subtypes and they can be considered as low-avidity or assisting collagen receptors [54]. Furthermore, denatured collagen (e.g. collagen I) can be recognized by the fibronectin receptor integrins based on the cryptic RGD motifs in collagen α chains [34].

10.2 Collagens and Other ECM Ligands

Collagens are structural proteins of extracellular matrix that typically have triple helical domains of variable length [76]. Collagens form, for example, large fibrils in connective tissues and networks in basement membranes, while some collagens are transmembrane proteins. Metazoans from sea sponges [2] to mammals express collagens and in man, 28 structurally and functionally different collagen subtypes have been published [76]. The collagen subtypes are named from I to XXVIII based on the order in which they have been found. The collagen family is composed of several subgroups. The fibril-forming collagens have a long, continuous triple helix that gives to the molecule a rigid, rod-like structure. These collagens form large fibrils, which is essential for the structural integrity and the tensile strength of the tissues. The network-forming collagens have interruptions in the triple helix. Basement membrane collagen IV belongs to this subfamily. Two collagen subtypes have unique functions and they are the only members of the corresponding subgroups: beaded-filaments are built from collagen VI and anchoring fibrils from collagen VII. Fibril-associated collagens with interruptions in triple-helices (FACIT) form a large subgroup. Collagens IX and XII are typical FACITs. All FACIT collagens may not, however, be able to bind to fibrils. Collagens XIII and XVII were the first subtypes shown to be transmembrane proteins. Collagen XVII is a structural component of hemidesmosomes, whereas collagen XIII is found, for example, in muscle, bone and skin. Multiplexins are collagens that are associated to basement membranes. Collagens XV and XVIII belong to this subfamily. Their C-terminal cleavage products have become known as angiogenesis-blocking endostatins [71].

Specific collagenous motifs are recognized by integrin αI domains. Best known is the GFOGER (O = hydroxyproline) sequence in triple-helical conformation [52], which is a binding site for all four collagen receptor integrins. GLOGER, GASGER, GROGER, and GLOGEN represent other similar motifs [58]. However, many collagen subtypes are not homotrimers, but the triple helix is formed by two or three different α chains. In these cases the integrin binding mechanism may be different. For example in collagen IV α1β1 integrin may recognize one arginine and two aspartic acid residues all coming from a different α chain [76]. Many reports have named the fibril-forming collagens as high-avidity ligands of α2β1 and α11β1 integrins, whereas α1β1 seems to the best receptor for collagens IV and XIII [35, Table 1]. Receptor for collagen XIV is CD44 and it may not be an optimal ligand for the integrins [48]. Recognition of distinct collagen subtypes may also be dependent on the activation stage of the integrin [77, 12]. The αI domain has at least two activation stages. In nonactivated integrins the αI domain is in the closed conformation, that is able to recognize the ligands, but the interaction is weaker than with the activated, open αI domain. Activation may also diminish selectivity between high and low avidity binding motifs [12] and ligand proteins [77] and between hetero and homotrimeric collagen subtypes [12].

The collagen receptors also have non-collagenous ligands. Integrin α1β1 is a receptor for laminins, collagen IV derived antiangiogenic degradation product called arresten [77] and semaphorin protein, Sema7A [72].

Integrin α2β1 has numerous ligands, including ECM proteins tenascin C [76] and chondroadherin [10] as well as different laminins. This receptor can also bind to proteoglycans, such as decorin [32] and endorepellin (C-terminal domain of perlecan; [6]), and collectin family members, namely C1q complement protein, mannose-binding lectin (MBL) and surfactant protein A (SP-A) [77].

10.3 Regulation of Cellular Signaling Pathways by the Collagen Receptor Integrins

During the canonical integrin outside-in signaling, the ligand binding induces rapid increase in the levels of phosphatidylinositol-4,5-bidphosphate and phosphatidylinositol-3,4,5-triphosphate and promotes the tyrosine phosphorylation of proteins such as focal adhesion kinase (FAK), p130Cas and Src [57]. Soon after that small GTPases belonging to the Rho-family are activated [57]. Finally, the integrins regulate many pathways controlling cell survival, proliferation, differentiation, migration and metabolism. There is no reason to believe that the collagen receptors would act in a different manner. Indeed, there are numerous papers using different experimental models and demonstrating the regulation of FAK, Src, p130Cas, mitogen activated protein kinases (MAPKs), including extracellular signal regulated kinase (ERK), c-Jun N-terminal kinase (JNK) and p38, as well as phosphoinositide

3-kinase (PI3K) and Akt (Protein kinase B) by $\alpha1\beta1$ and $\alpha2\beta1$ integrins. Activation of Rac-1 GTPase has been connected to $\alpha1\beta1$ mediated cell migration in vitro [76] and invasion in vivo [77]. Integrin $\alpha2\beta1$ has been reported to activate RhoA and slow down cell locomotion [77]. Integrin $\alpha2\beta1$ may also regulate Rac-1 by orchestrating the membrane anchorage of Rac [9].

The functions of $\alpha2\beta1$ have been investigated in detail in platelets. These studies have revealed that Rac and/or p21-activated kinase (PAK) activation, but not Rho, participate in integrin $\alpha2\beta1$ mediated platelet spreading on collagen and that Src family kinases and PI3K are also involved [76]. Another study indicates that Pyk2 (a FAK-family protein tyrosine kinase) regulates PI3K and Akt downstream of integrin $\alpha2\beta1$ [20]. In platelets $\alpha2\beta1$ has also been connected to the regulation of Spleen tyrosine kinase (Syk), SH2 domain-containing leukocyte phosphoprotein (SLP-76), and phospholipase Cγ2 (LPCγ2) as well as plasma membrane calcium ATPase and FAK [33, 39]. Many reports indicate that both $\alpha1\beta1$ [17, 18] and $\alpha2\beta1$ [38] can regulate the formation reactive oxygen species (ROS), which partially explains their effects on cellular functions, e.g. p38 MAPK phosphorylation and cyclin expression [38].

The general signaling events are often mediated by the β subunit and therefore they are not specific to any individual $\beta1$ containing heterodimer. Still, the collagen receptors may have different, even opposite, effects on gene expression. For example, $\alpha1\beta1$ is a negative regulator of collagen I synthesis [55, 77], whereas $\alpha2\beta1$ increases the expression [40, 77]. Some regulatory functions of the collagen receptors seem to be dependent on the cytoplasmic tails of the α subunits. For example the effect of $\alpha1\beta1$ on Rac-1 activation and consequently on cell migration is dependent on the $\alpha1$ cytoplasmic domain [76]. The mechanism of signaling through $\alpha2\beta1$ integrin has also been studied by deletions, mutations and swaps of the $\alpha2$ subunit cytoplasmic domain [13, 40, 45, 49, 51]. Deletion of the entire tail targets the integrin to the focal adhesion sites, even in the absence of collagen [45]. The essential role of the $\alpha2$ cytoplasmic tail in other functions of $\alpha2\beta1$ integrin, such as contraction of collagen gels, has also been well characterized [13]. Replacement of the $\alpha2$ cytoplasmic tail with one from the $\alpha1$ integrin renders the integrin unable to signal normally [40, 49, 51]. This also supports the idea that different collagen receptors have distinct signaling functions. Furthermore, these observations suggest that the cytoplasmic domains of the collagen receptor α subunits may be directly connected to cellular signaling proteins, or that they modify the molecular interactions of the $\beta1$ cytoplasmic domain. Still very little is known about α subunit binding proteins or the mechanisms of α specific signaling. The cytoplasmic tail of $\alpha1$ integrin selectively interacts with a ubiquitously expressed protein tyrosine phosphatase TCPTP (T-cell protein tyrosine phosphatase) and activates it after cell adhesion to collagen [61]. Other α subunit binding proteins include Rab21 and SHARPIN, but they are not selective for any single α subunit. Rab21 regulates integrin trafficking [73] and SHARPIN is an inhibitor of integrin ligand binding function [76].

Many integrins may also act as cellular receptors for viruses. Human pathogen, echovirus-1 (EV-1) binds to the α2I domain in $\alpha2\beta1$ integrin and is rapidly internalized into the host cell. Interestingly, EV-1 seems to recognize the closed conformation of α2I domain and keep the integrin in the nonactivated stage [43]. Still the virus is able to activate protein kinase C-α (PKC-α) and Rac-1, which is also requited for the macropinocytosis-like entry of the virus-integrin complex [44, 77]. In this case the clustering of integrins seems to be the activating factor rather than the conformational change in the receptors. The nonactivated stage of $\alpha2\beta1$ may also have other biological functions, e.g. during platelet collagen binding under shear stress [67]. Actually, activated $\alpha2\beta1$ integrins are very rapidly internalized in a ligand dependent manner, which leads to remarkable decrease in the number of collagen receptors on platelets [76].

10.4 Interplay of the Collagen Receptors with Growth Factor Receptors

Recent studies have indicated that integrins have numerous mechanisms to interact with growth factor receptors [42]. Cell adhesion is often required for the establishment of molecular platforms that enable the signaling by growth factor receptors. Integrins may even activate the growth factor receptors in the absence of the growth factor. Integrins can also orchestrate the trafficking of other receptors and in that way regulate the copy number of growth factor receptors on cell surface [42].

The interplay between collagen receptors and epidermal growth factor receptor (EGFR) has been studied in different cellular model systems. Integrin $\alpha1\beta1$ is reported to negatively regulate EGFR. This may be related to the ability of $\alpha1\beta1$ to increase caveolin-1 levels and to activate protein tyrosine phosphatase TCPTP [18, 61]. Integrin $\alpha2\beta1$ is known to modify EGFR signaling, too. EGFR may also reduce the levels of $\alpha2\beta1$ on cell surface by increasing its internalization [66]. Published findings have also suggested that crosstalk between hepatocyte growth factor receptor (HGFR/c-met) and $\alpha2\beta1$ integrin is requited for mast-cell activation [62].

In addition to ECM proteins many integrin can also directly bind to growth factors [42]. Accordingly, integrin $\alpha1\beta1$ is not only a receptor for collagens, but it has also been reported to recognize a semaphorin protein, Sema7A, that enhances axon growth [72]. Sema7A binding to $\alpha1\beta1$ integrin can activate several signaling proteins, including FAK, ERK MAPKs [72], Abelson (Abl), and Abl-related gene (Arg) tyrosine kinases [64].

10.5 Crosstalk Between Integrins and Discoidin Domain Type Collagen Receptors

The collagen receptor integrins are not the only cellular receptors that can recognize collagenous triple-helical motifs [58]. Platelet glycoprotein VI (GPVI) on platelets is critical for response to collagen. Discoidin domain receptors 1 and 2 (DDR1 and DDR2) are tyrosine kinases that bind to collagens and regulate e.g. cell proliferation. Leukocyte associated immunoglobulin-like receptor 1 (LAIR-1) is an inhibitory receptor on leukocytes. The triple helical motifs formed by peptides harboring GVMGFO sequences have been described as binding sites for DDR1 and DDR2. The minimum functional binding site for GPVI contains two GPO triplets in collagenous triple helix. Similarly, LAIR-1 binds to peptides containing multiple GPOs [58]. Thus, these receptors do not compete with the integrins in collagen binding.

Activation and autophosphorylation of DDRs are independent of the integrins [76]. However, several studies indicate that the signaling functions of DDRs and collagen receptor integrins are linked together. DDRs can for example regulate integrin activity [1, 77]. The interplay between the two receptor systems is also obvious in studies focused on collagen I induced epithelial-mesenchymal-transition in pancreatic cancer cells. During the process N-cadherin is upregulated by JNK-dependent mechanism. Both receptors are needed for p130Cas-dependent activation of Rap1, but they act in a different manner. DDR1 regulates Pyk2, while $\alpha2\beta1$ integrin activates FAK [76].

In another study DDR1 was shown to inhibit collagen-induced tyrosine phosphorylation of Stat 1/3 and cell migration triggered by $\alpha2\beta1$ integrin via SHP-2. SHP-2 is a phosphotyrosol phosphatase (PTP) that via SH2 domain binds to phosphorylated tyrosine residues in DDR1 [77].

10.6 Regulation of Cell Proliferation and Survival by the Collagen Receptors

Analysis of $\alpha1$ integrin deficient knock-out mice has supported the idea that $\alpha1$ might promote cell proliferation, since the dermis of the animals seems to be hypocellular [50] and their bone marrow derived mesenchymal stem cells are less proliferative than those of the control animals [27]. Integrin $\alpha 1\beta 1$ is among the integrins that can activate Shc (a SH2 domain containing adaptor protein) in a process requiring caveolin-1 and Fyn (a protein tyrosine kinase) [50, 76]. Shc activation subsequently leads to activation of Ras and the growth promoting MAPKs. However, the fact that $\alpha 1\beta 1$ negatively regulates EGFR [61] and caveolin-1 phosphorylation [7] also connects this integrin to growth inhibiting mechanisms. The idea that $\alpha 1\beta 1$ can generate different or even opposite signals is also supported by the observation that the binding of this receptor to two different domains in laminin has distinct effects [22].

Integrin $\alpha 2$ null mice are viable, fertile and without defects that could suggest general dysregulation of cell growth [14, 37]. However, in cell culture assays $\alpha 2\beta 1$ is capable of generating negative growth signals. Its interaction with laminin, a low affinity ligand, results in growth arrest in endothelial cells [63]. It also increases cell commitment toward quiesence by a mechanism involving changes in the anchorage of Ras to membranes and a tetraspanin CD9 [9]. The role of $\alpha 2\beta 1$ as a negative growth regulator is also in agreement with the observation that the expression level of $\alpha 2$ is often very low in breast cancer cells [77]. The effect of collagen—$\alpha 2\beta 1$ interaction on cell proliferation seems to be dependent on the organization of the collagenous matrix. In mesenchymal cells, including smooth muscle cells and fibroblasts, fibrillar collagen prevents proliferation [28, 53]. Similar results have been reported with melanoma cells [36]. In fibroblasts growth arrest may require that the cells are inside floating and contracting collagen gels [28]. In smooth muscle and melanoma cells as well as in fibroblasts, growth arrest has been connected to $\alpha 2\beta 1$ function and the accumulation of cyclin/cyclin-dependent kinase inhibitor, p27kip [28, 36, 53]. Recent studies have also indicated that the crosslinking and therefore the stiffness of collagenous matrix is an important regulator of cell behavior [56].

On a contrary in murine mammary gland-derived epithelial cells, $\alpha 2\beta 1$ has been reported to increase proliferation, when tested in monolayer cultures on non-fibrillar collagen [76]. In these cells the intracellular tail of $\alpha 2$ subunit has been analyzed by targeted mutations, and two distinct sites have been identified which regulate p38 and ERK pathways [49, 51], connected to cell migration and proliferation, respectively. Similarly, in human adenocarcinoma cells (Caco-2) interaction of $\alpha 2\beta 1$ with collagen IV promotes G1/S transition [38].

The conclusion is that the organization of the collagenous matrix is critical for the action of $\alpha 2\beta 1$. It has been speculated that the clustering of $\alpha 2\beta 1$ by antibodies or non-fibrillar collagen may actually promote proliferation, while fibrillar collagen prevents $\alpha 2\beta 1$ clustering and therefore inhibits proliferation [36]. Similarly, physical forces, such as the stiffness of the tissue [56] or the shear stress in blood stream [67], may be important modulators of integrin action and signal transduction.

A recent study indicates that integrins $\alpha 2\beta 1$ and $\alpha 11\beta 1$ promote the survival of mesenchymal stem cells [74]. Integrin $\alpha 2\beta 1$ is also known to be a marker protein of prostate epithelial stem cells [19]. Anti-apoptotic effect of $\alpha 2\beta 1$ has been reported with human mammary epithelial cells [4], Jurkat T cells [3], A431 cells [77] and Madin-Darby canine kidney cells [76]. In an acute liver injury model in mice $\alpha 1\beta 1$ has been shown to mediate survival promoting signals after contact to collagen XVIII [24].

In certain experimental models the collagen receptors also promote apoptosis. Release of mechanical tension in a three-dimensional collagen gel model triggers apoptosis in fibroblasts [31, 65]. In these conditions $\alpha 2\beta 1$ mediates cell adhesion to collagen and is essential for contraction. Function blocking antibodies against $\alpha 2\beta 1$ and $\alpha 1\beta 1$ integrin can reduce the number of apoptotic cells, and $\alpha 2$-negative rhabdomyosarcoma cells undergo apoptosis only if they are cDNA-transfected to express the $\alpha 2$ subunit [65].

Apoptosis may be partially regulated by the same signaling pathways as proliferation. In addition to the ERK pathway, Akt seems to be involved. In fibroblasts and osteosarcoma cells, protein phosphatase 2A (PP2A) is activated in a process that requires the presence of α2 cytoplasmic domain and Cdc42 activity [41]. Activation of PP2A leads consequently to dephosphorylation of Akt, a well-known promoter of cell survival. PP2A can also inhibit the cell cycle in several different ways, and α2β1 dependent activation of PP2A may be one mechanism leading to growth arrest. Dephosphorylation of Akt inside collagen gel can be prevented by α2 integrin antibodies [41]. Another study indicates that apoptosis of fibroblasts inside contracting collagen gels can be prevented by β1 integrin antibodies which also prevent Akt dephosphorylation [76].

Integrin α1β1 is a receptor for arresten, a 26 kDa non-collagenous domain of α1-chain in collagen IV. Arresten promotes apoptosis of endothelial by decreasing the amount of anti-apoptotic molecules of the Bcl-family, namely Bcl-2 and Bcl-xL [68].

10.7 Regulation of Matrix Gene Expression by the Collagen Receptors

Several lines of evidence support the idea that α1β1 integrin is a negative regulator of collagen synthesis, especially in cells that are surrounded by three-dimensional collagenous matrix. Early observations using osteosarcoma cells with a low expression level of α1β1 showed that these cells do not down-regulate collagen synthesis inside collagen [77]. This was later confirmed by experiments performed with cells derived from α1 integrin deficient mice [29]. Furthermore, experiments utilizing functional integrin antibodies or mutant α1 integrins have led to the same conclusion [55, 77]. Similar regulatory mechanisms may also function in tissues since α1 knock-out mice have increased collagen synthesis rate in their dermis while a concomitant increase in matrix metalloproteinase (MMP) expression prevents the accumulation of collagen [29]. Integrin α1 null mice are also more sensitive to glomerulosclerosis than their wild type littermates [16].

Overexpression of α2β1 integrin in cells increases collagen synthesis, suggesting that α2β1 is a positive regulator of collagen gene expression [40, 77]. Increased collagen synthesis can be prevented using selective inhibitors of the α isoform of p38 MAPK. p38 is activated by α2β1 after contact with collagen [40]. Activation of the p38 pathway has been frequently observed after α2β1—collagen interaction in several different cell lines and experiment models [40, 49, 51, 77, 76]. The mechanism of collagen gene suppression by α1β1 is not clear but the receptor can activate ERK [50] that is known to be a negative regulator of collagen synthesis [77]. It is also possible to speculate that the opposite effects of α1β1 and α2β1 on many cellular functions may partially be due to the alterations in the balance between p38 and ERK pathways.

Cell contact to three-dimensional collagen activates the expression of collagenase-1 (matrix metalloproteinase-1, MMP-1) [30]. This phenomenon has been linked to signaling through the α2β1 integrin [77]. Other MMPs regulated by either α1β1 or α2β1 include stromalysin-1 (MMP-3) [60] and collagenase-3 (MMP-13) [77]. In skin fibroblasts inside collagen collagenase-1 (MMP-1) expression seems to be activated by a pathway involving PKC-ζ and nuclear factor κB (NF-κB) [76, 77]. However, the above described p38 pathway may also participate in the process [76]. Interestingly, α2β1 can also regulate its own expression by a positive signaling loop involving PKC-ζ/NF-κB [77]. The p38 pathway seems to mediate the upregulation of MMP-13 by α2β1 integrin [77].

Signaling by α11β1 integrin has recently been shown to induce MMP-13 expression [5]. Integrin α1β1 has been reported to regulate MMP-2, MMP-9 and MMP-14 in a p38-dependent manner in mesengial cells [21] and MMP-2 and MMP-9 expression in colon cancer cells via p130Cas and JNK [77]. In general, there is a strong link between the collagen receptor integrins and the expression of MMPs. This connection may play an important role in the maintenance of tissue homeostasis, in the regulation of wound healing and tissue repair processes and during cell invasion.

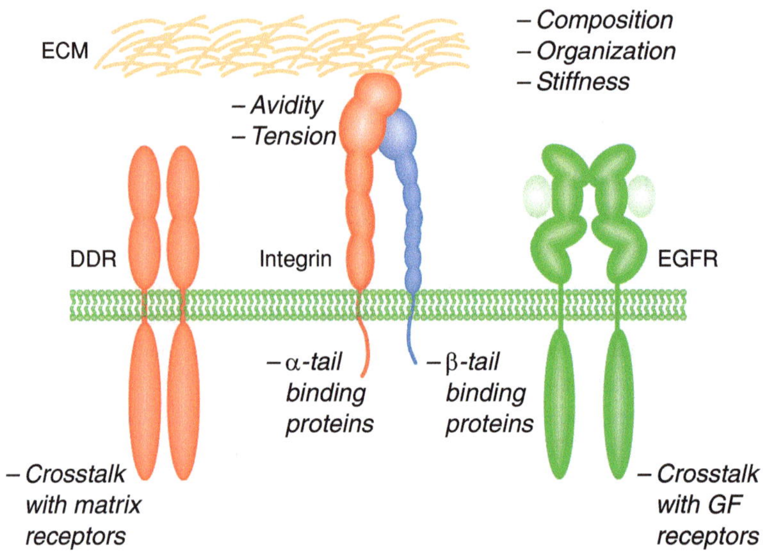

Fig. 10.1 Numerous factors influence on the cellular signals generated by the collagen receptor integrins. Composition, organization and stiffness of collagenous ECM are all important factors. Activation stage of integrins regulates the avidity and specificity of ligand binding. Crosstalk with other adhesion and growth factor receptors is involved. There are also cell type specific differences in the expression of integrin cytoplasmic domain binding proteins

10.8 Perspectives

Cellular signaling after adhesion to collagen is influenced by numerous factors (Fig. 10.1). Different collagen subtypes are recognized by different receptors and also the organization of the ECM is critical. Recent observations have also stressed the important role of ECM stiffness. Activation stage of integrins regulates the avidity and specificity of ligand binding. Integrin signaling is also influenced by other adhesion receptors, e.g. DDR-type collagen receptors, and growth factor receptors. Finally, cell type specific differences, e.g. in the expression of integrin cytoplasmic domain binding proteins, may modify the activation of distinct cellular pathways.

Acknowledgments Original integrin-related studies in the author's research group have been supported by the Academy of Finland, the Sigrid Jusélius Foundation, the Finnish Cancer Association and the European Union (Metafight and CAFFEIN in the 7th Framework Programme).

References

1. Abbonante V, Gruppi C, Rubel D, Gross O, Moratti R, Balduini A (2013) Discoidin domain receptor 1 protein is a novel modulator of megakaryocyte-collagen interactions. J Biol Chem 288:16738–16746
2. Aho S, Turakainen H, Onnela ML, Boedtker H (1993) Characterization of an intronless collagen gene family in the marine sponge Microciona prolifera. Proc Natl Acad Sci USA 90:7288–7292
3. Aoudjit F, Vuori K (2000) Engagement of the $\alpha2\beta1$ integrin inhibits Fas ligand expression and activation-induced cell death in T cells in a focal adhesion kinase dependent manner. Blood 95:2044–2051
4. Baeckstrom D, Lu PJ, Taylor-Papadimitriou J (2000) Activation of the $\alpha2\beta1$ integrin prevents c-erbB2-induced scattering and apoptosis of human mammary epithelial cells in collagen. Oncogene 19:4592–4603
5. Barczyk MM, Lu N, Popova SN, Bolstad AI, Gullberg D (2013) $\alpha11\beta1$ integrin-mediated MMP-13-dependent collagen lattice contraction by fibroblasts: evidence for integrin-coordinated collagen proteolysis. J Cell Physiol 228:1108–1119
6. Bix G, Fu J, Gonzalez EM, Macro L, Barker A, Campbell S, Zutter MM, Santoro SA, Kim JK, Höök M, Reed CC, Iozzo RV (2004) Endorepellin causes endothelial cell disassembly of actin

cytoskeleton and focal adhesions through $\alpha2\beta1$ integrin. J Cell Biol 166:97–109
7. Borza CM, Chen X, Mathew S, Mont S, Sanders CR, Zent R, Pozzi A (2010) Integrin $\alpha1\beta1$ promotes caveolin-1 dephosphorylation by activating T cell protein-tyrosine phosphatase. J Biol Chem 285:40114–40124
8. Briesewitz R, Epstein MR, Marcantonio EE (1993) Expression of native and truncated forms of the human integrin $\alpha1$ subunit. J Biol Chem 268:2989–2996
9. Cailleteau L, Estrach S, Thyss R, Boyer L, Doye A, Domange B, Johnsson N, Rubinstein E, Boucheix C, Ebrahimian T, Silvestre JS, Lemichez E, Meneguzzi G, Mettouchi A (2010) $\alpha2\beta1$ integrin controls association of Rac with the membrane and triggers quiescence of endothelial cells. J Cell Sci 123:2491–2501
10. Camper L, Heinegârd D, Lundgren-Akerlund E (1997) Integrin $\alpha2\beta1$ is a receptor for the cartilage matrix protein chondroadherin. J Cell Biol 138:1159–1167
11. Camper L, Hellman U, Lundgren-Akerlund E (1998) Isolation, cloning, and sequence analysis of the integrin subunit $\alpha10$, a $\beta1$-associated collagen binding integrin expressed on chondrocytes. J Biol Chem 273:20383–20389
12. Carafoli F, Hamaia SW, Bihan D, Hohenester E, Farndale RW (2013) An activating mutation reveals a second binding mode of the integrin α2I domain to the GFOGER motif in collagens. PLoS ONE 8:e69833
13. Chan BM, Kassner PD, Schiro JA, Byers HR, Kupper TS, Hemler ME (1992) Distinct cellular functions mediated by different VLA integrin α subunit cytoplasmic domains. Cell 68:1051–1060
14. Chen J, Diacovo TG, Grenache DG, Santoro SA, Zutter MM (2002) The $\alpha2$ integrin subunit-deficient mouse: a multifaceted phenotype including defects of branching morphogenesis and hemostasis. Am J Pathol 161:337–344
15. Chen M, O'Toole EA, Li YY, Woodley DT (1999) $\alpha2\beta1$ integrin mediates dermal fibroblast attachment to type VII collagen via a 158-amino-acid segment of the NC1 domain. Exp Cell Res 249:231–239
16. Chen X, Moeckel G, Morrow JD, Cosgrove D, Harris RC, Fogo AB, Zent R, Pozzi A (2004) Lack of integrin $\alpha1\beta1$ leads to severe glomerulosclerosis after glomerular injury. Am J Pathol 165:617–630
17. Chen X, Abair TD, Ibanez MR, Su Y, Frey MR, Dise RS, Polk DB, Singh AB, Harris RC, Zent R, Pozzi A (2007) Integrin $\alpha1\beta1$ controls reactive oxygen species synthesis by negatively regulating epidermal growth factor receptor-mediated Rac activation. Mol Cell Biol 27:3313–3326
18. Chen X, Whiting C, Borza C, Hu W, Mont S, Bulus N, Zhang MZ, Harris RC, Zent R, Pozzi A (2010) Integrin $\alpha1\beta1$ regulates epidermal growth factor receptor activation by controlling peroxisome proliferator-activated receptor gamma-dependent caveolin-1 expression. Mol Cell Biol 30:3048–3058
19. Collins AT, Habib FK, Maitland NJ, Neal DE (2001) Identification and isolation of human prostate epithelial stem cells based on $\alpha2\beta1$-integrin expression. J Cell Sci 114:3865–3872
20. Consonni A, Cipolla L, Guidetti G, Canobbio I, Ciraolo E, Hirsch E, Falasca M, Okigaki M, Balduini C, Torti M (2012) Role and regulation of phosphatidylinositol 3-kinase β in platelet integrin $\alpha2\beta1$ signaling. Blood 119:847–856
21. Cosgrove D, Meehan DT, Delimont D, Pozzi A, Chen X, Rodgers KD, Tempero RM, Zallocchi M, Rao VH (2008) Integrin $\alpha1\beta1$ regulates matrix metalloproteinases via P38 mitogen-activated protein kinase in mesangial cells: implications for Alport syndrome. Am J Pathol 172:761–773
22. Desban N, Lissitzky JC, Rousselle P, Duband JL (2006) $\alpha1\beta1$-integrin engagement to distinct laminin-1 domains orchestrates spreading, migration and survival of neural crest cells through independent signaling pathways. J Cell Sci 119:3206–3218
23. DiPersio CM, Shah S, Hynes RO (1995) $\alpha3A\beta1$ integrin localizes to focal contacts in response to diverse extracellular matrix proteins. J Cell Sci 108:2321–2336
24. Duncan MB, Yang C, Tanjore H, Boyle PM, Keskin D, Sugimoto H, Zeisberg M, Olsen BR, Kalluri R (2013) Type XVIII collagen is essential for survival during acute liver injury in mice. Dis Model Mech 6:942–951
25. Eble JA, Kassner A, Niland S, Mörgelin M, Grifka J, Grässel S (2006) Collagen XVI harbors an integrin $\alpha1\beta1$ recognition site in its C-terminal domains. J Biol Chem 281:25745–25756
26. Ehnis T, Dieterich W, Bauer M, Lampe B, Schuppan D (1996) A chondroitin/dermatan sulfate form of CD44 is a receptor for collagen XIV (undulin). Exp Cell Res 229:388–397
27. Ekholm E, Hankenson KD, Uusitalo H, Hiltunen A, Gardner H, Heino J, Penttinen R (2002) Diminished callus size and cartilage synthesis in $\alpha1\beta1$ integrin deficient mice during bone fracture healing. Am J Pathol 160:1779–1785
28. Fringer J, Grinnell F (2001) Fibroblast quiescence in floating or released collagen matrices: contribution of the ERK signaling pathway and actin cytoskeletal organization. J Biol Chem 276:31047–31052
29. Gardner H, Broberg A, Pozzi A, Laato M, Heino J (1999) Absence of integrin $\alpha1\beta1$in the mouse causes loss of feedback regulation of collagen synthesis in normal and wounded dermis. J Cell Sci 112:263–272
30. Grinnell F (1994) Fibroblasts, myofibroblasts, and wound contraction. J Cell Biol 124:401–404
31. Grinnell F, Zhu M, Carlson MA, Abrams JM (1999) Release of mechanical tension triggers apoptosis of

human fibroblasts in a model of regressing granulation tissue. Exp Cell Res 248:608–619
32. Guidetti G, Bertoni A, Viola M, Tira E, Balduini C, Torti M (2002) The small proteoglycan decorin supports adhesion and activation of human platelets. Blood 100:1707–1714
33. Guidetti GF, Bernardi B, Consonni A, Rizzo P, Gruppi C, Balduini C, Torti M (2009) Integrin $\alpha 2\beta 1$ induces phosphorylation-dependent and phosphorylation-independent activation of phospholipase Cγ2 in platelets: role of Src kinase and Rac GTPase. J Thromb Haemost 7:1200–1206
34. Gullberg D, Gehlsen KR, Turner DC, Ahlén K, Zijenah LS, Barnes MJ, Rubin K (1992) Analysis of $\alpha 1\beta 1$, $\alpha 2\beta 1$ and $\alpha 3\beta 1$ integrins in cell–collagen interactions: identification of conformation dependent $\alpha 1\beta 1$ binding sites in collagen type I. EMBO J 11:3865–3873
35. Heino J (2007) The collagen family members as cell adhesion proteins. BioEssays 29:1001–1010
36. Henriet P, Zhong ZD, Brooks PC, Weinberg KI, DeClerck YA (2000) Contact with fibrillar collagen inhibits melanoma cell proliferation by up-regulating p27KIP1. Proc Natl Acad Sci USA 97:10026–10031
37. Holtkötter O, Nieswandt B, Smyth N, Müller W, Hafner M, Schulte V, Krieg T, Eckes B (2002) Integrin α2-deficient mice develop normally, are fertile, but display partially defective platelet interaction with collagen. J Biol Chem 277:10789–10794
38. Honoré S, Kovacic H, Pichard V, Briand C, Rognoni JB (2003) $\alpha 2\beta 1$-integrin signaling by itself controls G1/S transition in a human adenocarcinoma cell line (Caco-2): implication of NADPH oxidase-dependent production of ROS. Exp Cell Res 285:59–71
39. Inoue O, Suzuki-Inoue K, Dean WL, Frampton J, Watson SP (2003) Integrin $\alpha 2\beta 1$ mediates outside-in regulation of platelet spreading on collagen through activation of Src kinases and PLCgamma2. J Cell Biol 160:769–780
40. Ivaska J, Reunanen H, Westermarck J, Koivisto L, Kähäri VM, Heino J (1999) Integrin $\alpha 2\beta 1$ mediates isoform-specific activation of p38 and upregulation of collagen gene transcription by a mechanism involving the α2 cytoplasmic tail. J Cell Biol 147:401–416
41. Ivaska J, Nissinen L, Immonen N, Eriksson JE, Kähäri VM, Heino J (2002) Integrin $\alpha 2\beta 1$ promotes activation of protein phosphatase 2A and dephosphorylation of Akt and GSK3 β. Mol Cell Biol 22:1352–1359
42. Ivaska J, Heino J (2011) Cooperation between integrins and growth factor receptors in signaling and endocytosis. Annu Rev Cell Dev Biol 27:291–320
43. Jokinen J, White DJ, Salmela M, Huhtala M, Käpylä J, Sipilä K, Puranen JS, Nissinen L, Kankaanpää P, Marjomäki V, Hyypiä T, Johnson MS, Heino J (2010) Molecular mechanism of $\alpha 2\beta 1$ integrin interaction with human echovirus 1. EMBO J 29:196–208
44. Karjalainen M, Kakkonen E, Upla P, Paloranta H, Kankaanpää P, Liberali P, Renkema GH, Hyypiä T, Heino J, Marjomäki V (2008) A Raft-derived, Pak1-regulated entry participates in $\alpha 2\beta 1$ integrin-dependent sorting to caveosomes. Mol Biol Cell 19:2857–2869
45. Kawaguchi S, Bergelson JM, Finberg RW, Hemler ME (1994) Integrin α2 cytoplasmic domain deletion effects: loss of adhesive activity parallels ligand-independent recruitment into focal adhesions. Mol Biol Cell 5:977–988
46. Käpylä J, Jäälinoja J, Tulla M, Ylöstalo J, Nissinen L, Viitasalo T, Vehviläinen P, Marjomäki V, Nykvist P, Säämänen AM, Farndale RW, Birk DE, Ala-Kokko L, Heino J (2004) The fibril-associated collagen IX provides a novel mechanism for cell adhesion to cartilaginous matrix. J Biol Chem 279:51677–51687
47. Kern A, Eble J, Golbik R, Kuhn K (1993) Interaction of type IV collagen with the isolated integrins $\alpha 1\beta 1$ and $\alpha 2\beta 1$. Eur J Biochem 215:151–159
48. Klein G, Kibler C, Schermutzki F, Müller CA, Timpl R (1998) Cell binding properties of collagen type XIV for human hematopoietic cells. Matrix Biol 16:307–317
49. Klekotka PA, Santoro SA, Zutter MM (2001a) α2 integrin subunit cytoplasmic domain dependent cellular migration requires p38 MAPK. J Biol Chem 276:9503–9511
50. Pozzi A, Wary KK, Giancotti FG, Gardner HA (1998) Integrin $\alpha 1\beta 1$ mediates a unique collagen-dependent proliferation pathway in vivo. J Cell Biol 142:587–594
51. Klekotka PA, Santoro SA, Wang H, Zutter MM (2001b) Specific residues within the α2 integrin subunit cytoplasmic domain regulate migration and cell cycle progression via distinct MAPK pathways. J Biol Chem 276:32353–32361
52. Knight CG, Morton LF, Peachey AR, Tuckwell DS, Farndale RW, Barnes MJ (2000) The collagen-binding A-domains of integrins $\alpha 1\beta 1$ and $\alpha 2\beta 1$ recognize the same specific amino acid sequence, GFOGER, in native (triple-helical) collagens. J Biol Chem 275:35–40
53. Koyama H, Raines EW, Bornfeldt KE, Roberts JM, Ross R (1996) Fibrillar collagen inhibits arterial smooth muscle proliferation through regulation of Cdk2 inhibitors. Cell 87:1069–1078
54. Lahti M, Heino J, Käpylä J (2013) Leukocyte integrins $\alpha L\beta 2$, $\alpha M\beta 2$ and $\alpha X\beta 2$ as collagen receptors–receptor activation and recognition of GFOGER motif. Int J Biochem Cell Biol 45:1204–1211
55. Langholz O, Rockel D, Mauch C, Kozlowska E, Bank I, Krieg T, Eckes B (1995) Collagen and collagenase gene expression in three-dimensional

collagen lattices are differentially regulated by $\alpha 1\beta 1$ and $\alpha 2\beta 1$ integrins. J Cell Biol 131:1903–1915
56. Levental KR, Yu H, Kass L, Lakins JN, Egeblad M, Erler JT, Fong SF, Csiszar K, Giaccia A, Weninger W, Yamauchi M, Gasser DL, Weaver VM (2009) Matrix crosslinking forces tumor progression by enhancing integrin signaling. Cell 139:891–906
57. Legate KR, Wickström SA, Fässler R (2009) Genetic and cell biological analysis of integrin outside-in signaling. Genes Dev 23:397–418
58. Leitinger B (2011) Transmembrane collagen receptors. Annu Rev Cell Dev Biol 27:265–290
59. Leitinger B, Kwan AP (2006) The discoidin domain receptor DDR2 is a receptor for type X collagen. Matrix Biol 25:355–364
60. Lochter A, Navre M, Werb Z, Bissell MJ (1999) $\alpha 1$ and $\alpha 2$ integrins mediate invasive activity of mouse mammary carcinoma cells through regulation of stromelysin-1 expression. Mol Biol Cell 10:271–282
61. Mattila E, Pellinen T, Nevo J, Vuoriluoto K, Arjonen A, Ivaska J (2005) Negative regulation of EGFR signalling through integrin-$\alpha 1\beta 1$-mediated activation of protein tyrosine phosphatase TCPTP. Nat Cell Biol 7:78–85
62. McCall-Culbreath KD, Li Z, Zutter MM (2008) Crosstalk between the $\alpha 2\beta 1$ integrin and c-met/HGF-R regulates innate immunity. Blood 111:3562–3570
63. Mettouchi A, Klein S, Guo W, Lopez-Lago M, Lemichez E, Westwick JK, Giancotti FG (2001) Integrin-specific activation of Rac controls progression through the G(1) phase of the cell cycle. Mol Cell 8:115–127
64. Moresco EM, Donaldson S, Williamson A, Koleske AJ (2005) Integrin-mediated dendrite branch maintenance requires Abelson (Abl) family kinases. J Neurosci 25:6105–6118
65. Niland S, Cremer A, Fluck J, Eble JA, Krieg T, Sollberg S (2001) Contraction-dependent apoptosis of normal dermal fibroblasts. J Invest Dermatol 116:686–692
66. Ning Y, Zeineldin R, Liu Y, Rosenberg M, Stack MS, Hudson LG (2005) Down-regulation of integrin $\alpha 2$ surface expression by mutant epidermal growth factor receptor (EGFRvIII) induces aberrant cell spreading and focal adhesion formation. Cancer Res 65:9280–9286
67. Nissinen L, Koivunen J, Käpylä J, Salmela M, Nieminen J, Jokinen J, Sipilä K, Pihlavisto M, Pentikäinen OT, Marjamäki A, Heino J (2012) Novel $\alpha 2\beta 1$ integrin inhibitors reveal that integrin binding to collagen under shear stress conditions does not require receptor preactivation. J Biol Chem 287:44694–44702
68. Nyberg P, Xie L, Sugimoto H, Colorado P, Sund M, Holthaus K, Sudhakar A, Salo T, Kalluri R (2008) Characterization of the anti-angiogenic properties of arresten, an $\alpha 1\beta 1$ integrin-dependent collagen-derived tumor suppressor. Exp Cell Res 314:3292–3305
69. Nykvist P, Tu H, Ivaska J, Kapyla J, Pihlajaniemi T, Heino J (2000) Distinct recognition of collagen subtypes by $\alpha 1\beta 1$ and $\alpha 2\beta 1$ integrins. $\alpha 1\beta 1$ mediates cell adhesion to type XIII collagen. J Biol Chem 275:8255–8261
70. Nykvist P, Tasanen K, Viitasalo T, Käpylä J, Jokinen J, Bruckner-Tuderman L, Heino J (2001) The cell adhesion domain of type XVII collagen promotes integrin-mediated cell spreading by a novel mechanism. J Biol Chem 276:38673–38679
71. O'Reilly MS, Boehm T, Shing Y, Fukai N, Vasios G, Lane WS, Flynn E, Birkhead JR, Olsen BR, Folkman J (1997) Endostatin: an endogenous inhibitor of angiogenesis and tumor growth. Cell 88:277–285
72. Pasterkamp RJ, Peschon JJ, Spriggs MK, Kolodkin AL (2003) Semaphorin 7A promotes axon outgrowth through integrins and MAPKs. Nature 424:398–405
73. Pellinen T, Arjonen A, Vuoriluoto K, Kallio K, Fransen JA, Ivaska J (2006) Small GTPase Rab21 regulates cell adhesion and controls endosomal traffic of $\beta 1$-integrins. J Cell Biol 173:767–780
74. Popov C, Radic T, Haasters F, Prall WC, Aszodi A, Gullberg D, Schieker M, Docheva D. Integrins $\alpha 2\beta 1$ and $\alpha 11\beta 1$ regulate the survival of mesenchymal stem cells on collagen I (2011) Cell Death Dis 2:e186
75. Ramirez NE, Zhang Z, Madamanchi A, Boyd KL, O'Rear LD, Nashabi A, Li Z, Dupont WD, Zijlstra A, Zutter MM (2011) The $\alpha 2\beta 1$integrin is a metastasis suppressor in mouse models and human cancer. J Clin Invest 121:226–237
76. Rantala JK, Pouwels J, Pellinen T, Veltel S, Laasola P, Mattila E, Potter CS, Duffy T, Sundberg JP, Kallioniemi O, Askari JA, Humphries MJ, Parsons M, Salmi M, Ivaska J (2011) SHARPIN is an endogenous inhibitor of $\beta 1$-integrin activation. Nat Cell Biol 13:1315–1324
77. Ravanti L, Heino J, Lopez-Otin C, Kahari VM (1999) Induction of collagenase-3 (MMP-13) expression in human skin fibroblasts by three-dimensional collagen is mediated by p38 mitogen-activated protein kinase. J Biol Chem 274:2446–2455
78. Rehn M, Veikkola T, Kukk-Valdre E, Nakamura H, Ilmonen M, Lombardo C, Pihlajaniemi T, Alitalo K, Vuori K (2001) Interaction of endostatin with integrins implicated in angiogenesis. Proc Natl Acad Sci USA 98:1024–1029
79. Reunanen N, Foschi M, Han J, Kahari VM (2000) Activation of extracellular signal regulated kinase 1/2 inhibits type I collagen expression by human skin fibroblasts. J Biol Chem 275:34634–34639
80. Ricard-Blum S (2011) The collagen family. Cold Spring Harb Perspect Biol 3:a004978
81. Riikonen T, Westermarck J, Koivisto L, Broberg A, Kahari VM, Heino J (1995) Integrin $\alpha 2\beta 1$ is a positive regulator of collagenase (MMP-1) and collagen $\alpha 1$(I) gene expression. J Biol Chem 270:13548–13552

82. Sacca B, Fiori S, Moroder L (2003) Studies of the local conformational properties of the cell-adhesion domain of collagen type IV in synthetic heterotrimeric peptides. Biochemistry 42:3429–3436
83. Saelman EU, Nieuwenhuis HK, Hese KM, de Groot PG, Heijnen HF, Sage EH, Williams S, McKeown L, Gralnick HR, Sixma JJ (1994) Platelet adhesion to collagen types I through VIII under conditions of stasis and flow is mediated by GPIa/IIa ($\alpha2\beta1$-integrin). Blood 83:1244–1250
84. Saelman EU, Keely PJ, Santoro SA (1995) Loss of MDCK cell $\alpha2\beta1$ integrin expression results in reduced cyst formation, failure of hepatocyte growth factor/scatter factor-induced branching morphogenesis, and increased apoptosis. J Cell Sci 108:3531–3540
85. Shi M, Pedchenko V, Greer BH, Van Horn WD, Santoro SA, Sanders CR, Hudson BG, Eichman BF, Zent R, Pozzi A (2012) Enhancing integrin $\alpha1$ inserted (I) domain affinity to ligand potentiates integrin $\alpha1\beta1$-mediated down-regulation of collagen synthesis. J Biol Chem 287:35139–35152
86. Shintani Y, Fukumoto Y, Chaika N, Svoboda R, Wheelock MJ, Johnson KR (2008) Collagen I mediated up-regulation of N-cadherin requires cooperative signals from integrins and discoidin domain receptor 1. J Cell Biol 180:1277–1289
87. van Slambrouck S, Grijelmo C, De Wever O, Bruyneel E, Emami S, Gespach C, Steelant WF (2007) Activation of the FAK-src molecular scaffolds and p130Cas-JNK signaling cascades by $\alpha1$-integrins during colon cancer cell invasion. Int J Oncol 31:1501–1508
88. Smerling C, Tang K, Hofmann W, Danker K (2007) Role of the $\alpha1$ integrin cytoplasmic tail in the formation of focal complexes, actin organization, and in the control of cell migration. Exp Cell Res 313:3153–3165
89. Smida Rezgui S, Honore S, Rognoni JB, Martin PM, Penel C (2000) Up-regulation of $\alpha2\beta1$ integrin cell-surface expression protects A431 cells from epidermal growth factor-induced apoptosis. Int J Cancer 87:360–367
90. Sriramarao P, Mendler M, Bourdon MA (1993) Endothelial cell attachment and spreading on human tenascin is mediated by $\alpha2\beta1$ and $\alpha V\beta3$ integrins. J Cell Sci 105:1001–1012
91. Sudhakar A, Nyberg P, Keshamouni VG, Mannam AP, Li J, Sugimoto H, Cosgrove D, Kalluri R (2005) Human $\alpha1$ type IV collagen NC1 domain exhibits distinct antiangiogenic activity mediated by $\alpha1\beta1$ integrin. J Clin Invest 115:2801–2810
92. Suzuki-Inoue K, Yatomi Y, Asazuma N, Kainoh M, Tanaka T, Satoh K, Ozaki Y (2001) Rac, a small guanosine triphosphate-binding protein, and p21-activated kinase are activated during platelet spreading on collagen-coated surfaces: roles of integrin $\alpha2\beta1$. Blood 98:3708–3716
93. Takada Y, Hemler ME (1989) The primary structure of the VLA-2/collagen receptor $\alpha2$ subunit (platelet GPIa): homology to other integrins and the presence of a possible collagen-binding domain. J Cell Biol 109:397–407
94. Tian B, Lessan K, Kahm J, Kleidon J, Henke C (2002) $\beta1$ integrin regulates fibroblast viability during collagen matrix contraction through a phosphatidylinositol 3-kinase AKT/protein kinase B signaling pathway. J Biol Chem 277:24667–24675
95. Tuckwell DS, Reid KB, Barnes MJ, Humphries MJ (1996) The A-domain of integrin $\alpha2$ binds specifically to a range of collagens but is not a general receptor for the collagenous motif. Eur J Biochem 241:732–739
96. Tulla M, Pentikinen OT, Viitasalo T, Kapyla J, Impola U, Nykvist P, Nissinen L, Johnson MS, Heino J (2001) Selective binding of collagen subtypes by integrin $\alpha1$I, $\alpha2$I, and $\alpha10$I domains. J Biol Chem 276:48206–48612
97. Tulla M, Lahti M, Puranen JS, Brandt AM, Käpylä J, Domogatskaya A, Salminen TA, Tryggvason K, Johnson MS, Heino J (2008) Effects of conformational activation of integrin $\alpha1$I and $\alpha2$I domains on selective recognition of laminin and collagen subtypes. Exp Cell Res 314:1734–1743
98. Tulla M, Pentikäinen O, Viitasalo T, Käpylä J, Impola U, Nykvist P, Nissinen L, Johnson M, Heino J (2001) Selective binding of collagen subtypes by integrin $\alpha1$I, $\alpha2$I, and $\alpha10$I domains. J Biol Chem 276:48206–48212
99. Upla P, Marjomäki V, Kankaanpää P, Ivaska J, Hyypiä T, Van Der Goot FG, Heino J (2004) Clustering induces a lateral redistribution of $\alpha2\beta1$ integrin from membrane rafts to caveolae and subsequent protein kinase C-dependent internalization. Mol Biol Cell 15:625–636
100. Veit G, Zwolanek D, Eckes B, Niland S, Käpylä J, Zweers MC, Ishada-Yamamoto A, Krieg T, Heino J, Eble JA, Collagen Koch M (2011) XXIII, novel ligand for integrin $\alpha2\beta1$ in the epidermis. J Biol Chem 286:27804–27813
101. Velling T, Kusche-Gullberg M, Sejersen T, Gullberg D (1999) cDNA cloning and chromosomal localization of human $\alpha11$ integrin. A collagen-binding, I domain-containing, $\beta1$-associated integrin α-chain present in muscle tissues. J Biol Chem 274:25735–25742
102. Vogel W, Brakebusch C, Fässler R, Alves F, Ruggiero F, Pawson T (2000) Discoidin domain receptor 1 is activated independently of $\beta1$ integrin. J Biol Chem 275:5779–5784
103. Wang CZ, Su HW, Hsu YC, Shen MR, Tang MJ (2006) A discoidin domain receptor 1/SHP-2 signaling complex inhibits $\alpha2\beta1$-integrin-mediated signal transducers and activators of transcription 1/3 activation and cell migration. Mol Biol Cell 17:2839–2852
104. Wary KK, Mariotti A, Zurzolo C, Giancotti FG (1998) A requirement for caveolin-1 and associated kinase Fyn in integrin signaling and anchorage-dependent cell growth. Cell 94:625–634

105. Xu H, Bihan D, Chang F, Huang PH, Farndale RW, Leitinger B (2012) Discoidin domain receptors promote $\alpha1\beta1$- and $\alpha2\beta1$-integrin mediated cell adhesion to collagen by enhancing integrin activation. PLoS ONE 7:e52209
106. Xu J, Clark RA (1997) A three-dimensional collagen lattice induces protein kinase C-ζ activity: role in $\alpha2$ integrin and collagenase mRNA expression. J Cell Biol 136:473–483
107. Xu J, Zutter MM, Santoro SA, Clark RA (1998) A three-dimensional collagen lattice activates NF-κB in human fibroblasts: role in integrin $\alpha2$ gene expression and tissue remodeling. J Cell Biol 140:709–719
108. Xu J, Clark RA, Parks WC (2001) p38 mitogen-activated kinase is a bidirectional regulator of human fibroblast collagenase-1 induction by three-dimensional collagen lattices. Biochem J 355:437–447
109. Zallocchi M, Johnson BM, Meehan DT, Delimont D, Cosgrove D (2013) $\alpha1\beta1$ integrin/Rac1-dependent mesangial invasion of glomerular capillaries in Alport syndrome. Am J Pathol 183:1269–1280
110. Zhang WM, Kapyla J, Puranen JS, Knight CG, Tiger CF, Pentikainen OT, Johnson MS, Farndale RW, Heino J, Gullberg D (2003) $\alpha11\beta1$ integrin recognizes the GFOGER sequence in interstitial collagens. J Biol Chem 278:7270–7277
111. Zhou H, Kramer RH (2005) Integrin engagement differentially modulates epithelial cell motility by RhoA/ROCK and PAK1. J Biol Chem 280:10624–10635
112. Zou Z, Schmaier AA, Cheng L, Mericko P, Dickeson SK, Stricker TP, Santoro SA, Kahn ML (2009) Negative regulation of activated $\alpha2$ integrins during thrombopoiesis. Blood 113:6428–6439
113. Zutter MM, Edelson BT (2007) The $\alpha2\beta1$ integrin: a novel collectin/C1q receptor. Immunobiology 212:343–353
114. Zutter MM, Santoro SA, Wu JE, Wakatsuki T, Dickeson SK, Elson EL (1999) Collagen receptor control of epithelial morphogenesis and cell cycle progression. Am J Pathol 155:927–940

The Therapeutic Potential of I-Domain Integrins 11

Marian Brennan and Dermot Cox

Abstract

Due to their role in processes central to cancer and autoimmune disease I-domain integrins are an attractive drug target. Both antibodies and small molecule antagonists have been discovered and tested in the clinic. Much of the effort has focused on $\alpha L\beta 2$ antagonists. Maybe the most successful was the monoclonal antibody efalizumab, which was approved for the treatment of psoriasis but subsequently withdrawn from the market due to the occurrence of a serious adverse effect (progressive multifocal leukoencephalopathy). Other monoclonal antibodies were tested for the treatment of reperfusion injury, post-myocardial infarction, but failed to progress due to lack of efficacy. New potent small molecule inhibitors of αv integrins are promising reagents for treating fibrotic disease. Small molecule inhibitors targeting collagen-binding integrins have been discovered and future work will focus on identifying molecules selectively targeting each of the collagen receptors and identifying appropriate target diseases for future clinical studies.

Keywords

I-domain integrins · Therapeutics · Integrin antagonists · Integrin structure

11.1 Introduction

The discovery of the first integrin [129] and the realization that this was only a member of a large family of cell adhesion molecules [108] opened up possibilities for novel therapeutics for diseases such as cancer, inflammation and thrombosis. Prior to this, drug discovery was very much chemistry-led where novel chemicals were screened for potential biological activities.

M. Brennan · D. Cox (✉)
Molecular and Cellular Therapeutics, Royal College of Surgeons in Ireland, 123 St. Stephen's Green, Dublin 2, Ireland
e-mail: dcox@rcsi.ie

D. Gullberg (ed.), *I Domain Integrins*, Advances in Experimental Medicine and Biology 819,
DOI: 10.1007/978-94-017-9153-3_11,

The discovery of the integrin provided an opportunity for target-led drug discovery where chemicals could be screened for their ability to bind to specific receptors. The potential for this new paradigm was confirmed with the approval of the αIIbβ3 antagonist abciximab in 1993. However, since then, aside from two other αIIbβ3 antagonists the only other integrins with approved antagonists are the α4 integrins.

Why the poor success in developing anti-integrin agents? It is certainly not due to the lack of a clinical potential for integrin antagonists as integrins are clearly involved in many of the big diseases such as cardiovascular disease, autoimmune disease and cancer. It is not due to a difficulty in discovering potent antagonists as many antibody and small molecule antagonists have been discovered. Most of the difficulties have arisen due to the complexity of integrins and a poor understanding of the pharmacology of these agents. Despite the poor record of developing anti-integrin agents there is still great potential for the development of this class of drugs. This chapter focuses on the history of integrin antagonists with a specific focus on the development of I-domain antagonists.

11.2 Integrin Families

Integrins are cell adhesion molecules that are found on virtually every cell in the body where they mediate cell–cell and cell-substrate interactions, which are essential for regulating cell growth and cell function. However, they do not simply act as "glue-like" molecules as they are true receptors, generating intracellular signals. Their importance is reflected in the diverse range of diseases in which integrins play key roles including cancer, thrombosis, autoimmune diseases and infection.

Integrins are heterodimers formed from the combination of an α and a β subunit. As there are only eight distinct β subunits and eighteen α subunits combining to form twenty-four unique receptors many β subunits must complex with more than one α subunit (see Preface). Since there are more α subunits than β subunits the integrins were originally described as a superfamily composed of families defined by their β subunit. Table 1 in the Preface chapter lists the integrin families as defined by their β subunits and their associated ligands.

Initially defining integrin families based on their β subunit made sense, as it appeared that the α subunits only associated with a specific β subunit. However, it soon became clear that this is not so as αV can associate with many different β subunits and the α subunit appears to be important in defining the ligand binding properties of an integrin. Thus, among the twelve β1 integrins there is a wide range of ligands with little in common between them other than their shared β subunit. On the other hand, all five αV integrins have similar ligand-binding properties and all bind vitronectin. The β2 integrins are unique in that they do not share their α subunits with any other β subunit and their ligand-binding properties are very similar. Integrins can also be categorized according to their amino acid recognition sequences. The most common recognition sequence is Arg-Gly-Asp (RGD), which is recognized by many integrins. However, this is complicated by the fact that some integrins can bind some proteins in an RGD-dependent manner and others in an RGD-independent manner. For instance αIIbβ3 can bind to fibrinogen via the 2-RGD sequences and via the γ-chain dodecapeptide [123]. As a result defining integrin families requires a more flexible system. Integrins can also be classified by structure. In particular 9 integrin α-subunits contain an I-domain, which is important for ligand binding (see Fig. 11.1). Thus, integrins can be defined as I-domain and non-I-domain-containing integrins. This review will focus on the I-domain-containing integrins and their properties are described below after being loosely grouped into families based on their ligand-binding properties and α and β-subunits.

11.2.1 β2 Integrins

The β2 integrins are a well-defined and distinct group of integrins and their α subunits exclusively associate with the β2 subunit. They are

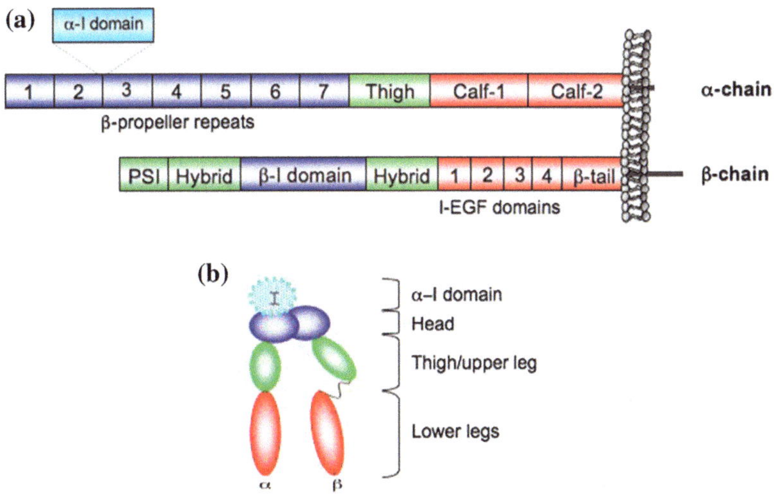

Fig. 11.1 Schematic of the domain structure of I-domain integrins [25, 119, 148]. **a** The I-domain is inserted between the 2nd and 3rd beta propeller subunit on the alpha chain. **b** Outline of the domain structure of an activated I-domain containing integrin

primarily found on leucocytes and are important in normal immune function. The β2 subunit was originally identified as CD18 and the α subunit as CD11. There are four α subunits: αL (LFA-1), αM (Mac-1), αX and αD. All four receptors bind at least one member of the ICAM family, two are complement receptors (αMβ2 (CR3) and αXβ2 (CR4)) and two are fibrinogen receptors (αXβ2 and αDβ2).

11.2.2 Collagen-Binding Integrins

There are 5 integrins that bind collagen and all are β1 integrins (α1, α2, α3, α10 and α11) with differing collagen selectivity [154] (see Preface chapter).All except α3β1 are I-domain containing integrins. α3β1 is a high-affinity receptor for laminin but it also binds collagen type IV and VI through the collagen NC1 domain [2, 11, 16], which is distinct to the binding of the other collagen receptors. α1β1 is found on many cell types including endothelial cells, fibroblasts, astrocytes, T-cells, natural killer cells and macrophages not B-cells. Expression levels usually increase with cytokine stimulation. α2β1 is expressed on platelets, epithelial cells, and many mesenchymal cell types. α10β1 is expressed on chondrocytes [19] while α11β1 is expressed on fibroblasts [100].

11.2.3 αEβ7

The β7 integrins are found on lymphocytes. αEβ7 is the E-Cadherin receptor primarily found on intraepithelial T-lymphocytes while α4β7 is found on lymphocytes in gut-associated lymphoid tissue. αEβ7 facilitates lymphocyte homing to the lamina propria resulting in increased expression of αEβ7, which facilitates lymphocyte extravasation [41]. It is found on T-lymphocytes especially in the gut and in high potency regulatory T-cells [73].

11.3 Integrin Ligands

Integrins bind to a diverse collection of ligands that are large molecules. They are either subendothelial matrix proteins such as fibronectin, vitronectin and collagens or plasma proteins such as complement factors, C-reactive protein [14] and fibrinogen. There is also a group of secreted proteins known as small integrin-binding ligand N-linked glycoproteins (SIBLINGs), which include osteopontin and bone sialoprotein [8]. An interesting ligand for a number of integrins is latent TGFβ1 (LAP-TGF-β). This is an inactive complex of TGF that requires activation by binding to integrins before it becomes biologically active [91].

Many integrins recognize the amino acid sequence Arg-Gly-Asp (RGD) in their ligands. However, RGD is a relatively common motif in proteins many of which are not known to be integrin ligands. Also many RGD-containing proteins can be shown to bind to an integrin that they are unlikely to encounter in vivo or they may require unnatural conformations to expose the RGD motif and thus are not genuine ligands. NGR is also another integrin-recognition motif used by some integrins [68].

11.4 Integrin Physiology

Cell–cell and cell-substrate interactions are critical for every cell of the body, even those in the circulation. Contact with other cells and extracellular matrix components regulates the activity of all cells and since integrins are an important family of receptors that mediate these interactions, they play essential roles in the function of most cells in the body. Areas where integrins are especially important are those that involve growth of tissue or where cell attachment is necessary for function. Thus, embryonic development and the growth of new blood vessels (angiogenesis) [121] are critically dependent on integrins as is the immune system [122] where immune cell attachment is necessary for normal function.

11.5 Integrin Pathology

While integrins have many different physiological roles, identifying a role for integrins in pathology, especially with respect to identifying drug targets, is difficult. This is because many diseases are multi-factorial and while integrins may play a role in the disease process they are only one of many receptors involved. Many cells have multiple integrins with similar binding properties, which can compensate for the inhibited integrin and as a result targeting specific integrins often does not provide therapeutic benefit despite a role for that integrin in the disease process.

11.5.1 Thrombosis

The first disease to be clearly identified with integrins was thrombosis and this was the first therapeutic target for anti-integrin therapy. Thrombosis occurs when platelets adhere to damaged blood vessels and become activated. These activated platelets recruit other platelets resulting in the formation of a haemostatic plug. This is the essential mechanism for preventing blood loss but inappropriate thrombus formation can lead to a stroke or myocardial infarction.

One of the earliest events in haemostasis is the interaction between platelets and exposed collagen in the damaged endothelium. While there are 5 collagen-binding integrins only one ($\alpha 2\beta 1$) is expressed on the platelet. The $\alpha 2\beta 1$ interaction with exposed collagen in conjunction with a second, non-integrin collagen receptor (GPVI) leads to platelet activation [23]. The platelet-platelet interaction that mediates thrombus formation is facilitated by fibrinogen binding to the integrin αIIb$\beta 3$, which becomes activated as a result of platelet activation. Currently the only approved anti-integrin inhibitors for thrombosis are the αIIb$\beta 3$ antagonists although there are also inhibitors of $\alpha 2\beta 1$ under development [84].

11.5.2 Cancer

During carcinogenesis the growth of the tumour and its subsequent metastasis is highly dependent on the cell being able to regulate its attachment to the extracellular matrix and adjacent cells. As integrins play an important role in cell attachment the ability to up-regulate and down-regulate these receptors is critical in carcinogenesis [90]. Integrins play important roles in cell attachment, survival, migration and invasion [59]. Integrins are also essential in the process of angiogenesis, which is also critical for cell growth [6, 121]. As a result there has been a focus on inhibiting integrins to disrupt tumour growth, cancer-associated fibroblasts and metastasis [30].

Arresten is an angiogenesis inhibitor that is derived from collagen. It binds to $\alpha 1\beta 1$ and inhibits invasion of a squamous cell carcinoma in vitro and in vivo [3]. Endorepellin is a anti-angiogenic fragment of perlecan [87] and binds to both $\alpha 2\beta 1$ and VEGF receptor triggering their down regulation [50]. $\alpha 1\beta 1$ is implicated in peritoneal dissemination of gastric cancer cells [44]. $\alpha 2\beta 1$ acts to inhibit metastasis in mouse models of cancer [103].

11.5.3 Immunology

Both $\beta 1$ and $\beta 2$ integrins are important in immune function where they play an essential role in localizing the immune response to the site of inflammation. Engagement of the T-cell receptor and subsequent inside-out signalling leads to activation of T-cell integrins [122]. While inhibition of these integrins can modulate the immune response the specific integrin to target or immune disorder to treat have yet to be determined.

Much of the work has focused on the role of $\alpha 4$, which is implicated in multiple sclerosis [9, 47, 106, 126] and Crohn's disease [131]. Both $\alpha 4\beta 1$ and $\alpha L\beta 2$ mediate leucocyte adhesion in an animal model of epilepsy and anti-$\alpha 4$ antibodies reduced seizure activity [39]. $\alpha L\beta 2$ antagonists have shown potential benefit in lupus [135], renal transplant [138], psoriasis [34, 75] and experimental autoimmune encephalomyelitis [139]. $\alpha 2\beta 1$ is important co-stimulatory molecule on Th17 cells [10].

11.5.4 Infection

A number of infectious agents have developed the ability to interact with integrins and subsequently become internalized allowing access to the intracellular milieu. There are three general mechanisms used to achieve this: binding of integrin ligands that mediate the interaction, a direct interaction with the integrin or binding of a secreted product. The $\alpha M\beta 2$ and $\alpha 4\beta 1$ ligand mindin [58] also acts as a pattern recognition molecule for microbes and thus plays a role in both adaptive and innate immunity [76]. $\alpha M\beta 2$ and $\alpha X\beta 2$ are complement receptors [83, 111].

Streptococcus agalactiae and *Staphylococcus aureus* have both been shown to bind osteopontin, which triggers phagocytosis by binding to monocyte $\alpha X\beta 2$ in an RGD-independent manner [113]. *Shigella flexneri* interacts directly with integrin-linked kinase to enhance adhesion to cells [66].

LFA-1 binds RTX (repeat in toxin) family of cytotoxins from a number of different species [28, 31, 89]. *Helicobacter pylori* Vac A toxin binds to $\beta 2$ integrins enabling it to enter T-lymphocytes [115]. Peptides derived from $\beta 2$ integrin bind LPS and were found to reduce mortality in a mouse model of sepsis [145]. Viruses have also been shown to bind to integrins, which can facilitate cell entry. Rotavirus binds to $\alpha 2\beta 1$ and $\alpha 4\beta 1$ [43, 51, 52]. The interactions of viruses with integrins has been reviewed by Stewart and Nemerow [127].

11.5.5 Osteoporosis and Kidney Disease

Osteoporosis occurs when the balance between bone formation and degradation is disturbed. Integrins play an important role in the function of osteoclasts, which are responsible for degradation of bone. Osteoclast $\alpha 1\beta 1$ is responsible for adhesion of osteoclasts to collagen and polymorphisms in this receptor are related to bone mineral density and fractures [71]. Integrin $\alpha 1$ expression is increased in mesangial cells in Alport disease [125]. $\alpha 1\beta 1$ is involved in diabetic neuropathy [152]. $\alpha 1\beta 1$ binding to collagen IV down-regulates collagen IV synthesis highlighting the importance of this integrin for basement membrane dynamics [116].

11.6 Structural Analysis of I-Domain Integrins

I-domain integrins are a subgroup of integrins containing an inserted I-domain between the β propeller domains 2 and 3 on the α subunit

(Fig. 11.1). I-domain containing integrins bind to their ligands through the metal ion-dependent adhesion site (MIDAS) on their I-domain. Collagen is represented bound to α2 as an example of this mechanism of action in Fig. 11.2. Ligand binding leads to integrin activation through movements of the α1 and α7 helices in the I-domain, which translate to allosteric movements in the integrin and subsequent signaling events. The design of integrin antagonists has presented some difficulties as competitive antagonists often lead to allosteric changes and therefore act as partial agonists. I-domain containing integrins also have a second I-like domain on their β chains. With activation, an interaction occurs between the two I-domains leading to signaling events. This has led to the development of inhibitors to the linker site (Fig. 11.3) [118].

11.6.1 Inhibitors

Inhibitors for I-domain containing integrins fall into two categories, competitive antagonists, and allosteric or non-competitive antagonists. Competitive antagonists bind in the region of the MIDAS domain on the α-chain. Only a few competitive inhibitors of I-domains have been identified. One such example is the AQC2 antibody which binds to the a1b1 I-domain [22, 61], competitively inhibiting collagen binding while maintaining the closed conformation of the I-domain. The snake venom of *Echis multisquamatus* (EMS16) binds competitively to the ligand-binding site of the α2 I-domain also maintaining the closed conformation of the MIDAS domain. Although it is a competitive antagonist, it does not interact directly with the MIDAS domain itself, however stearically blocks ligand binding [55]. The antibody AL-57 binds specifically to the activated/open αL MIDAS [117]. These inhibitors demonstrate that it is possible to develop competitive activation specific inhibitors for the I-domain integrins.

A group of small molecule αL ligand mimetics were designed based on the stucture of ICAM-1

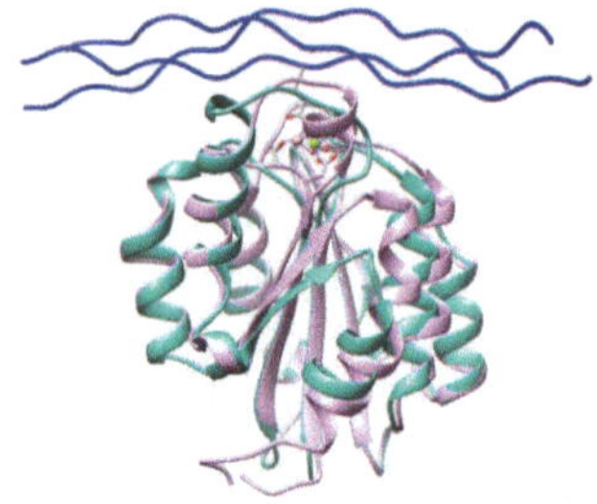

Fig. 11.2 Model of ligand binding and activation for I-domain integrins. The crystal structure of α_2 in its unbound form is represented in pink, PDB 1AOX [36]. This is overlaid onto the bound form of α2 which is represented in cyan PDB 1DZI [37]. Movements of the helices are highlighted with the arrows

as competitive antagonists to the MIDAS site [45, 62]. There is however some evidence to suggest that they bind to the ligand-binding site between the I-like domain MIDAS and the β-propeller (Fig. 11.3), also known as the linker site, or the α/β I-domain allosteric site [109, 119, 144, 151].

11.6.2 Allosteric Inhibition

The interaction of αLβ2 with ICAM provides a model for true allosteric inhibition of I-domain integrins. The carboxyl group of Glu^{34} on ICAM-1 co-ordinates directly with the Mg^{2+} in the MIDAS domain of αL [119]. Displacement of helices α1 and α7 occurs with ligand binding and provides the possibility of allosteric inhibition at the linker site. The I-domain of the αL subunit has been co-crystalized with a number of small molecule inhibitors [27, 32, 60, 77, 101, 140, 143] identifying a major site for allosteric inhibition of the I-domain (Fig. 11.4). Small molecules bind to a pocket in contact with the α7 helix on the opposite side of the molecule from the MIDAS domain [27, 32, 60, 101, 140, 143]. This region has been shown to be important for ligand binding and receptor function [20, 82]. Binding of small molecules do not cause any significant changes to the I-domain, suggesting that they do not activate the receptor. Kallen et al. suggest

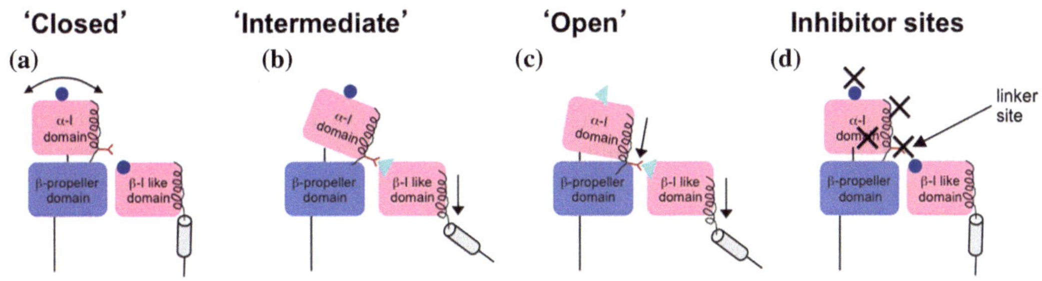

Fig. 11.3 I-domain integrin activation states and inhibitor sites. Figures a to c represent integrin activation states based on the model proposed for the complement receptor 4 [149]. The closed MIDAS site is represented as a blue circle. The open high affinity MIDAS site is represented as a cyan triangle.The α-I domain is thought to have a great deal of freedom in the closed state (**a**). Activation of the β-I domain MIDAS site and interaction between the MIDAS and the acidic residue (at the '*linker site*') locks the α-I domain at an angle (**b**). The α_7 helix moves downwards due to the activation of the α-I MIDAS site and reduces the angle of the α-I domain to the β-I domain (**c**). The sites that have been used to develop inhibitors are presented as black crosses in figure (**d**). This figure is modified from Cox et al. [25]

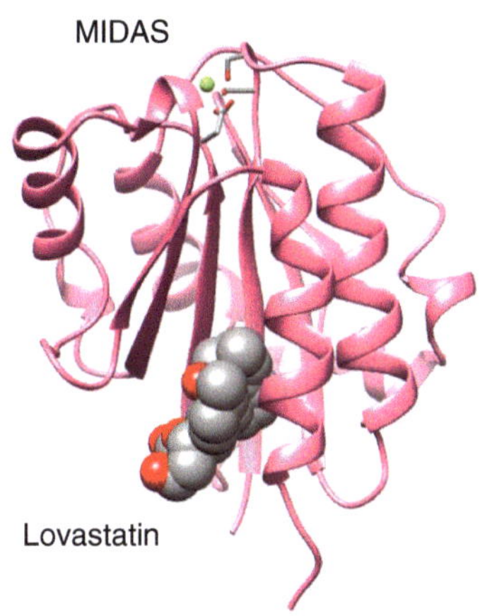

Fig. 11.4 Allosteric binding site on αL. Lovastatin is depicted in space fill in contact with the α7 helix. The binding site is on the opposite side of the molecule to the MIDAS, PDB 1CQP [60]

that they may inhibit signalling by locking the molecule in an inactive, low-affinity state by stabilizing the C-terminal α7 helix. Lovastatin and isoflurane have been co-crystallized in this position [60, 155]. Recently propofol has also been demonstrated to bind to this site [153]. These small molecules are proof of concept for the development of allosteric inhibitors for I-domain integrins that have no agonist activity.

Efalizumab binds to another region of αL which is also distal from the MIDAS [77]. It is largely in contact with the α1 helix, however its binding site is close enough to the MIDAS to cause stearic hindrence of ICAM-1 binding. Therefore it is possible that it acts as an allosteric inhibitor, however it is also possible that it is acting by sterically inhibiting ICAM-1 binding. The structure of efalizumab-bound αL adopts the unliganded, resting conformation. Efalizumab interacts with the α1 and α3 helices, and is thus thought to also act by stabilizing the closed, low affinity conformation. Therefore, there is structural evidence that competitive and non-competitive allosteric inhibitors can be developed for I-domain integrins.

11.7 Therapeutically Targeting I-Domain Integrins

The original anti-integrin drug discovery strategy was to develop monoclonal antibodies to the receptors and this has been commercially successful. A second strategy was to develop peptide antagonists usually based on the peptide sequence from the natural ligand (e.g., RGD) or snake venom peptides. Finally small molecule non-peptide antagonists have also been developed. As our understanding of the nature of the drug-integrin interaction has grown it has created the opportunity for developing different types of integrin antagonists.

11.7.1 Monoclonal Antibodies

Monoclonal antibodies provide an effective source of anti-integrins and were the initial strategy used. There are two types of antibody possible: complex-specific antibodies and subunit-specific antibodies. The advantage of complex-specific antibodies is that they are very specific and this can reduce the level of adverse effects. The specificity of subunit-specific antibodies depends on the target subunit. While the less specific antibodies may have increased adverse effects they may also be more effective as they target a complete family of receptors. The first anti-integrin antibody to be commercialized was the anti-$\beta 3$ abciximab (ReoPro®). It is a potent inhibitor of platelet aggregation and has been extensively tested in clinical studies resulting in approval for use during percutaneous coronary intervention (PCI) [132, 133]. Natalizumab (Tysabri®) is an anti-$\alpha 4\beta 1/\alpha 4\beta 7$ antibody approved for multiple sclerosis [54, 99, 107] and has been shown to be beneficial in the treatment of Crohn's disease [131]. Efalizumab (Raptiva®) is an anti-$\alpha L\beta 2$ antibody approved for the treatment of plaque psoriasis [48, 70, 95] and proved to be effective and safe for long-term use [74] although it has recently been withdrawn from the market.

11.7.2 Peptide-Based Inhibitors

Peptide-based antagonists of integrins have been very attractive for a number of reasons primarily because short peptide sequences that mediate integrin binding were identified from integrin-binding proteins, e.g. RGD, and the effectiveness of phage display libraries at identifying novel integrin-binding sequences. In most cases cyclic peptides are used due to their enhanced stability and potency. Snake venoms are a rich source of bioactive molecules. Snake C-type lectins (snaclecs) have been found to modulate haemostasis. Rhinocetin is a snaclec isolated from the venom of *Bitis gabonica rhinoceros* and is an $\alpha 2\beta 1$ antagonist. It blocks collagen-induced platelet activation [137]. Rhodocetin from *Calloselasma rhodostoma* [13, 35], EMS16 from *Echis multisquamatus* [80] and VP12 from *Vipera palestina* [124] are all snaclecs that inhibit $\alpha 2\beta 1$. Del-1 is a 52 kDa natural inhibitor of $\alpha L\beta 2$ and administration of it has anti-inflammatory effects in diseases such as periodontitis [38].

11.7.3 Non-peptide Small Molecule Antagonists

The ideal drug is an orally active non-peptide small molecule and this has been the goal in anti-integrin therapy. While there are many small molecule inhibitors in pre-clinical development tirofiban (Aggrastat®) is still the only approved non-peptide inhibitor. It is an $\alpha IIb\beta 3$ antagonist although it has no oral activity and like eptifibatide [57] its development was based on a viper venom peptide (echistatin). It was approved for use in PCI and acute coronary syndromes [134]. $\alpha L\beta 2$ has attracted a lot of interest with the discovery of non-peptide inhibitors with nM IC_{50} values [32, 45, 46, 63, 78, 101, 102, 140, 141, 147]. BMS-587101 a small molecule $\alpha L\beta 2$ antagonist reduces symptoms in a mouse model of RA [128]. A few of the compounds showed efficacy in mouse models of inflammation but it is still unclear what their target disease is likely to be. SAR 1118 [45] blocks $\alpha L\beta 2$ and is undergoing clinical trials in diabetic macular oedema [97] and dry eye [114]. $\alpha M\beta 2$ small molecule agonists enhance cell adhesion and thus reduce chemotaxis [79]. See Fig. 11.5 for sample structures.

11.7.4 Collagen and Cadherin Receptor Antagonists

$\alpha 1\beta 1$ is a collagen and laminin receptor and has been implicated in angiogenesis and fibrosis and diabetic neuropathy. Jerdostatin [112], viperistatin [67], lebestatin [94] and obtustatin [88] are snake venom disintegrins, which have low micromolar potency and high selectivity against $\alpha 1\beta 1$. Their selectivity is due to the presence of

Fig. 11.5 Some structures of small molecule αLβ2 antagonists. Sample structures of small molecule αLβ2 antagonists

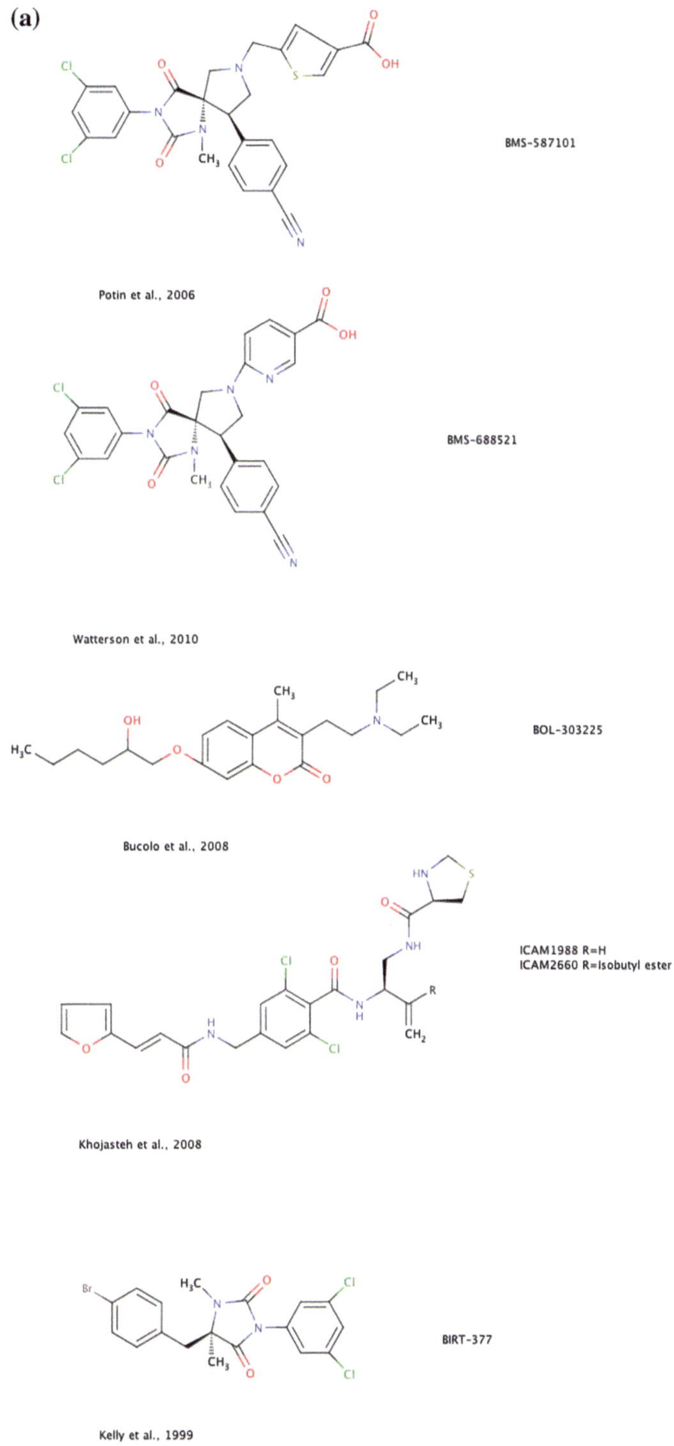

Fig. 11.5 continued

(b)

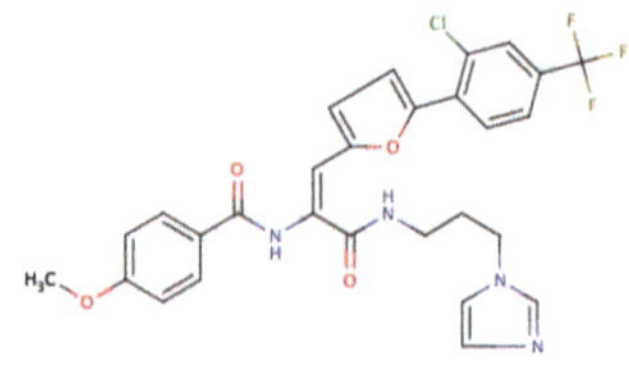

Shoda et al., 2007

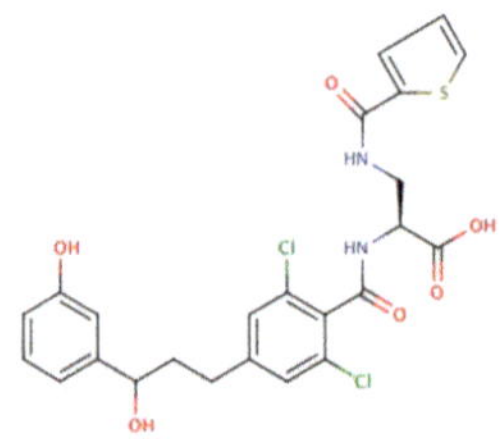

Gadek et al., 2002
Yang et al., 2006 (agonist like properties

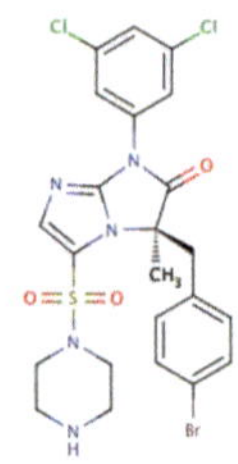

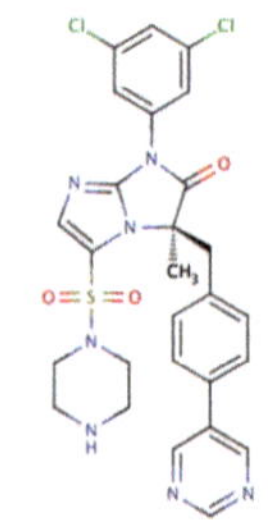

Wu et al., 2004

Panzerbeck et al., 2006

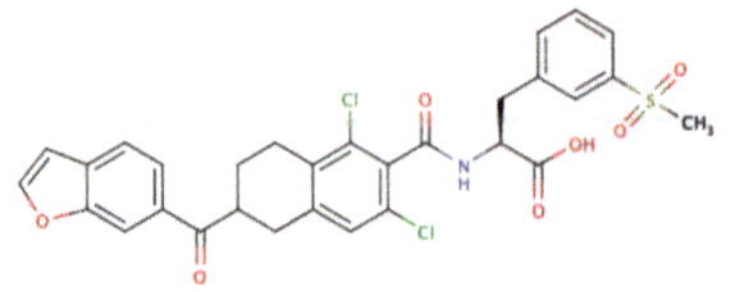

SAR 1118

Zhong et al., 2012

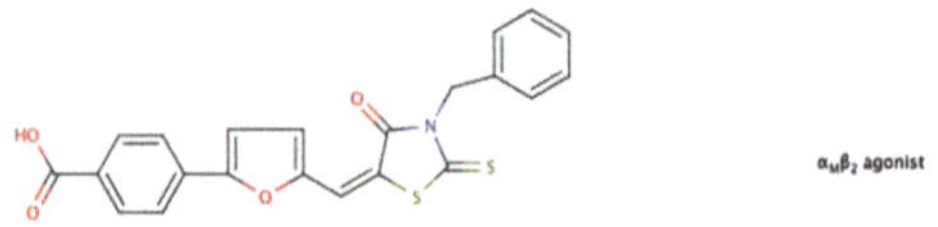

$\alpha_M\beta_2$ agonist

Faridi et al, 2013

the KTS sequence and the flanking residues further enhance their potency. Thus, viperistatin, which contains KTSR, is more potent than obtustatin, which contains KTSL [15, 67]. Jerdiostatin contains the sequence RTS rather than KTS [112]. These have been shown to be effective in models of angiogenesis [81] and melanoma metastasis [18, 124]. However, there is no evidence of small molecule $\alpha 1\beta 1$ antagonists in development probably due to a lack of a clear target disease for an antagonist.

$\alpha 2\beta 1$ is also a collagen and laminin receptor but its presence on platelets where it is an important collagen receptor mediating thrombus formation has made it an attractive target for drug discovery. As with many integrins there are snake venom C-type lectin related proteins that specifically target it. These include rhinocetin [137] rhodocetin [37], vixapatin (VP12) [124], EMS16 [80], flavocetin [4] and VP-I [5]. Potent small molecule inhibitors have also been discovered with sub-micromolar IC_{50} values [84, 92, 93] (see Fig. 11.6) for sample structures. These compounds have been tested in a number of different disease models. Vixapatin was shown to be effective in a model of angiogenesis [86, 110], melanoma metastasis [124] and thrombosis [93]

$\alpha 10\beta 1$ and $\alpha 11\beta 1$ are collagen receptors and αE is an E-cadherin receptor. However, these are newly discovered receptors and there is no evidence of any antagonists under development.

11.7.5 $\beta 2$ Antagonists

$\alpha L\beta 2$ is an ICAM receptor and is an active target for drug discovery. One of the first $\alpha L\beta 2$ antagonists was the humanized monoclonal antibody efalizumab [29, 49]. This was investigated for use in psoriasis [48, 75] and was ultimately approved for this indication. It was also investigated for use in renal transplant [138]. While it was shown to be effective in plaque psoriasis [74] it was ultimately withdrawn from the market due to an increase in the incidence of progressive multifocal leukoencephalopathy (PML) [69]. PML is a serious disease due to reactivation of a JC virus infection in the brain. The original infection is usually asymptomatic but re-activation of the infection leads to demyelination and is associated with a high level of mortality. Once primarily associated with HIV it has more recently been associated with biological immunosuppressive therapy especially natalizumab, efalizumab and rituximab [130]. While the response of Élan/Biogen Idec to the occurrence of PML with natalizumab was to develop a Risk Evaluation and Mitigation Strategy (REMS) in conjunction with the FDA along with a black-box warning, Genentech opted to withdraw efalizumab from the market.

Other antibodies to $\beta 2$ integrins have been described. Hu23F2G [42] and rhuMAb CD18 [7] are anti-$\beta 2$ antibodies that failed to reduce infarct size after angioplasty or thrombolysis respectively. There were also small phase II studies with rhuMAb CD18 in haemorrhagic stroke [105] and Hu23F2G in multiple sclerosis [12]. AL-57 is a monoclonal antibody that specifically recognizes activated $\alpha L\beta 2$ [117]. UK-279,276 is a recombinant glycoprotein also known as Neutrophil Inhibitory Factor [136] that selectively binds to $\alpha M\beta 2$ [72] and investigated for use in stroke although its clinical development appears to be ended. Phage display has been used to identify a peptide-based inhibitor of $\alpha M\beta 2$ [56].

BMS-587101 is a small molecule $\alpha L\beta 2$ antagonist that entered clinical trials for transplant rejection [101]. It also showed benefit in an animal model of rheumatoid arthritis [128]. BMS-688521 is a more potent follow-on compound [141]. BOL-303225-A is a coumarin derivative that has inhibitory activity against both $\alpha L\beta 2$ and $\alpha M\beta 2$ [17]. ICAM1988 is the active metabolite of the small molecule prodrug ICAM2660 that inhibits $\alpha L\beta 2$ [65]. Virtual screening has also identified potential small molecule $\alpha L\beta 2$ antagonists [120]. A number of other groups have also discovered small molecule $\alpha L\beta 2$ antagonists [45, 62, 63, 96, 98, 147] (see Fig. 11.5).

Lifitegrast (SAR 1118) is an $\alpha L\beta 2$ antagonist [156] that is undergoing clinical development in the area of ocular inflammation. It has undergone both phase I [97] and phase II [114] trials. It is

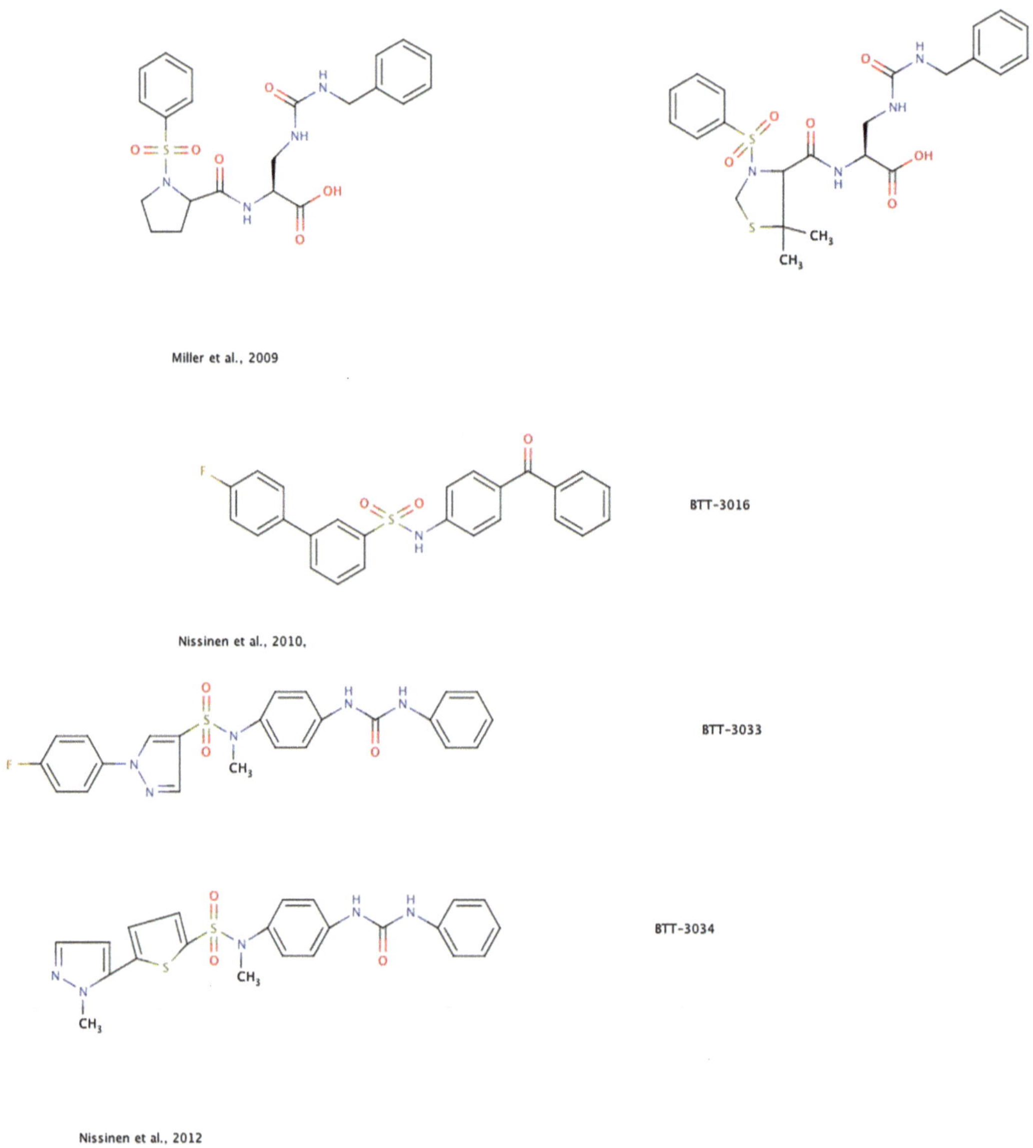

Fig. 11.6 Some structures of small molecule $\alpha 2\beta 1$ antagonists. Sample structures of small molecule $\alpha 2\beta 1$ antagonists

being investigated for dry eye [114] and diabetic macular oedema [97]. Currently this is likely to be the first I-domain antagonist to be commercialised. $\alpha L\beta 2$ antagonists have also shown potential benefit in lupus [135], renal transplant [138], psoriasis [34, 75] and experimental autoimmune encephalomyelitis [139]. However, there has been a very high failure rate in the clinical development programmes for $\beta 2$ integrin antagonists [33, 53], although it is worth noting that the failed trials tended to focus on cardiovascular indications such as reperfusion injury, myocardial infarction, and stroke and thus may only reflect these indications. The $\beta 2$ integrin chain binds LPS and the region between amino acids 266–318 in the A domain has been identified as the LPS binding site [146]. This peptide has been shown to be effective in a mouse model of sepsis [145].

The $\alpha IIb\beta 3$ discovery programme was severely impacted by the nature of the interaction between the drugs and the receptor [24, 25]. Rather than being pure antagonists many of the inhibitors had agonist activity [26]. This is also true of $\alpha L\beta 2$ as compounds that were previously identified as allosteric inhibitors have been shown to be $\alpha L\beta 2$ agonists [150] and a number of compounds have been found to have some agonist like properties similar to that seen with $\alpha IIb\beta 3$ antagonists [119]. Pure agonists have also been discovered [79] and are known as leukadherins [40]. The explanation for this lies in the concept of permissive antagonism [64]. The conventional view of integrin function is that the agonist binds to the receptor, induces conformational changes leading to receptor clustering and finally to outside-in signalling and that this happens in a linear fashion. Leukadherins appear to bind to the receptor at an allosteric site that facilitates ligand binding. However, the leukadherins also block receptor clustering and outside-in signalling and thus are antagonists. So leukadherins are best described as permissive antagonists.

11.7.6 Problems with Integrin Antagonists

As many integrin antagonists are designed around the natural ligand for the receptors such as RGD it is not surprising that the resulting antagonists often display agonist-like activity [142]. RGD-based $\alpha V\beta 3$ and $\alpha V\beta 5$ inhibitors were found to stimulate angiogenesis at low doses [104]. This has also been seen with oral $\alpha IIb\beta 3$ antagonists where low doses were shown to induce platelet aggregation while higher doses were inhibitory [26]. In both cases the problems appear to arise during trough periods. In the case of $\alpha IIb\beta 3$ antagonists this is not a problem with the intravenous agents, as these are maintained at high plasma concentrations using an infusion. However, it was a bigger problem for the oral compounds. A similar situation exists with the I-domain integrin antagonists where compounds have significant agonist-like activity [40, 79, 150]. It is not yet clear whether this will prove to be a problem for the development programme. Another problem identified with I-domain integrin antagonists is PML, which led to the withdrawal of efalizumab from the market [69]. It is not clear whether this is unique to efalizumab or is only associated with the use of biological agents or could happen with any antagonist of $\alpha L\beta 2$. Clearly this will be an issue that will have to be addressed in the development programme for any $\alpha L\beta 2$ antagonist and if it is an issue companies will need to decide whether they will stop the development cycle or implement a REMS.

11.7.7 Modulation of Integrin Expression

In many cases integrin expression on the cell surface is dynamic and is regulated to modulate cell function. This is important in processes such as tumour metastasis where tumour cells must loose their adhesive properties to metastasise and must gain new adhesive properties to colonise the target organ. The expression of some integrins especially the $\beta 2$-integrins is controlled by micro RNAs and this creates the potential for using specific micro RNA to influence the expression of individual integrins [21]. The tumour suppressor genes tuberous sclerosis complex (TSC) regulates $\alpha 1\beta 1$ expression and thus cell migration [85].

11.8 Conclusions

Integrins were discovered almost 30-years ago at a time when the pharmaceutical industry was undergoing a paradigm-shift from chemistry-led drug discovery to target-led drug discovery. Prior to this, drug discovery projects typically involved screening a library of compounds for activity in a disease model. The discovery of integrins and their recognition sequence RGD allowed for a different approach to drug discovery where pharmacologists screen chemical libraries for activity on a specific receptor. The big advantage of a chemistry-led approach is that it produces a drug with desirable activity even if its mechanism of action is unknown. On the other hand a target-led approach makes high-throughput screening a possibility and allows for the discovery of drugs with a known mechanism of action. However, the success of a target-led approach is dependent on the biology of the target being well understood, i.e., a validated target, integrins have proven to be a difficult drug target to commercialize, despite their importance in thrombosis, autoimmune disease and cancer only 4 agents are on the market and three of these are in the field of thrombosis.

There are a few key factors necessary for a successful drug discovery programme. These include a validated target for a specific disease, highly potent and specific ligands, a good ADME (Absorption, Distribution, Metabolism, Excretion) profile and low toxicity. As integrins are large receptors binding to equally large ligands the possibility of developing small molecule antagonists would be expected to be the major challenge; however small molecule antagonists have been developed for many of the integrins. This was facilitated by the identification of short binding motifs such as RGD. Another obvious problem is the development of specific inhibitors, which is a challenge as many of the integrins bind the same motifs such as RGD. Yet highly specific small molecule inhibitors were developed for many of the integrins.

A poor ADME profile is one of the major reasons for a drug failing in the development pipeline. The use of monoclonal antibodies overcomes many of the ADME problems with the exception of absorption. However, they are often dosed on a monthly basis, which mitigates the lack of oral activity. Monoclonal antibodies were approved for all of the successfully targeted integrins (αIIbβ3, α4β1 and αLβ2). Small molecule antagonists were also approved for αIIbβ3 although they were not orally active. Orally active antagonists were also investigated but these all failed in part due to poor ADME profile.

Toxicity is another major reason for failure of drugs in development. This has been a problem for many integrin antagonists. All of the orally active αIIbβ3 antagonists failed due to increased cardiovascular mortality. This was in part due to a poor ADME profile but it was also due to the presence of significant agonist-like activity with the drugs. Both natalizumab (anti-α4 antibody) and efalizumab (anti-αLβ2 antibody) were associated with PML, a very serious adverse effect that ultimately led to the withdrawal of efalizumab from the market. However, this is due to their immune suppressive effects, which is a function of the targets themselves.

The lack of validated targets appears to be the major hurdle in developing anti-integrin antagonists. It is not surprising that the most successful integrin target was αIIbβ3 as it is specific for platelets and was found to play a unique and critical role in platelet function. It was also clear that those patients who lacked αIIbβ3 (Glanzmann's thrombasthenia) had a complete lack of platelet function. Yet despite the presence of a platelet-specific, validated target αIIbβ3 antagonists did not live up to the expectation that they would be the next generation "super-aspirin" and instead are restricted to high-risk patients. It was the $P2Y_{12}$ ADP receptor antagonists such as clopidogrel and prasugrel, which were to become the anti-platelet agent of choice and ultimately become one of the biggest selling drugs today. While α4 integrins were known to play a role in lymphocyte function they were only one of many integrins found on lymphocytes so it was by no means certain that the anti-α4 antibody natalizumab would be successful in multiple sclerosis and Crohn's disease.

There is a lack of validated targets for I-domain integrins. The most obvious target disease for the collagen receptors is thrombosis where collagen-induced platelet activation is important and Biotie have been developing small molecule α2β1 antagonists for thrombosis. However, the success of αIIbβ3 antagonists and $P2Y_{12}$ receptor antagonists such as clopidogrel and prasugrel suggest that there is no market for a platelet collagen receptor antagonist even if it was effective or at least it will be difficult to convince a pharmaceutical company to enter this field. Currently the α2β1 project is not listed as an active project with Biotie. Collagen receptors, like many other integrins, also suffer from the problem of redundancy as many cell types (including platelets) contain multiple collagen receptors suggesting that blocking any one specific receptor may not produce a strong pharmacological effect. Furthermore the widespread distribution of collagen receptors also suggests a potential for adverse effects. There is interest in $\alpha 10\beta 1$ as a chondrocyte biomarker and important in cartilage production although it is not clear if there is a role for drugs that target $\alpha 10\beta 1$. Fibrosis is an important therapeutic area where collagen-binding integrins are very important [57]. However, there are also other integrins involved such as αV integrins and thus the integrins that should be specifically targeted has yet to be elucidated.

$\alpha L\beta 2$ is probably the most investigated I-domain integrin however; the lack of a validated target has proven a problem. Initial targets focused on reperfusion injury post-MI but this did not produce significant clinical results. Typical of the strategy used by pharmaceutical companies they all pursued the same target with the same result. Ultimately the problem was not a lack of potent inhibitors but a lack of understanding of reperfusion injury. However, when a drug fails in a clinical study history has shown that a company is more likely to drop the drug and target entirely rather than investigating other potential uses of the drug.

So what is the future for integrin antagonists and more specifically I-domain antagonists? $\beta 2$ integrins are important in the immune response but $\alpha 4$ integrin antagonists are the first to market in this space. It will be important to identify a disease in which $\beta 2$ integrins are more significant than $\alpha 4$ integrins. Certainly efalizumab was effective in psoriasis as it was clinically approved and also showed benefit in renal transplant. Ocular inflammation appears to be the most advanced therapeutic area for $\alpha L\beta 2$ inhibitors. Ultimately the problem of toxicity, especially PML will need to be addressed. Was this unique to efalizumab or will it be a problem for all $\beta 2$ antagonists? At least in ocular inflammation this is unlikely to be an issue as there is low systemic exposure to the drug. Cancer is another important area for anti-integrins and there has been a lot of interest in this for decades however, an anti-integrin has yet to be approved in this area. Currently the major focus is on $\alpha V\beta 3$ inhibitors but I-domain collagen-binding integrins appear to play a significant role in angiogenesis, which may make them potential targets in cancer. Infection is another potential target disease as a number of I-domain integrins are involved in infection of cells by bacteria and viruses. This is the new area of targeting host factors such as adhesion receptors to supplement anti-microbial therapy.

Ultimately the future of anti-integrin pharmacology lies in further work on the role of integrins in disease. Advances in our understanding of integrin structure means that we can now develop better antagonists that are activation-specific and that are not partial agonists. This needs to be supplemented by a better understanding of the role of integrins in health and disease if successful therapeutics are to be developed.

11.9 Perspectives

Potent antibodies and small molecule inhibitors for I-domain integrins have been discovered and a number of the antibodies have entered into clinical development programmes. However, none have proven to be successful either due to lack of efficacy or adverse effects. Since I-domain integrins are known to be important in cancer, autoimmune disease and fibrosis, all areas where there are significant unmet needs, there will remain interest in these receptors as potential drug targets. A key issue to be addressed with I-domain integrins is identification of validated targets. As multiple integrins are often involved in these target diseases further research on specific integrins and their roles in the disease process will be required to ensure that the appropriate integrin is being targeted in each disease. This in conjunction with recent studies that elucidated the molecular interaction between antagonists and their target integrins will allow new generations of potent and specific antagonists to be tested in theses diseases.

References

1. IMPACT investigators (1997) Randomised placebo-controlled trial of effect of eptifibatide on complications of percutaneous coronary intervention: IMPACT-II. Integrilin to minimise platelet aggregation and coronary thrombosis-II. *Lancet* 349:1422–1428.
2. Aggeli AS, Kitsiou PV, Tzinia AK, Boutaud A, Hudson BG, Tsilibary EC (2009) Selective binding of integrins from different renal cell types to the NC1 domain of alpha3 and alpha1 chains of type IV collagen. J Nephrol 22:130–136
3. Aikio M, Alahuhta I, Nurmenniemi S, Suojanen J, Palovuori R, Teppo S et al (2012) Arresten, a collagen-derived angiogenesis inhibitor, suppresses invasion of squamous cell carcinoma. PLoS ONE 7:e51044
4. Arlinghaus FT, Eble JA (2013) The Collagen-binding integrin $\alpha 2\beta 1$ is a novel interaction partner of the *Trimeresurus flavoviridis* venom protein flavocetin-A. J Biol Chem 288:947–955
5. Arlinghaus FT, Momic T, Ammar NA, Shai E, Spectre G, Varon D et al (2013) Identification of $\alpha 2\beta 1$ integrin inhibitor VP-i with anti-platelet properties in the venom of *Vipera palaestinae*. Toxicon 64:96–105
6. Avraamides CJ, Garmy-Susini B, Varner JA (2008) Integrins in angiogenesis and lymphangiogenesis. Nat Rev Cancer 8:604–617
7. Baran KW, Nguyen M, McKendall GR, Lambrew CT, Dykstra G, Palmeri ST et al (2001) Double-blind, randomized trial of an anti-CD18 antibody in conjunction with recombinant tissue plasminogen activator for acute myocardial infarction: limitation of myocardial infarction following thrombolysis in acute myocardial infarction (LIMIT AMI) study. Circulation 104:2778–2783
8. Bellahcene A, Castronovo V, Ogbureke KUE, Fisher LW, Fedarko NS (2008) Small integrin-binding ligand N-linked glycoproteins (SIBLINGs): multifunctional proteins in cancer. Nat Rev Cancer 8:212–226
9. Bennett J (2006) Natalizumab and progressive multifocal leukoencephalopathy: migrating towards safe adhesion molecule therapy in multiple sclerosis. Neurol Res 28:291–298
10. Boisvert M, Chetoui N, Gendron S, Aoudjit F (2010) $\alpha 2\beta 1$ integrin is the major collagen-binding integrin expressed on human Th17 cells. Eur J Immunol 40:2710–2719
11. Borza CM, Pozzi A, Borza D-B, Pedchenko V, Hellmark T, Hudson BG et al (2006) Integrin a_3b_1, a novel receptor for α3(IV) noncollagenous domain and a trans-dominant inhibitor for integrin $\alpha v\beta 3$. J Biol Chem 281:20932–20939
12. Bowen JD, Petersdorf SH, Richards TL, Maravilla KR, Dale DC, Price TH et al (1998) Phase I study of a humanized anti-CD11/CD18 monoclonal antibody in multiple sclerosis[ast]. Clin Pharmacol Ther 64:339–346
13. Bracht T, Figueiredo de Rezende F, Stetefeld J, Sorokin LM, Eble JA (2011) Monoclonal antibodies reveal the alteration of the rhodocetin structure upon $\alpha 2\beta 1$ integrin binding. Biochem J 440:1–11
14. Brennan MP, Moriarty RD, Grennan S, Chubb AJ, Cox D (2008) C-reactive protein binds to αIIbβ3. J Thromb Haemost 6:1239–1241
15. Brown MC, Eble JA, Calvete JJ, Marcinkiewicz C (2009) Structural requirements of KTS-disintegrins for inhibition of $\alpha 1\beta 1$ integrin. Biochem J 417:95–101
16. Bryant JE, Shamhart PE, Luther DJ, Olson ER, Koshy JC, Costic DJ et al (2009) Cardiac myofibroblast differentiation is attenuated by α3 integrin blockade: potential role in post-MI remodeling. J Mol Cell Cardiol 46:186–192
17. Bucolo C, Maltese A, Maugeri F, Ward K, Baiula M, Spartà A et al (2008) New coumarin-based anti-inflammatory drug: putative antagonist of the integrins $\alpha L\beta 2$ and $\alpha M\beta 2$. J Pharm Pharmacol 60:1473–1479
18. Calvete J, Marcinkiewicz C, Sanz L (2007) KTS and RTS-disintegrins: anti-angiogenic viper venom peptides specifically targeting the $\alpha 1\beta$ 1 integrin. Curr Pharm Des 13:2853–2859
19. Camper L, Holmvall K, Wängnerud C, Aszódi A, Lundgren-Akerlund E (2001) Distribution of the collagen-binding integrin $\alpha 10\beta 1$ during mouse development. Cell Tissue Res 306:107–116
20. Champe M, McIntyre BW, Berman PW (1995) Monoclonal antibodies that block the activity of leukocyte function-associated antigen 1 recognize three discrete epitopes in the inserted domain of CD11a. J Biol Chem 270:1388–1394
21. Chen W, Harbeck MC, Zhang W, Jacobson JR (2013) MicroRNA regulation of integrins. Transl Res 162:133–143
22. Clark LA, Boriack-Sjodin PA, Eldredge J, Fitch C, Friedman B, Hanf KJ et al (2006) Affinity enhancement of an in vivo matured therapeutic antibody using structure-based computational design. Protein Sci 15:949–960
23. Clemetson K, Clemetson J (2007) Collagen receptors as potential targets for novel anti-platelet agents. Curr Pharm Des 13:2673–2683
24. Cox D (2004) Oral GPIIb/IIIa antagonists: what went wrong? Curr Pharm Des 10:1587–1596
25. Cox D, Brennan M, Moran N (2010) Integrins as therapeutic targets: lessons and opportunities. Nat Rev Drug Discov 9:804–820
26. Cox D, Smith R, Quinn M, Theroux P, Crean P, Fitzgerald DJ (2000) Evidence of platelet activation during treatment with a GPIIb/IIIa antagonist in patients presenting with acute coronary syndromes. J Am Coll Cardiol 36:1514–1519
27. Crump MP, Ceska TA, Spyracopoulos L, Henry A, Archibald SC, Alexander R et al (2004) Structure of an allosteric inhibitor of LFA-1 bound to the I-

domain studied by crystallography, NMR, and calorimetry. Biochemistry 43:2394–2404
28. Dassanayake RP, Maheswaran SK, Srikumaran S (2007) Monomeric expression of bovine b_2-integrin subunits reveals their role in *Mannheimia haemolytica* leukotoxin-induced biological effects. Infect Immun 75:5004–5010
29. Dedrick RL, Walicke P, Garovoy M (2002) Anti-adhesion antibodies: efalizumab, a humanized anti-CD11a monoclonal antibody. Transpl Immunol 9:181–186
30. Desgrosellier JS, Cheresh DA (2010) Integrins in cancer: biological implications and therapeutic opportunities. Nat Rev Cancer 10:9–22
31. Dileepan T, Kachlany SC, Balashova NV, Patel J, Maheswaran SK (2007) Human CD18 is the functional receptor for *Aggregatibacter actinomycetemcomitans* leukotoxin. Infect Immun 75:4851–4856
32. Dodd DS, Sheriff S, Chang CJ, Stetsko DK, Phillips LM, Zhang Y et al (2007) Design of LFA-1 antagonists based on a 2,3-dihydro-1H-pyrrolizin-5(7aH)-one scaffold. Bioorgan Med Chem Lett 17:1908–1911
33. Dove A (2000) CD18 trials disappoint again. Nat Biotech 18:817–818
34. Dubertret L, Sterry W, Bos JD, Chimenti S, Shumack S, Larsen CG et al (2006) CLinical experience acquired with the efalizumab (Raptiva) (CLEAR) trial in patients with moderate-to-severe plaque psoriasis: results from a phase III international randomized, placebo-controlled trial. Br J Dermatol 155:170–181
35. Eble JA, Niland S, Bracht T, Mormann M, Peter-Katalinic J, Pohlentz G et al. (2009) The alpha2beta1 integrin-specific antagonist rhodocetin is a cruciform, heterotetrameric molecule. FASEB J 23: 2917–2927
36. Emsley J, King SL, Bergelson JM, Liddington RC (1997) Crystal structure of the I domain from integrin $\alpha 2\beta 1$. J Biol Chem 272:28512–28517
37. Emsley J, Knight CG, Farndale RW, Barnes MJ, Liddington RC (2000) Structural basis of collagen recognition by integrin $\alpha 2\beta 1$. Cell 101:47–56
38. Eskan MA, Jotwani R, Abe T, Chmelar J, Lim J-H, Liang S et al (2012) The leukocyte integrin antagonist Del-1 inhibits IL-17-mediated inflammatory bone loss. Nat Immunol 13:465–473
39. Fabene PF, Mora GN, Martinello M, Rossi B, Merigo F, Ottoboni L et al (2008) A role for leukocyte-endothelial adhesion mechanisms in epilepsy. Nat Med 14:1377–1383
40. Faridi MH, Altintas MM, Gomez C, Duque JC, Vazquez-Padron RI, Gupta V (2013) Small molecule agonists of integrin CD11b/CD18 do not induce global conformational changes and are significantly better than activating antibodies in reducing vascular injury. Biochimica et Biophysica Acta (BBA) - General Subjects 1830:3696–3710
41. Farstad I, Halstensen T, Lien B, Kilshaw P, Lazarovits A, Brandtzaeg P et al (1996) Distribution of $\beta 7$ integrins in human intestinal mucosa and organized gut-associated lymphoid tissue. Immunology 89:227–237
42. Faxon DP, Gibbons RJ, Chronos NAF, Gurbel PA, Sheehan F (2002) The effect of blockade of the CD11/CD18 integrin receptor on infarct size in patients with acute myocardial infarction treated with direct angioplasty: the results of the HALT-MI study. J Am Coll Cardiol 40:1199–1204
43. Fleming FE, Graham KL, Taniguchi K, Takada Y, Coulson BS (2007) Rotavirus-neutralizing antibodies inhibit virus binding to integrins $\alpha 2\beta 1$ and $\alpha 4\beta 1$. Arch Virol 152:1087
44. Fukuda K, Saikawa Y, Yagi H, Wada N, Takahashi T, Kitagawa Y (2012) Role of integrin $\alpha 1$ subunits in gastric cancer patients with peritoneal dissemination. Mol Med Rep 5:336–340
45. Gadek TR, Burdick DJ, McDowell RS, Stanley MS, Marsters JC Jr, Paris KJ et al (2002) Generation of an LFA-1 antagonist by the transfer of the ICAM-1 immunoregulatory epitope to a small molecule. Science 295:1086–1089
46. Gadek TR, McDowell RS (2003) Discovery of small molecule leads in a biotechnology datastream. Drug Discovery Today 8:545–550
47. Gonzalez-Amaro R, Mittelbrunn M, Sanchez-Madrid F (2005) Therapeutic anti-integrin ($\alpha 4$ and αL) monoclonal antibodies: two-edged swords? Immunology 116:289–296
48. Gordon KB, Papp KA, Hamilton TK, Walicke PA, Dummer W, Li N et al (2003) Efalizumab for patients with moderate to severe plaque psoriasis: a randomized controlled trial. J Am Med Assoc 290:3073–3080
49. Gottlieb AB, Krueger JG, Wittkowski K, Dedrick R, Walicke P, Garovoy M (2002) Psoriasis as a model for t-cell–mediated disease: immunobiologic and clinical effects of treatment with multiple doses of efalizumab, an anti–cd11a antibody. Arch Dermatol 138:591–600
50. Goyal A, Pal N, Concannon M, Paul M, Doran M, Poluzzi C et al (2011) Endorepellin, the angiostatic module of perlecan, interacts with both the $\alpha 2\beta 1$ integrin and vascular endothelial growth factor receptor 2 (VEGFR2): a dual receptor antagonism. J Biol Chem 286:25947–25962
51. Graham KL, Fleming FE, Halasz P, Hewish MJ, Nagesha HS, Holmes IH et al (2005) Rotaviruses interact with $\alpha 4\beta 7$ and $\alpha 4\beta 1$ integrins by binding the same integrin domains as natural ligands. J Gen Virol 86:3397–3408
52. Graham KL, Halasz P, Tan Y, Hewish MJ, Takada Y, Mackow ER et al (2003) Integrin-using rotaviruses bind $\alpha 2\beta 1$ integrin $\alpha 2$ I domain via VP4 DGE sequence and recognize $\alpha X\beta 2$ and $\alpha V\beta 3$ by using VP7 during cell entry. J Virol 77:9969–9978

53. Harlan JM, Winn RK (2002) Leukocyte-endothelial interactions: clinical trials of anti-adhesion therapy. Crit Care Med 30:S214–S219
54. Havrdova E, Galetta S, Hutchinson M, Stefoski D, Bates D, Polman CH et al (2009) Effect of natalizumab on clinical and radiological disease activity in multiple sclerosis: a retrospective analysis of the natalizumab safety and efficacy in relapsing-remitting multiple sclerosis (AFFIRM) study. Lancet Neurol 8:254–260
55. Horii K, Okuda D, Morita T, Mizuno H (2004) Crystal structure of EMS16 in complex with the integrin alpha2-I domain. J Mol Biol 341:519–527
56. Houimel M, Mazzucchelli L (2012) Random phage-epitope library based identification of a peptide antagonist of Mac-1 $\beta 2$ integrin ligand binding. Matrix Biol 31:66–77
57. Huang C, Ogawa R (2012) Fibroproliferative disorders and their mechanobiology. Connect Tissue Res 53:187–196
58. Jia W, Li H, He Y-W (2005) The extracellular matrix protein mindin serves as an integrin ligand and is critical for inflammatory cell recruitment. Blood 106:3854–3859
59. Jin H, Varner J (2004) Integrins: roles in cancer development and as treatment targets. Br J Cancer 90:561–565
60. Kallen J, Welzenbach K, Ramage P, Geyl D, Kriwacki R, Legge G et al (1999) Structural basis for LFA-1 inhibition upon lovastatin binding to the CD11a I-domain. J Mol Biol 292:1–9
61. Karpusas M, Ferrant J, Weinreb PH, Carmillo A, Taylor FR, Garber EA (2003) Crystal structure of the $\alpha 1\beta 1$ integrin I domain in complex with an antibody Fab fragment. J Mol Biol 327:1031–1041
62. Keating SM, Clark KR, Stefanich LD, Arellano F, Edwards CP, Bodary SC et al (2006) Competition between intercellular adhesion molecule-1 and a small-molecule antagonist for a common binding site on the a_L subunit of lymphocyte function-associated antigen-1. Protein Sci 15:290–303
63. Kelly TA, Jeanfavre DD, McNeil DW, Woska JR Jr, Reilly PL, Mainolfi EA et al (1999) Cutting edge: a small molecule antagonist of LFA-1-mediated cell adhesion. J Immunol 163:5173–5177
64. Kenakin T (2005) New concepts in drug discovery: collateral efficacy and permissive antagonism. Nat Rev Drug Discov 4:919–927
65. Khojasteh SC, Leipold DD, Lai F, La H, Baumgardner MJ, Desino KE et al (2008) Preclinical absorption, distribution, metabolism and excretion (ADME) characterization of ICAM1988, an LFA-1/ICAM antagonist, and its prodrug. Xenobiotica 38:340–352
66. Kim M, Ogawa M, Fujita Y, Yoshikawa Y, Nagai T, Koyama T et al (2009) Bacteria hijack integrin-linked kinase to stabilize focal adhesions and block cell detachment. Nature 459:578–582
67. Kisiel DG, Calvete JJ, Katzhendler J, Fertala A, Lazarovici P, Marcinkiewicz C (2004) Structural determinants of the selectivity of KTS-disintegrins for the $\alpha 1\beta 1$ integrin. FEBS Lett 577:478–482
68. Koivunen E, Wang B, Ruoslahti E (1994) Isolation of a highly specific ligand for the $\alpha 5\beta 1$ integrin from a phage display library. J Cell Biol 124:373–380
69. Korman BD, Tyler KL, Korman NJ (2009) Progressive multifocal leukoencephalopathy, efalizumab, and immunosuppression: a cautionary tale for dermatologists. Arch Dermatol 145:937–942
70. Lebwohl M, Tyring SK, Hamilton TK, Toth D, Glazer S, Tawfik NH et al (2003) A novel targeted T-cell modulator, efalizumab, for plaque psoriasis. N Engl J Med 349:2004–2013
71. Lee H-J, Kim S-Y, Koh J-M, Bok J, Kim K-J, Kim K-S et al (2007) Polymorphisms and haplotypes of integrin a_1 (ITGA1) are associated with bone mineral density and fracture risk in postmenopausal Koreans. Bone 41:979–986
72. Lees KR, Diener H-C, Asplund K, Krams M, UK-279 -SI (2003) UK-279,276, a neutrophil inhibitory glycoprotein, in acute stroke: tolerability and pharmacokinetics. Stroke 34:1704–1709
73. Lehmann J, Huehn J, de la Rosa M, Maszyna F, Kretschmer U, Krenn V et al (2002) Expression of the integrin $\alpha E\beta 7$ identifies unique subsets of CD25 + as well as CD25- regulatory T cells. PNAS 99:13031–13036
74. Leonardi C, Menter A, Hamilton T, Caro I, Xing B, Gottlieb AB (2008) Efalizumab: results of a 3-year continuous dosing study for the long-term control of psoriasis. Br J Dermatol 158:1107–1116
75. Leonardi CL, Papp KA, Gordon KB, Menter A, Feldman SR, Caro I et al (2005) Extended efalizumab therapy improves chronic plaque psoriasis: results from a randomized phase III trial. J Am Acad Dermatol 52:425–433
76. Li Y, Cao C, Jia W, Yu L, Mo M, Wang Q et al (2009) Structure of the F-spondin domain of mindin, an integrin ligand and pattern recognition molecule. EMBO J 28:286–297
77. Li S, Wang H, Peng B, Zhang M, Zhang D, Hou S et al (2009) Efalizumab binding to the LFA-1 αL I domain blocks ICAM-1 binding via steric hindrance. Proc Natl Acad Sci U S A 106:4349–4354
78. Liu G, Huth JR, Olejniczak ET, Mendoza R, DeVries P, Leitza S et al (2001) Novel p-arylthio cinnamides as antagonists of leukocyte function-associated antigen-1/intracellular adhesion molecule-1. Interaction. 2. Mechanism of inhibition and structure-based improvement of pharmaceutical properties. J Med Chem 44:1202–1210

79. Maiguel D, Faridi MH, Wei C, Kuwano Y, Balla KM, Hernandez D et al (2011) Small Molecule-mediated activation of the integrin CD11b/CD18 Reduces inflammatory disease. Sci. Signal 4:ra57
80. Marcinkiewicz C, Lobb RR, Marcinkiewicz MM, Daniel JL, Smith JB, Dangelmaier C et al (2000) Isolation and characterization of EMS16, a C-lectin type protein from Echis multisquamatus venom, a potent and selective inhibitor of the $\alpha 2\beta 1$ integrin†. Biochemistry 39:9859–9867
81. Marcinkiewicz C, Weinreb PH, Calvete JJ, Kisiel DG, Mousa SA, Tuszynski GP et al (2003) Obtustatin: a potent selective inhibitor of $\alpha 1\beta 1$ integrin in vitro and angiogenesis in vivo. Cancer Res 63:2020–2023
82. McDowall A, Leitinger B, Stanley P, Bates PA, Randi AM, Hogg N (1998) The I domain of integrin leukocyte function-associated antigen-1 is involved in a conformational change leading to high affinity binding to ligand intercellular adhesion molecule 1 (ICAM-1). J Biol Chem 273:27396–27403
83. Micklem K, Sim R (1985) Isolation of complement-fragment-iC3b-binding proteins by affinity chromatography. The identification of p150,95 as an iC3b-binding protein. Biochem J 231:233–236
84. Miller MW, Basra S, Kulp DW, Billings PC, Choi S, Beavers MP et al (2009) Small-molecule inhibitors of integrin a_2b_1 that prevent pathological thrombus formation via an allosteric mechanism. Proc Natl Acad Sci 106:719–724
85. Moir LM, Black JL, Krymskaya VP (2012) TSC2 modulates cell adhesion and migration via integrin-$\alpha 1\beta 1$. Am J Physiol Lung Cell Mol Physiol 303:L703–L710
86. Momic T, Cohen G, Reich R, Arlinghaus FT, Eble JA, Marcinkiewicz C et al (2012) Vixapatin (VP12), a C-type lectin-protein from Vipera xantina palestinae venom: characterization as a novel anti-angiogenic compound. Toxins 4:862–877
87. Mongiat M, Sweeney SM, San Antonio JD, Fu J, Iozzo RV (2003) Endorepellin, a novel inhibitor of angiogenesis derived from the c terminus of perlecan. J Biol Chem 278:4238–4249
88. Moreno-Murciano MP, Monleón D, Calvete JJ, Celda B, Marcinkiewicz C (2003) Amino acid sequence and homology modeling of obtustatin, a novel non-RGD-containing short disintegrin isolated from the venom of *Vipera lebetina obtusa*. Protein Sci 12:366–371
89. Morova J, Osicka R, Masin J, Sebo P (2008) RTX cytotoxins recognize b_2 integrin receptors through N-linked oligosaccharides. Proc Nat Acad Sci 105:5355–5360
90. Moschos S, Drogowski L, Reppert S, Kirkwood J (2007) Integrins and cancer. Oncology (Williston Park) 21:13–20
91. Munger JS, Harpel JG, Giancotti FG, Rifkin DB (1998) Interactions between growth factors and integrins: latent forms of transforming growth factor-beta are ligands for the integrin $\alpha V\beta 1$. Mol Biol Cell 9:2627–2638
92. Nissinen L, Koivunen J, Käpylä J, Salmela M, Nieminen J, Jokinen J et al (2012) Novel $\alpha 2\beta 1$ integrin inhibitors reveal that integrin binding to collagen under shear stress conditions does not require receptor preactivation. J Biol Chem 287:44694–44702
93. Nissinen L, Pentikäinen O, Jouppila A, Käpylä J, Ojala M, Nieminen J et al (2010) A small-molecule inhibitor of integrin alpha2 beta1 introduces a new strategy for antithrombotic therapy. Thromb Haemost 103:387–397
94. Olfa K-Z, Jose L, Salma D, Amine B, Najet SA, Nicolas A et al (2005) Lebestatin, a disintegrin from *Macrovipera venom*, inhibits integrin-mediated cell adhesion, migration and angiogenesis. Lab Invest 85:1507–1516
95. Ortonne JP, Shear N, Shumack S, Henninger E (2005) Impact of efalizumab on patient-reported outcomes in high-need psoriasis patients: results of the international, randomized, placebo-controlled Phase III clinical experience acquired with raptiva (CLEAR) trial [NCT00256139]. BMC Dermatol 5:13
96. Panzenbeck MJ, Jeanfavre DD, Kelly TA, Lemieux R, Nabozny G, Reilly PL et al (2006) An orally active, primate selective antagonist of LFA-1 inhibits delayed-type hypersensitivity in a humanized-mouse model. Eur J Pharmacol 534:233–240
97. Paskowitz DM, Nguyen QD, Gehlbach P, Handa JT, Solomon S, Stark W et al (2012) Safety, tolerability, and bioavailability of topical SAR 1118, a novel antagonist of lymphocyte function-associated antigen-1: a phase 1b study. Eye 26:944–949
98. Patwardhan AP, Pulgam VR, Zhang Y, Wulff WD (2005) Highly diastereoselective alkylation of aziridine-2-carboxylate esters: enantioselective synthesis of LFA-1 antagonist BIRT-377. Angew Chem Int Ed 44:6169–6172
99. Polman CH, O'Connor PW, Havrdova E, Hutchinson M, Kappos L, Miller DH et al (2006) A randomized, placebo-controlled trial of natalizumab for relapsing multiple sclerosis. N Engl J Med 354:899–910
100. Popova SN, Rodriguez-Sánchez B, Lidén Å, Betsholtz C, van den Bos T, Gullberg D (2004) The mesenchymal $\alpha 11\beta 1$ integrin attenuates PDGF-BB-stimulated chemotaxis of embryonic fibroblasts on collagens. Dev Biol 270:427–442
101. Potin D, Launay M, Monatlik F, Malabre P, Fabreguettes M, Fouquet A et al (2006) Discovery and development of 5-[(5S,9R)-9- (4-Cyanophenyl)-3-(3,5-dichlorophenyl)-1- methyl-2,4-dioxo-1,3,7-triazaspiro[4.4]non- 7-yl-methyl]-3-thiophenecarboxylic Acid (BMS-587101)-a small molecule antagonist leukocyte function associated antigen-1. J Med Chem 49:6946–6949

102. Potin D, Launay M, Nicolai E, Fabreguette M, Malabre P, Caussade F et al (2005) De novo design, synthesis, and in vitro activity of LFA-1 antagonists based on a bicyclic[5.5] hydantoin scaffold. Bioorg Med Chem Lett 15:1161–1164
103. Ramirez NE, Zhang Z, Madamanchi A, Boyd KL, x, Rear LD et al (2011) The $\alpha2\beta1$ integrin is a metastasis suppressor in mouse models and human cancer. J Clin Invest 121:226–237
104. Reynolds AR, Hart IR, Watson AR, Welti JC, Silva RG, Robinson SD et al (2009) Stimulation of tumor growth and angiogenesis by low concentrations of RGD-mimetic integrin inhibitors. Nat Med 15:392–400
105. Rhee P, Morris J, Durham R, Hauser C, Cipolle M, Wilson R et al (2000) Recombinant humanized monoclonal antibody against CD18 (rhuMAb CD18) in traumatic hemorrhagic shock: results of a phase II clinical trial. Traumatic shock group. J Trauma 49:611–619
106. Rice GPA, Hartung H-P, Calabresi PA (2005) Anti-$\alpha4$ integrin therapy for multiple sclerosis: mechanisms and rationale. Neurology 64:1336–1342
107. Rudick RA, Stuart WH, Calabresi PA, Confavreux C, Galetta SL, Radue EW et al (2006) Natalizumab plus interferon $\beta1$a for relapsing multiple sclerosis. N Engl J Med 354:911–923
108. Ruoslahti E, Pierschbacher MD (1987) New perspectives in cell adhesion: RGD and integrins. Science 238:491–497
109. Salas A, Shimaoka M, Kogan AN, Harwood C, von Andrian UH, Springer TA (2004) Rolling adhesion through an extended conformation of integrin $\alpha L\beta2$ and relation to α I and β I-like domain interaction. Immunity 20:393–406
110. San Antonio JD, Zoeller JJ, Habursky K, Turner K, Pimtong W, Burrows M et al (2009) A key role for the integrin $\alpha2\beta1$ in experimental and developmental angiogenesis. Am J Pathol 175:1338–1347
111. Sanchez-Madrid F, Nagy J, Robbins E, Simon P, Springer T (1983) A human leukocyte differentiation antigen family with distinct alpha-subunits and a common beta-subunit: the lymphocyte function-associated antigen (LFA-1), the C3bi complement receptor (OKM1/Mac-1), and the p150,95 molecule. J Exp Med 158:1785–1803
112. Sanz L, Chen R-Q, Pérez A, Hilario R, Juárez P, Marcinkiewicz C et al (2005) cDNA cloning and functional expression of jerdostatin, a novel RTS-disintegrin from trimeresurus jerdonii and a specific antagonist of the $\alpha1\beta1$ integrin. J Biol Chem 280:40714–40722
113. Schack L, Stapulionis R, Christensen B, Kofod-Olsen E, Skov Sorensen UB, Vorup-Jensen T et al (2009) Osteopontin enhances phagocytosis through a novel osteopontin receptor, the $\alpha X\beta2$ integrin. J Immunol 182:6943–6950
114. Semba CP, Torkildsen GL, Lonsdale JD, McLaurin EB, Geffin JA, Mundorf TK et al (2012) A phase 2 randomized, double-masked, placebo-controlled study of a novel integrin antagonist (SAR 1118) for the treatment of dry eye. Am J Ophthalmol 153(1050–60):e1
115. Sewald X, Gebert-Vogl B, Prassl S, Barwig I, Weiss E, Fabbri M et al (2008) Integrin subunit CD18 is the T-lymphocyte receptor for the *Helicobacter pylori* vacuolating cytotoxin. Cell Host Microbe 3:20–29
116. Shi M, Pedchenko V, Greer BH, Van Horn WD, Santoro SA, Sanders CR et al (2012) Enhancing integrin $\alpha1$ inserted (I) domain affinity to ligand potentiates integrin $\alpha1\beta1$-mediated down-regulation of collagen synthesis. J Biol Chem 287:35139–35152
117. Shimaoka M, Kim M, Cohen EH, Yang W, Astrof N, Peer D et al (2006) AL-57, a ligand-mimetic antibody to integrin LFA-1, reveals chemokine-induced affinity up-regulation in lymphocytes. Proc Natl Acad Sci U S A 103:13991–13996
118. Shimaoka M, Salas A, Yang W, Weitz-Schmidt G, Springer TA (2003) Small molecule integrin antagonists that bind to the $\beta2$ subunit I-like domain and activate signals in one direction and block them in the other. Immunity 19:391–402
119. Shimaoka M, Springer TA (2003) Therapeutic antagonists and conformational regulation of integrin function. Nat Rev Drug Discov 2:703–716
120. Shoda M, Harada T, Yano K, Stahura FL, Himeno T, Shiojiri S et al (2007) Virtual screening leads to the discovery of an effective antagonist of lymphocyte function-associated antigen-1. ChemMedChem 2:515–521
121. Silva R, D'Amico G, Hodivala-Dilke KM, Reynolds LE (2008) Integrins: the keys to unlocking angiogenesis. Arterioscler Thromb Vasc Biol 28:1703–1713
122. Smith-Garvin JE, Koretzky GA, Jordan MS (2009) T cell activation. Ann Rev Immunol 27:591
123. Springer TA, Zhu J, Xiao T (2008) Structural basis for distinctive recognition of fibrinogen γC peptide by the platelet integrin αIIb$\beta3$. J Cell Biol 182:791–800
124. Staniszewska I, Walsh EM, Rothman VL, Gaathon A, Tuszynski GP, Calvete JJ et al (2009) Effect of VP12 and viperistatin on inhibition of collagen receptors: dependent melanoma metastasis. Cancer Biol Ther 8:1507–1516
125. Steenhard BM, Vanacore R, Friedman D, Zelenchuk A, Stroganova L, Isom K et al (2012) Upregulated expression of integrin $\alpha1$ in mesangial cells and integrin $\alpha3$ and vimentin in podocytes of *Col4a3*-Null (Alport) mice. PLoS ONE 7:e50745
126. Steinman L (2005) Blocking adhesion molecules as therapy for multiple sclerosis: natalizumab. Nat Rev Drug Discovery 4:510

127. Stewart PL, Nemerow GR (2007) Cell integrins: commonly used receptors for diverse viral pathogens. Trends Microbiol 15:500–507
128. Suchard SJ, Stetsko DK, Davis PM, Skala S, Potin D, Launay M et al (2010) An LFA-1 ({alpha}L{beta}2) small-molecule antagonist reduces inflammation and joint destruction in murine models of arthritis. J Immunol 184:3917–3926
129. Tamkun J, DeSimone D, Fonda D, Patel R, Buck C, Horwitz A et al (1986) Structure of integrin, a glycoprotein involved in the transmembrane linkage between fibronectin and actin. Cell 46:271–282
130. Tan CS, Koralnik IJ (2010) Progressive multifocal leukoencephalopathy and other disorders caused by JC virus: clinical features and pathogenesis. Lancet Neurol 9:425–437
131. Targan SR, Feagan BG, Fedorak RN, Lashner BA, Panaccione R, Present DH et al (2007) Natalizumab for the treatment of active Crohn's disease: results of the ENCORE trial. Gastroenterol 132:1672–1683
132. The Epilog Investigators (1997) Platelet glycoprotein IIb/IIIa receptor blockade and low-dose heparin during percutaneous coronary revascularization. N Engl J Med 336:1689–1697
133. The EPISTENT Investigators (1998) Randomised placebo-controlled and balloon-angioplasty-controlled trial to assess safety of coronary stenting with use of platelet glycoprotein-IIb/IIIa blockade. Evaluation of Platelet IIb/IIIa inhibitor for stenting. Lancet 352:87–92
134. The RESTORE Investigators (1997) Effects of platelet glycoprotein IIb/IIIa blockade with tirofiban on adverse cardiac events in patients with unstable angina or acute myocardial infarction undergoing coronary angioplasty. Circulation 96:1445–1453
135. Usmani N, Goodfield M (2007) Efalizumab in the treatment of discoid lupus erythematosus. Arch Dermatol 143:873–877
136. Ustinov VA, Plow EF (2002) Delineation of the key amino acids involved in neutrophil inhibitory factor binding to the I-domain supports a mosaic model for the capacity of integrin $\alpha M\beta 2$ to recognize multiple ligands. J Biol Chem 277:18769–18776
137. Vaiyapuri S, Hutchinson EG, Ali MS, Dannoura A, Stanley RG, Harrison RA et al (2012) Rhinocetin, a venom-derived integrin-specific antagonist inhibits collagen-induced platelet and endothelial cell functions. J Biol Chem 287:26235–26244
138. Vincenti F, Mendez R, Pescovitz M, Rajagopalan PR, Wilkinson AH, Butt K et al (2007) A phase I/II randomized open-label multicenter trial of efalizumab, a humanized anti-CD11a, anti-LFA-1 in renal transplantation. Am J Transplant 7:1770–1777
139. Wang Y, Kai H, Chang F, Shibata K, Tahara-Hanaoka S, S-i Honda et al (2007) A critical role of LFA-1 in the development of Th17 cells and induction of experimental autoimmune encephalomyelytis. Biochem Biophys Res Commun 353:857–862
140. Wattanasin S, Kallen J, Myers S, Guo Q, Sabio M, Ehrhardt C et al (2005) 1,4-diazepane-2,5-diones as novel inhibitors of LFA-1. Bioorg Med Chem Lett 15:1217–1220
141. Watterson SH, Xiao Z, Dodd DS, Tortolani DR, Vaccaro W, Potin D et al (2010) Small molecule antagonist of leukocyte function associated antigen-1 (LFA-1): structure-activity Relationships leading to the identification of 6-((5S,9R)-9-(4-cyanophenyl)-3-(3,5-dichlorophenyl)-1-methyl-2,4-dioxo-1,3,7-triazaspiro[4.4]nonan-7-yl)nicotinic acid (BMS-688521). J Med Chem 53:3814–3830
142. Weis SM, Stupack DG, Cheresh DA (2009) Agonizing integrin antagonists? Cancer Cell 15:359–361
143. Weitz-Schmidt G, Welzenbach K, Dawson J, Kallen J (2004) Improved lymphocyte function-associated antigen-1 (LFA-1) inhibition by statin derivatives: molecular basis determined by x-ray analysis and monitoring of LFA-1 conformational changes in vitro and ex vivo. J Biol Chem 279:46764–46771
144. Welzenbach K, Hommel U, Weitz-Schmidt G (2002) Small molecule inhibitors induce conformational changes in the I domain and the I-like domain of lymphocyte function-associated antigen-1. Molecular insights into integrin inhibition. J Biol Chem 277:10590–10598
145. Wong K, Wo J, Ho D, Poon R, Casasnovas J, Luk J (2010) Prophylactic uses of integrin CD18-βA peptide in a murine polymicrobial peritonitis model. World J Gastroenterol 16:2648–2656
146. Wong KF, Luk JM, Cheng RH, Klickstein LB, and Fan S-T (2007) Characterization of two novel LPS-binding sites in leukocyte integrin bA domain. FASEB J 21: 3231–3239
147. Wu JP, Emeigh J, Gao DA, Goldberg DR, Kuzmich D, Miao C et al (2004) Second-generation lymphocyte function-associated antigen-1 Inhibitors: 1H-Imidazo[1,2-]imidazol-2-one derivatives. J Med Chem 47:5356–5366
148. Xiao T, Takagi J, Coller BS, Wang J-H, Springer TA (2004) Structural basis for allostery in integrins and binding to fibrinogen-mimetic therapeutics. Nature 432:59–67
149. Xie C, Zhu J, Chen X, Mi L, Nishida N, Springer TA (2010) Structure of an integrin with an αI domain, complement receptor type 4. EMBO J 29:666–679
150. Yang W, Carman CV, Kim M, Salas A, Shimaoka M, Springer TA (2006) A small molecule agonist of an integrin, $\alpha L\beta 2$. J Biol Chem 281:37904–37912
151. Yang W, Shimaoka M, Salas A, Takagi J, Springer TA (2004) Intersubunit signal transmission in

integrins by a receptor-like interaction with a pull spring. Proc Natl Acad Sci U S A 101:2906–2911
152. Yu L, Su Y, Paueksakon P, Cheng H, Chen X, Wang H et al (2012) Integrin [alpha]1/Akita double-knockout mice on a Balb/c background develop advanced features of human diabetic nephropathy. Kidney Int 81:1086–1097
153. Yuki K, Bu W, Xi J, Shimaoka M, Eckenhoff R (2013) Propofol shares the binding site with isoflurane and sevoflurane on leukocyte function-associated antigen-1. Anesth Analg 117:803–811
154. Zeltz C, Orgel J, and Gullberg D (2013) Molecular composition and function of integrin-based collagen glues—introducing COLINBRIs. Biochimica et Biophysica Acta (BBA)—General Subjects (in press}
155. Zhang H, Astrof NS, Liu JH, Wang JH, and Shimaoka M (2009) Crystal structure of isoflurane bound to integrin LFA-1 supports a unified mechanism of volatile anesthetic action in the immune and central nervous systems. FASEB J 23(8):2735–2740
156. Zhong M, Gadek TR, Bui M, Shen W, Burnier J, Barr KJ et al (2012) Discovery and development of potent LFA-1/ICAM-1 antagonist SAR 1118 as an ophthalmic solution for treating dry eye. ACS Med Chem Lett 3:203–206

Index

D. Gullberg (ed.), *I Domain Integrins*, Advances in Experimental Medicine and Biology 819,
DOI: 10.1007/978-94-017-9153-3, © Springer Science+Business Media Dordrecht 2014

MIX
Papier aus verantwortungsvollen Quellen
Paper from responsible sources
FSC® C105338

If you have any concerns about our products,
you can contact us on
ProductSafety@springernature.com

In case Publisher is established outside the EU,
the EU authorized representative is:
Springer Nature Customer Service Center GmbH
Europaplatz 3, 69115 Heidelberg, Germany

Printed by Libri Plureos GmbH
in Hamburg, Germany